Introduction to Quantum Fields on a Lattice

Quantum field theory, our description of the fundamental forces in nature, was originally formulated in continuous space–time, where it leads to embarrassing infinities which have to be eliminated by a process called renormalization. A simple but rigorous formulation can be obtained by replacing continuous space–time by a discrete set of points on a lattice. This clarifies the essentials of quantum fields using concepts such as universality of critical phenomena and the renormalization group.

This book provides a clear and pedagogical introduction to quantum fields on a lattice. The path integral on the lattice is explained in concrete examples using weak- and strong-coupling expansions. Fundamental concepts, such as 'triviality' of Higgs fields and confinement of quarks and gluons into hadrons, are described and illustrated with the results of numerical simulations. The book also provides an introduction to chiral symmetry and chiral gauge theory. Based on the lecture notes of a course given by the author, this book contains many explanatory examples and exercises, and is suitable as a textbook for advanced undergraduate and graduate courses. This title, first published in 2002, has been reissued as an Open Access publication on Cambridge Core.

JAN SMIT holds a position at the Institute of Theoretical Physics of the University of Amsterdam and, since 1991, he has been Professor of Theoretical Physics at Utrecht University. He is well known for his fundamental contributions to lattice gauge theory. His current interests are lattice methods for quantum gravity, applications to cosmology and the creation of the quark–gluon plasma in the laboratory.

CAMBRIDGE LECTURE NOTES IN PHYSICS 15
General Editors: P. Goddard, J. Yeomans

1. Clarke: The Analysis of Space–Time Singularities

2. Dorey: Exact s-Matrices in Two Dimensional Quantum Field Theory

3. Sciama: Modern Cosmology and the Dark Matter Problem

4. Veltman: Diagrammatica – The Path to Feynman Rules

5. Cardy: Scaling and Renormalization in Statistical Physics

6. Heusler: Black Hole Uniqueness Theorems

7. Coles and Ellis: Is the Universe Open or Closed?

8. Razumov and Saveliev: Lie Algebras, Geometry, and Toda-type Systems

9. Forshaw and Ross: Quantum Chromodynamics and the Pomeron

10. Jensen: Self-organised Criticality

11. Vandezande: Lattice Models of Polymers

12. Esposito: Dirac Operators and Spectral Geometry

13. Kreimer: Knots and Feynman Diagrams

14. Dorfman: An Introduction to Chaos in Nonequilibrium Statistical Mechanics

15. Smit: Introduction to Quantum Fields on a Lattice

Introduction to Quantum Fields on a Lattice

'a robust mate'

JAN SMIT

University of Amsterdam

CAMBRIDGE
UNIVERSITY PRESS

CAMBRIDGE
UNIVERSITY PRESS

Shaftesbury Road, Cambridge CB2 8EA, United Kingdom

One Liberty Plaza, 20th Floor, New York, NY 10006, USA

477 Williamstown Road, Port Melbourne, VIC 3207, Australia

314–321, 3rd Floor, Plot 3, Splendor Forum, Jasola District Centre, New Delhi – 110025, India

103 Penang Road, #05-06/07, Visioncrest Commercial, Singapore 238467

Cambridge University Press is part of Cambridge University Press & Assessment, a department of the University of Cambridge.

We share the University's mission to contribute to society through the pursuit of education, learning and research at the highest international levels of excellence.

www.cambridge.org
Information on this title: www.cambridge.org/9781009402743

DOI: 10.1017/9781009402705

First published 2002
Reissued as OA 2023

A catalogue record for this publication is available from the British Library.

ISBN 978-1-009-40274-3 Hardback
ISBN 978-1-009-40275-0 Paperback

Cambridge University Press & Assessment has no responsibility for the persistence or accuracy of URLs for external or third-party internet websites referred to in this publication and does not guarantee that any content on such websites is, or will remain, accurate or appropriate.

Contents

Preface   *page* xi

1   **Introduction**   1
1.1   QED, QCD, and confinement   1
1.2   Scalar field   5

2   **Path-integral and lattice regularization**   8
2.1   Path integral in quantum mechanics   8
2.2   Regularization by discretization   10
2.3   Analytic continuation to imaginary time   12
2.4   Spectrum of the transfer operator   13
2.5   Latticization of the scalar field   15
2.6   Transfer operator for the scalar field   18
2.7   Fourier transformation on the lattice   20
2.8   Free scalar field   22
2.9   Particle interpretation   25
2.10   Back to real time   26
2.11   Problems   28

3 $O(n)$ **models**   32
3.1   Goldstone bosons   32
3.2 $O(n)$ models as spin models   34
3.3   Phase diagram and critical line   36
3.4   Weak-coupling expansion   39
3.5   Renormalization   46
3.6   Renormalization-group beta functions   48
3.7   Hopping expansion   51
3.8   Lüscher–Weisz solution   55
3.9   Numerical simulation   60

3.10	Real-space renormalization group and universality	67
3.11	Universality at weak coupling	71
3.12	Triviality and the Standard Model	74
3.13	Problems	79
4	**Gauge field on the lattice**	**83**
4.1	QED action	83
4.2	QCD action	85
4.3	Lattice gauge field	90
4.4	Gauge-invariant lattice path integral	95
4.5	Compact and non-compact Abelian gauge theory	97
4.6	Hilbert space and transfer operator	99
4.7	The kinetic-energy operator	102
4.8	Hamiltonian for continuous time	105
4.9	Wilson loop and Polyakov line	107
4.10	Problems	112
5	**$U(1)$ and $SU(n)$ gauge theory**	**115**
5.1	Potential at weak coupling	115
5.2	Asymptotic freedom	121
5.3	Strong-coupling expansion	125
5.4	Potential at strong coupling	129
5.5	Confinement versus screening	132
5.6	Glueballs	135
5.7	Coulomb phase, confinement phase	136
5.8	Mechanisms of confinement	138
5.9	Scaling and asymptotic scaling, numerical results	140
5.10	Problems	144
6	**Fermions on the lattice**	**149**
6.1	Naive discretization of the Dirac action	149
6.2	Species doubling	151
6.3	Wilson's fermion method	156
6.4	Staggered fermions	160
6.5	Transfer operator for Wilson fermions	161
6.6	Problems	165
7	**Low-mass hadrons in QCD**	**170**
7.1	Integrating over the fermion fields	170
7.2	Hopping expansion for the fermion propagator	171
7.3	Meson and baryon propagators	173
7.4	Hadron masses at strong coupling	177

7.5	Numerical results	179
7.6	The parameters of QCD	188
7.7	Computing the gauge coupling from the masses	190
7.8	Problems	190
8	**Chiral symmetry**	**193**
8.1	Chiral symmetry and effective action in QCD	193
8.2	Pseudoscalar masses and the $U(1)$ problem	199
8.3	Chiral anomalies	202
8.4	Chiral symmetry and the lattice	204
8.5	Spontaneous breaking of chiral symmetry	212
8.6	Chiral gauge theory	217
8.7	Outlook	223
8.8	Problems	223
Appendix A	$SU(n)$	**229**
A.1	Fundamental representation of $SU(n)$	229
A.2	Adjoint representation of $SU(n)$	231
A.3	Left and right translations in $SU(n)$	234
A.4	Tensor method for $SU(n)$	236
Appendix B	**Quantization in the temporal gauge**	**239**
Appendix C	**Fermionic coherent states**	**242**
Appendix D	**Spinor fields**	**253**
Notes		258
References		261
Index		267

Preface

She [field theory] *is not a robust mate ready to pitch in and lend a helping hand. She is a haunting mistress, refined, and much too beautiful for hard work. She is at her best in formal dress, and thus displayed in this book, where rigor will be found to be absolutely absent.* Bryce S. DeWitt

Since the above characterization appeared [1] in 1965 we have witnessed great progress in quantum field theory, our description of fundamental particles and their interactions. This book displays her in informal dress, robust and ready to give results, rigorous, while at a pedestrian mathematical level. By approximating space–time by a collection of points on a lattice we get a number of benefits:

- it serves as a precise but simple definition of quantum fields, which has its own beauty;
- it brings to the fore and clarifies essential aspects such as renormalization, scaling, universality, and the role of topology;
- it makes a fruitful connection to statistical physics;
- it allows numerical simulations on a computer, giving truly non-perturbative results as well as new physical intuition into the behavior of the system.

This book is based on notes of a lecture course given to advanced undergraduate students during the period 1984–1995. An effort was made to accomodate those without prior knowledge of field theory. In the present version, examples from numerical simulations have been replaced by more recent results, and a few sections (8.3–8.6) on lattice aspects of chiral symmetry have been added. The latter notoriously complicated topic was not dealt with in the lectures, but for this book it seemed appropriate to give an introduction.

An overview of the research area in this book is given by the proceedings of the yearly symposia 'Lattice XX', which contain excellent reviews in which the authors tried hard to make the material accessible. These meetings tend to be dominated by QCD, which is understandable, as many of the physical applications are in the sphere of the strong interactions, but a lot of exciting developments usually take place 'on the fringe', in the parallel sessions. In fact, Lattice XX may be considered as *the* arena for non-perturbative field theory. The appropriate papers can be retrieved from the e-print archive http://arXiv.org/ and its mirrors, or the SPIRES website http://www.slac.stanford.edu/spires/hep/

I would like to thank my students, who stumbled over my mistakes, for their perseverance and enthusiasm, and my colleagues for collaborations and for sharing their insight into this ever-surprising research field.

Amsterdam, November 2001

1

Introduction

We introduce here quarks and gluons. The analogy with electrodynamics
at short distances disappears at larger distances with the emergence
of the string tension, the force that confines the quarks and gluons
permanently into bound states called hadrons.

Subsequently we introduce the simplest relativistic field theory, the
classical scalar field.

1.1 QED, QCD, and confinement

Quantum electrodynamics (QED) is the quantum theory of photons
(γ) and charged particles such as electrons ($e^\pm$), muons ($\mu^\pm$), protons
(p), pions ($\pi^\pm$), etc. Typical phenomena that can be described by
perturbation theory are Compton scattering ($\gamma + e^- \rightarrow \gamma + e^-$), and
pair annihilation/production such as $e^+ + e^- \rightarrow \mu^+ + \mu^-$. Examples of
non-perturbative phenomena are the formation of atoms and molecules.
The expansion parameter of perturbation theory is the fine-structure
constant[1] $\alpha = e^2/4\pi$.

Quantum chromodynamics (QCD) is the quantum theory of quarks
(q) and gluons (g). The quarks u, d, c, s, t and b ('up', 'down', 'charm',
'strange', 'top' and 'bottom') are analogous to the charged leptons ν_e, e,
ν_μ, μ, ν_τ, and τ. In addition to electric charge they also carry 'color
charges', which are the sources of the gluon fields. The gluons are
analogous to photons, except that they are self-interacting because they
also carry color charges. The strength of these interactions is measured
by $\alpha_s = g^2/4\pi$ (alpha strong), with g analogous to the electromagnetic
charge e. The 'atoms' of QCD are $q\bar{q}$ ($\bar{q}$ denotes the antiparticle of q)

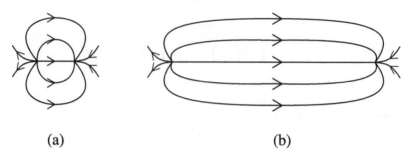

(a)   (b)

Fig. 1.1. Intuitive representation of chromoelectric field lines between a static quark–antiquark source pair in QCD: (a) Coulomb-like at short distances; (b) string-like at large distances, at which the energy content per unit length becomes constant.

bound states called mesons† (π, K, η, η', ρ, K^*, ω, ϕ, ...) and $3q$ bound states called baryons (the nucleon N, and furthermore Σ, Λ, Ξ, Δ, Σ^*, Λ^*,...). The mesons are bosons and the baryons are fermions. There may be also multi-quark states analogous to molecules. Furthermore, there are expected to be glueballs consisting mainly of gluons. These bound states are called 'hadrons' and their properties as determined by experiment are recorded in the tables of the Particle Data Group [2].

The way that the gluons interact among themselves has dramatic effects. At distances of the order of the hadron size, the interactions are strong and α_s effectively becomes arbitrarily large as the distance scale increases. Because of the increasing potential energy between quarks at large distances, it is not possible to have single quarks in the theory: they are permanently confined in bound states.

For a precise characterization of confinement one considers the theory with gluons only (no dynamical quarks) in which static external sources are inserted with quark quantum numbers, a distance r apart. The energy of this configuration is the quark–antiquark potential $V(r)$. In QCD confinement is realized such that $V(r)$ increases linearly with r as $r \to \infty$,

$$V(r) \approx \sigma r, \quad r \to \infty. \tag{1.1}$$

The coefficient σ is called the string tension, because there are effective string models for $V(r)$. Such models are very useful for grasping some of the physics involved (figure 1.1).

Because of confinement, quarks and gluons cannot exist as free parti-

† The quark content of these particles is given in table 7.1 in section 7.5.

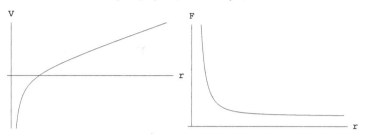

Fig. 1.2. Shape of the static $q\bar{q}$ potential and the force $F = \partial V/\partial r$.

cles. No such free particles have been found. However, scattering experiments at high momentum transfers (corresponding to short distances) have led to the conclusion that there are quarks and gluons inside the hadrons. The effective interaction strength α_s is *small* at short distances. Because of this, perturbation theory is applicable at short distances or large momentum transfers. This can also be seen from the force derived from the $q\bar{q}$ potential, $F = \partial V/\partial r$. See figure 1.2. Writing conventionally

$$F(r) = \frac{4}{3}\frac{\alpha_s(r)}{r^2}, \tag{1.2}$$

we know that $\alpha_s \to 0$ very slowly as the distance decreases,

$$\alpha_s(r) \approx \frac{4\pi}{11\ln(1/\Lambda^2 r^2)}. \tag{1.3}$$

This is called *asymptotic freedom*. The parameter Λ has the dimension of a mass and may be taken to set the dimension scale in quark-less 'QCD'. For the glueball mass m or string tension σ we can then write

$$m = C_m\Lambda, \qquad \sqrt{\sigma} = C_\sigma\Lambda. \tag{1.4}$$

Constants like C_m and C_σ, which relate short-distance to long-distance properties, are non-perturbative quantities. They are pure numbers whose computation is a challenge to be met by the theory developed in the following chapters.

The value of the string tension σ is known to be approximately $(400\text{ MeV})^2$. This information comes from a remarkable property of the hadronic mass spectrum, the fact that, for the leading spin states, the spin J is approximately linear in the squared mass m^2,

$$J = \alpha_0 + \alpha' m^2. \tag{1.5}$$

See figure 1.3. Such approximately straight 'Regge trajectories' can be

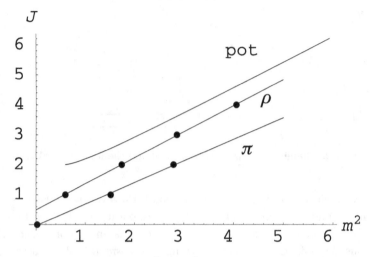

Fig. 1.3. Plot of spin J versus m^2 (GeV^2) for ρ- and π-like particles. The dots give the positions of particles, the straight lines are fits to the data, labeled by their particles with lowest spin. The line labeled 'pot' is L versus H^2 for the solution (1.10), for clarity shifted upward by two units, for $m_q = m_\rho/2$, $\sigma = 1/8\alpha'_\rho$.

understood from the following simple effective Hamiltonian for binding of a $q\bar{q}$ pair,

$$H = 2\sqrt{m_q^2 + p^2} + \sigma r. \tag{1.6}$$

Here m_q is the mass of the constituent quarks, taken to be equal for simplicity, $p = |\mathbf{p}|$ is the relative momentum, $r = |\mathbf{r}|$ is the relative separation, and the spin of the quarks is ignored. The potential is taken to be purely linear, because we are interested in the large-mass bound states with large relative angular momentum L, for which one expects that only the long-distance part of $V(r)$ is important.

For such states with large quantum number L the classical approximation should be reasonable. Hence, consider the classical Hamilton equations,

$$\frac{dr_k}{dt} = \frac{\partial H}{\partial p_k}, \quad \frac{dp_k}{dt} = -\frac{\partial H}{\partial r_k}. \tag{1.7}$$

and the following Ansatz for a circular solution:

$$r_1 = a\cos(\omega t), \quad r_2 = a\sin(\omega t), \quad r_3 = 0,$$
$$p_1 = -b\sin(\omega t), \quad p_2 = b\cos(\omega t), \quad p_3 = 0. \tag{1.8}$$

Substituting (1.8) into (1.7) we get relations among ω, a, and b, and expressions for p and r, which can be written in the form

$$p = b = \sigma\omega^{-1}, \quad r = a = 2s^{-1}\sigma^{-1}p, \quad s \equiv \sqrt{1 + m_q^2/p^2}, \qquad (1.9)$$

such that L and H can be written as

$$L = rp = 2s^{-1}\sigma^{-1}p^2, \quad H = 2(s + s^{-1})p. \qquad (1.10)$$

For $p^2 \gg m_q^2$, $s \approx 1$, $L \propto p^2$ and $H \propto p$. Then $L \propto H^2$ and, because $H = m$ is the mass (rest energy) of the bound state, we see that

$$\alpha' \equiv \left[LH^{-2} \right]_{p/m_q \to \infty} = (8\sigma)^{-1}. \qquad (1.11)$$

It turns out that L is approximately linear in H^2 even for quite small p^2, such that $L < 1$, as shown in figure 1.3. Of course, the classical approximation is suspect for L *not* much larger than unity, but the same phenomenon appears to take place quantum mechanically in nature, where the lower spin states are also near the straight line fitting the higher spin states.[2]

With $\alpha' = 1/8\sigma$, the experimental value $\alpha' \approx 0.90 \text{ GeV}^{-2}$ gives $\sqrt{\sigma} \approx$ 370 MeV. The effective string model (see e.g. [3] section 10.5) leads approximately to the same answer: $\alpha' = 1/2\pi\sigma$, giving $\sqrt{\sigma} \approx 420$ MeV. The string model is perhaps closer to reality if most of the bound-state energy is in the string-like chromoelectric field, but it should be kept in mind that both the string model and the effective Hamiltonian give only an approximate representation of QCD.

1.2 Scalar field

We start our exploration of field theory with the scalar field. Scalar fields $\varphi(x)$ $(x = (\mathbf{x}, t), t \equiv x^0)$ are used to describe spinless particles. Particles appearing elementary on one distance scale may turn out to be be composite bound states on a smaller distance scale. For example, protons, pions, etc. appear elementary on the scale of centimeters, but composed of quarks and gluons on much shorter distance scales. Similarly, fields may also be elementary or composite. For example, for the description of pions we may use elementary scalar fields $\varphi(x)$, or composite scalar fields of the schematic form $\bar{\psi}(x)\gamma_5\psi(x)$, where $\psi(x)$ and $\bar{\psi}(x)$ are quark fields and γ_5 is a Dirac matrix. Such composite fields can still be approximately represented by elementary $\varphi(x)$, which are then called effective fields. This is useful for the description of effective interactions, which are the result of more fundamental interactions on a shorter distance scale.

A basic tool in the description is the action $S = \int dt\, L$, with L the Lagrangian. For a nonrelativistic particle described by coordinates q_k, $k = 1, 2, 3$, the Lagrangian has the form kinetic energy minus potential energy, $L = \dot{q}_k \dot{q}_k / 2m - V(q)$.† For the anharmonic oscillator in three dimensions the potential has the form $V(q) = \omega^2 q^2 / 2 + \lambda (q^2)^2 / 4$, $q^2 \equiv q_k q_k$. In field theory a simple example is the action for the φ^4 theory,

$$S = \int_M d^4x\, \mathcal{L}(x), \quad d^4x = dx^0\, dx^1\, dx^2\, dx^3, \tag{1.12}$$

$$\mathcal{L}(x) = \tfrac{1}{2}\partial_t\varphi(x)\partial_t\varphi(x) - \tfrac{1}{2}\nabla\varphi(x)\cdot\nabla\varphi(x) - \tfrac{1}{2}\mu^2\varphi(x)^2 - \tfrac{1}{4}\lambda\varphi(x)^4, \tag{1.13}$$

Here M is a domain in space–time, $\varphi(x)$ is a scalar field, $\mathcal{L}(x)$ is the action density or Lagrange function, and λ and μ^2 are constants (λ is dimensionless and μ^2 has dimension $(\text{mass})^2 = (\text{length})^{-2}$). Note that the index $\mathbf{x}$ is a continuous analog of the discrete index k: $\varphi(\mathbf{x}, t) \leftrightarrow q_k(t)$.

Requiring the action to be stationary under variations $\delta\varphi(x)$ of $\varphi(x)$, such that $\delta\varphi(x) = 0$ for x on the boundary of M, leads to the equation of motion:

$$\delta S = \int d^4x \left[-\partial_t^2\varphi(x) + \nabla^2\varphi(x) - \mu^2\varphi(x) - \lambda\varphi(x)^3 \right] \delta\varphi(x)$$

$$= 0 \quad \Rightarrow \quad (\partial_t^2 - \nabla^2 + \mu^2)\varphi + \lambda\varphi^3 = 0. \tag{1.14}$$

In the first step we made a partial integration. In classical field theory the equations of motion are very important (e.g. Maxwell theory). In quantum field theory their importance depends very much on the problem and method of solution. The action itself comes more to the foreground, especially in the path-integral description of quantum theory.

Various states of the system can be characterized by the energy $H = \int d^3x\, \mathcal{H}$. The energy density has the form kinetic energy plus potential energy, and is given by

$$\mathcal{H} = \tfrac{1}{2}\dot{\varphi}^2 + \tfrac{1}{2}(\nabla\varphi)^2 + U, \tag{1.15}$$

$$U = \tfrac{1}{2}\mu^2\varphi^2 + \tfrac{1}{4}\lambda\varphi^4. \tag{1.16}$$

The field configuration with lowest energy is called the ground state. It has $\dot{\varphi} = \nabla\varphi = 0$ and minimal U. We shall assume $\lambda > 0$, such that $\mathcal{H}$ is

---

† Unless indicated otherwise, summation over repeated indices is implied, $\dot{q}_k \dot{q}_k \equiv \sum_k \dot{q}_k \dot{q}_k$.

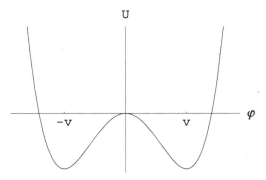

Fig. 1.4. The energy density for constant fields for $\mu^2 < 0$.

bounded from below for all φ. From a graph of $U(\varphi)$ (figure 1.4) we see that the cases $\mu^2 > 0$ and $\mu^2 < 0$ are qualitatively different:

$$\mu^2 > 0: \quad \varphi_g = 0, \quad U_g = 0;$$

$$\mu^2 < 0: \quad \varphi_g = \pm v, \quad v^2 = -\frac{\mu^2}{\lambda}, \quad U_g = -\frac{1}{4}\frac{\mu^2}{\lambda}. \tag{1.17}$$

So the case $\mu^2 < 0$ leads to a doubly degenerate ground state. In this case the symmetry of S or $\mathcal{H}$ under $\varphi(x) \to -\varphi(x)$ is broken, because a non-zero φ_g is not invariant, and one speaks of spontaneous (or dynamical) symmetry-breaking.

Small disturbances away from the ground state propagate and disperse in space and time in a characteristic way, which can be found by linearizing the equation of motion (1.14) around $\varphi = \varphi_g$. Writing $\varphi = \varphi_g + \varphi'$ and neglecting $O(\varphi'^2)$ gives

$$(\partial_t^2 - \nabla^2 + m^2)\varphi' = 0, \tag{1.18}$$

$$m^2 = U''(\varphi_g) = \begin{cases} \mu^2, & \mu^2 > 0; \\ \mu^2 + 3\lambda v^2 = -2\mu^2, & \mu^2 < 0. \end{cases} \tag{1.19}$$

Wavepacket solutions of (1.18) propagate with a group velocity $\mathbf{v} = \partial\omega/\partial\mathbf{k}$, where $\mathbf{k}$ is the average wave vector and $\omega = \sqrt{m^2 + \mathbf{k}^2}$. In the quantum theory these wavepackets are interpreted as particles with energy–momentum $(\omega, \mathbf{k})$ and mass m. The particles can scatter with an interaction strength characterized by the coupling constant λ. For $\lambda = 0$ there is no scattering and the field is called 'free'.

2

Path-integral and lattice regularization

In this chapter we introduce the path-integral method for quantum theory, make it precise with the lattice regularization and use it to quantize the scalar field. For a continuum treatment of path integrals in quantum field theory, see for example [8].

2.1 Path integral in quantum mechanics

To see how the path integral works, consider first a simple system with one degree of freedom described by the Lagrange function $L = L(q, \dot{q})$, or the corresponding Hamilton function $H = H(p, q)$,

$$L = \tfrac{1}{2}m\dot{q}^2 - V(q), \quad H = \frac{p^2}{2m} + V(q), \tag{2.1}$$

where p and q are related by $p = \partial L/\partial \dot{q} = m\dot{q}$. In the quantum theory p and q become operators $\hat{p}$ and $\hat{q}$ with $[\hat{q}, \hat{p}] = i\hbar$ (we indicate operators in Hilbert space by a caret $\hat{\ }$). The evolution in time is described by the operator

$$\hat{U}(t_1, t_2) = \exp[-i\hat{H}(t_1 - t_2)/\hbar], \tag{2.2}$$

with $\hat{H}$ the Hamilton operator, $\hat{H} = H(\hat{p}, \hat{q})$. Instead of working with q-numbers (operators) $\hat{p}$ and $\hat{q}$ we can also work with time dependent c-numbers (commuting numbers) $q(t)$, in the path-integral formalism. (Later we shall use anti-commuting numbers to incorporate Fermi–Dirac statistics.) In the coordinate basis $|q\rangle$ characterized by

$$\hat{q}|q\rangle = q|q\rangle, \tag{2.3}$$

$$\langle q'|q\rangle = \delta(q' - q), \quad \int dq\, |q\rangle\langle q| = 1, \tag{2.4}$$

8

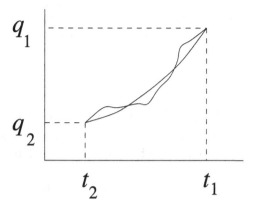

Fig. 2.1. Illustration of two functions $q(t)$ contributing to the path integral.

we can represent the matrix element of $\hat{U}(t_1, t_2)$ by a path integral

$$\langle q_1|\hat{U}(t_1, t_2)|q_2\rangle = \int Dq \, \exp[iS(q)/\hbar]. \tag{2.5}$$

Here S is the action functional of the system,

$$S(q) = \int_{t_2}^{t_1} dt \, L(q(t), \dot{q}(t)), \tag{2.6}$$

and $\int Dq$ symbolizes an integration over all functions $q(t)$ such that

$$q(t_1) = q_1, \quad q(t_2) = q_2, \tag{2.7}$$

as illustrated in figure 2.1. The path integral is a summation over all 'paths' ('trajectories', 'histories') $q(t)$ with given end points. The classical path, which satisfies the equation of motion $\delta S(q) = 0$, or

$$\frac{\partial L}{\partial q} - \frac{\partial}{\partial t}\frac{\partial L}{\partial \dot{q}} = 0, \tag{2.8}$$

is only one out of infinitely many possible paths. Each path has a 'weight' $\exp(iS/\hbar)$. If $\hbar$ is relatively small such that the phase $\exp(iS/\hbar)$ varies rapidly over the paths, then a stationary-phase approximation will be good, in which the classical path and its small neighborhood give the dominant contributions. The other extreme is when the variation of $S/\hbar$ is of order one. In the following we shall use again units in which $\hbar = 1$.

A formal definition of $\int Dq$ is given by

$$\int Dq = \prod_{t_2 < t < t_1} \int dq(t), \tag{2.9}$$

i.e. for every $t \in (t_2, t_1)$ we integrate over the domain of q, e.g. $-\infty < q < \infty$. The definition is formal because the continuous product $\prod_t$ still has to be defined. We shall give such a definition with the help of a discretization procedure.

2.2 Regularization by discretization

To define the path integral properly we discretize time in small units a, writing $t = na$, $q(t) = q_n$, with n integer. For a smooth function $q(t)$ the time derivative $\dot{q}(t)$ can be approximated by $\dot{q}(t) = (q_{n+1} - q_n)/a$, such that the discretized Lagrange function may be written as†

$$L(t) = \frac{m}{2a^2}(q_{n+1} - q_n)^2 - \frac{1}{2}V(q_{n+1}) - \frac{1}{2}V(q_n), \qquad (2.10)$$

where we have divided the potential term equally between q_n and q_{n+1}. We define a discretized evolution operator $\hat{T}$ by its matrix elements as follows:

$$\langle q_1 | \hat{T} | q_2 \rangle = c \exp\left\{ ia \left[\frac{m}{2a^2}(q_1 - q_2)^2 - \frac{1}{2}V(q_1) - \frac{1}{2}V(q_2) \right] \right\}, \qquad (2.11)$$

where c is a constant to be specified below. Note that the exponent is similar to the Lagrange function. The operator $\hat{T}$ is called the transfer operator, its matrix elements the transfer matrix. In terms of the transfer matrix we now give a precise definition of the discretized path integral:

$$\begin{aligned}
\langle q' | \hat{U}(t', t'') | q'' \rangle &= \int dq_1 \cdots dq_{N-1} \langle q' | \hat{T} | q_{N-1} \rangle \\
&\quad \times \langle q_{N-1} | \hat{T} | q_{N-2} \rangle \cdots \langle q_1 | \hat{T} | q'' \rangle \\
&= c \int \left(\prod c\, dq \right) \exp\left[\frac{im}{2a}(q' - q_{N-1})^2 \right. \\
&\quad - \frac{ia}{2}V(q') - iaV(q_{N-1}) + \frac{im}{2a}(q_{N-1} - q_{N-2})^2 \\
&\quad - iaV(q_{N-2}) + \cdots + \frac{im}{2a}(q_1 - q'')^2 - \left. \frac{ia}{2}V(q'') \right] \\
&\equiv \int Dq\, e^{iS}. \qquad (2.12)
\end{aligned}$$

Here the discretized action is defined by

$$S = a \sum_{n=0}^{N-1} L(na), \qquad (2.13)$$

† For notational simplicity we shall denote the discretized forms of L, S, ..., by the same symbols as their continuum counterparts.

where $q_N \equiv q'$ and $q_0 \equiv q''$. In the limit $N \to \infty$ this becomes equal to the continuum action, when we substitute smooth functions $q(t)$. Since the q_n are integrated over on every 'time slice' n, such smoothness is not typically present in the integrand of the path integral (typical paths q_n will look like having a very discontinuous derivative) and a continuum limit at this stage is formal.

It will now be shown that, with a suitable choice of the constant c, the transfer operator can be written in the form

$$\hat{T} = e^{-iaV(\hat{q})/2} \, e^{-ia\hat{p}^2/2m} \, e^{-iaV(\hat{q})/2}. \tag{2.14}$$

Taking matrix elements between $\langle q_1|$ and $|q_2\rangle$ we see that this formula is correct if

$$\langle q_1|e^{-ia\hat{p}^2/2m}|q_2\rangle = ce^{im(q_1-q_2)^2/2a}. \tag{2.15}$$

Inserting eigenstates $|p\rangle$ of the momentum operator $\hat{p}$ using

$$\langle q|p\rangle = e^{ipq}, \quad \int \frac{dp}{2\pi} \, |p\rangle\langle p| = 1, \tag{2.16}$$

we find that (2.15) is true provided that we choose

$$c = \sqrt{\frac{m}{2\pi ia}} = \sqrt{\frac{m}{2\pi a}} e^{-i\pi/4}. \tag{2.17}$$

The transfer operator $\hat{T}$ is the product of three unitary operators, so we may write

$$\hat{T} = e^{-ia\hat{H}}. \tag{2.18}$$

This equation defines a Hermitian Hamiltonian operator $\hat{H}$ modulo $2\pi/a$. For matrix elements between eigenstates with energy $E \ll 2\pi/a$ the expansion

$$\hat{T} = 1 - ia\hat{H} + O(a^2) \tag{2.19}$$

leads to the identification

$$\hat{H} = \frac{\hat{p}^2}{2m} + V(\hat{q}) + O(a^2), \tag{2.20}$$

in which we recognize the usual Hamilton operator. It should be kept in mind though that, as an operator equation, the expansion (2.19) is formal: because $\hat{p}^2$ is an unbounded operator there may be matrix elements for which the expansion does not converge.

2.3 Analytic continuation to imaginary time

It is very useful in practice to make an analytic continuation to imaginary time according to the substitution $t \to -it$. This can be justified if the potential $V(q)$ is bounded from below, as is the case, for example, for the anharmonic oscillator

$$V(q) = \tfrac{1}{2}m\omega^2 q^2 + \tfrac{1}{4}\lambda q^4. \tag{2.21}$$

Consider the discretized path integral (2.12). The integration over the variables q_n continues to converge if we rotate a in the complex plane according to

$$a = |a|e^{-i\varphi}, \quad \varphi\colon 0 \to \frac{\pi}{2}. \tag{2.22}$$

The reason is that, for all $\varphi \in (0, \pi/2]$, the real part of the exponent in (2.12) is negative:

$$\frac{i}{|a|e^{-i\varphi}} = \frac{1}{|a|}(-\sin\varphi + i\cos\varphi), \quad -i|a|e^{-i\varphi} = |a|(-\sin\varphi - i\cos\varphi). \tag{2.23}$$

The result of this analytic continuation in a is that the discretized path integral takes the form

$$\langle q'|\hat{U}_\Im(t', t'')|q''\rangle = |c| \int \left(\prod_n |c|\,dq_n\right) e^{S_\Im},$$

$$S_\Im = -|a| \sum_{n=0}^{N-1} \left[\frac{m}{2|a|^2}(q_{n+1} - q_n)^2 + \frac{1}{2}V(q_{n+1}) + \frac{1}{2}V(q_n)\right]. \tag{2.24}$$

Here the subscript $\Im$ denotes the imaginary-time versions of U and S.

The integrand in the imaginary-time path integral is real and bounded from above. This makes numerical calculations and theoretical analysis very much easier. Furthermore, in the generalization to field theory there is a direct connection to statistical physics, which has led to many fruitful developments. For most purposes the imaginary-time formulation is sufficient to extract the relevant physical information such as the energy spectrum of a theory. If necessary, one may analytically continue back to real time, by implementing the inverse of the rotation (2.22). (This can be done only in analytic calculations, since statistical errors in e.g. Monte Carlo computations have the tendency to blow up upon continuation.) In the following the subscript $\Im$ will be dropped and we will redefine $|a| \to a$, with a positive.

After transformation to imaginary time the transfer operator takes the Hermitian form

$$\hat{T} = e^{-aV(\hat{q})/2}\, e^{-a\hat{p}^2/2m}\, e^{-aV(\hat{q})/2}. \tag{2.25}$$

This is a positive operator, i.e. all its expectation values and hence all its eigenvalues are positive. We may therefore redefine the Hamiltonian operator $\hat{H}$ according to

$$\hat{T} = e^{-a\hat{H}}. \tag{2.26}$$

A natural object in the imaginary-time formalism is the partition function

$$Z = \text{Tr}\, e^{-\hat{H}(t_+ - t_-)} = \int dq\, \langle q| e^{-\hat{H}(t_+ - t_-)} |q\rangle = \text{Tr}\, \hat{T}^N, \tag{2.27}$$

where we think of t_+ (t_-) as the largest (smallest) time under consideration, with $t_+ - t_- = Na$. From quantum statistical mechanics we recognize that Z is the canonical partition function corresponding to the temperature

$$T = (t_+ - t_-)^{-1} \tag{2.28}$$

in units such that Boltzmann's constant $k_B = 1$. The path-integral representation of Z is obtained by setting in (2.24) $q_N = q_0 \equiv q$ ($q' = q'' \equiv q$) and integrating over q:

$$Z = \int_{\text{pbc}} Dq\, e^S. \tag{2.29}$$

Here 'pbc' indicates the fact that the integration is now over all discretized functions $q(t)$, $t_- < t < t_+$, with 'periodic boundary conditions' $q(t_+) = q(t_-)$.

2.4 Spectrum of the transfer operator

Creation and annihilation operators are familiar from the theory of the harmonic oscillator. Here we shall use them to derive the eigenvalue spectrum of the transfer operator of the harmonic oscillator, for which

$$V(q) = \tfrac{1}{2}m\omega^2 q^2. \tag{2.30}$$

For simplicity we shall use units in which $a = 1$ and $m = 1$, which may be obtained by transforming to variables $q' = q/a$, $p' = ap$, $m' = am$, and $\omega' = a\omega$, then to $q'' = q'\sqrt{m'}$ and $p'' = p'/\sqrt{m'}$, such that (omitting the primes) $[\hat{p}, \hat{q}] = -i$ and

$$\hat{T} = e^{-\omega^2\hat{q}^2/4}e^{-\hat{p}^2/2}e^{-\omega^2\hat{q}^2/4}. \tag{2.31}$$

Using the representation $\hat{q} \to q$, $\hat{p} \to -i\,\partial/\partial q$ or vice-versa one obtains the relation

$$\hat{T}\begin{pmatrix} \hat{p} \\ \hat{q} \end{pmatrix} = M\begin{pmatrix} \hat{p} \\ \hat{q} \end{pmatrix}\hat{T}, \tag{2.32}$$

where the matrix M is given by

$$M = \begin{pmatrix} 1 + \frac{1}{2}\omega^2 & i \\ -i(2 + \frac{1}{2}\omega^2)\frac{1}{2}\omega^2 & 1 + \frac{1}{2}\omega^2 \end{pmatrix}. \tag{2.33}$$

We want to find linear combinations $\kappa\hat{q} + \lambda\hat{p}$ such that

$$\hat{T}(\kappa\hat{q} + \lambda\hat{p}) = \mu(\kappa\hat{q} + \lambda\hat{p})\hat{T}, \tag{2.34}$$

from which it follows that (κ, λ) have to form an eigenvector of M^{T} (the transpose of M) with eigenvalue μ. The eigenvalues $\mu_\pm$ of M can be expressed as

$$\mu_\pm = e^{\pm\tilde{\omega}}, \quad \cosh\tilde{\omega} = 1 + \tfrac{1}{2}\omega^2, \tag{2.35}$$

and the linear combinations sought are given by

$$\hat{a} = \nu[\sinh(\tilde{\omega}\hat{q}) + i\hat{p}],$$
$$\hat{a}^\dagger = \nu[\sinh(\tilde{\omega}\hat{q}) - i\hat{p}], \tag{2.36}$$

where ν is a normalization constant. The $\hat{a}$ and $\hat{a}^\dagger$ are the annihilation and creation operators for the discretized harmonic oscillator. They satisfy the usual commutation relations

$$[\hat{a}, \hat{a}^\dagger] = 1, \quad [\hat{a}, \hat{a}] = [\hat{a}^\dagger, \hat{a}^\dagger] = 0, \tag{2.37}$$

provided that

$$\nu = \frac{1}{\sqrt{2\sinh\tilde{\omega}}}, \tag{2.38}$$

and furthermore

$$\hat{T}\hat{a} = e^{\tilde{\omega}}\,\hat{a}\hat{T}, \quad \hat{T}\hat{a}^\dagger = e^{-\tilde{\omega}}\,\hat{a}^\dagger\hat{T}. \tag{2.39}$$

The ground state $|0\rangle$ with the highest eigenvalue of $\hat{T}$ satisfies $\hat{a}|0\rangle = 0$, from which one finds (using for example the coordinate representation)

$$\langle q|0\rangle = e^{-\frac{1}{2}\sinh\tilde{\omega}\,q^2},$$
$$\hat{T}|0\rangle = e^{-E_0}|0\rangle,$$
$$E_0 = \tfrac{1}{2}\tilde{\omega}. \tag{2.40}$$

The ground-state energy is $E_0 = \tfrac{1}{2}\tilde{\omega}$ and using (2.39) one finds that the excitation energies occur in units of $\tilde{\omega}$, for example

$$\hat{T}\hat{a}^\dagger|0\rangle = e^{-\tilde{\omega}}\,\hat{a}^\dagger\hat{T}|0\rangle = e^{-(3/2)\,\tilde{\omega}}\hat{a}^\dagger|0\rangle. \tag{2.41}$$

Hence, the energy spectrum is given by

$$E_n = \left(n + \tfrac{1}{2}\right)\tilde{\omega}, \tag{2.42}$$

which looks familiar, except that $\tilde{\omega} \neq \omega$.

We now can take the continuum limit $a \to 0$ in the physical quantities E_n. Recalling that ω is really $a\omega$, and similarly for $\tilde{\omega}$, we see by expanding (2.35) in powers of a, i.e. $\cosh(a\tilde{\omega}) = 1 + a^2\tilde{\omega}^2/2 + a^4\tilde{\omega}^4/24 + \cdots = 1 + a^2\omega^2/2$, that

$$\tilde{\omega} = \omega + O(a^2). \tag{2.43}$$

Note that the corrections are $O(a^2)$, which is much better than $O(a)$ as might be expected naively. This is the reason for the symmetric division of the potential in (2.11).

2.5 Latticization of the scalar field

We now transcribe these ideas to field theory, taking the scalar field as the first example. The dynamical variables generalize as

$$q(t) \to \varphi(\mathbf{x}, t) \tag{2.44}$$

(i.e. there is a q for every $\mathbf{x}$). The coordinate representation is formally characterized by

$$\hat{\varphi}(\mathbf{x})|\varphi\rangle = \varphi(\mathbf{x})|\varphi\rangle, \tag{2.45}$$
$$|\varphi\rangle = \prod_{\mathbf{x}}|\varphi_{\mathbf{x}}\rangle, \tag{2.46}$$
$$\langle\varphi'|\varphi\rangle = \prod_{\mathbf{x}}\delta(\varphi'(\mathbf{x}) - \varphi(\mathbf{x})), \tag{2.47}$$
$$\prod_{\mathbf{x}}\int_{-\infty}^{\infty} d\varphi(\mathbf{x})\,|\varphi\rangle\langle\varphi| = 1. \tag{2.48}$$

The evolution operator is given by

$$\langle \varphi_1 | \hat{U}(t_1, t_2) | \varphi_2 \rangle = \int D\varphi \, e^{S(\varphi)}, \qquad (2.49)$$

where the integral is over all functions $\varphi(\mathbf{x}, t)$ with $\varphi(\mathbf{x}, t_{1,2}) = \varphi_{1,2}(\mathbf{x})$. The theory is specified furthermore by the choice of action S. For the standard φ^4 model

$$S(\varphi) = -\int_{t_2}^{t_1} dx_4 \int d^3x \left[\frac{1}{2} \partial_\mu \varphi(x) \partial_\mu \varphi(x) + \frac{\mu^2}{2} \varphi^2(x) + \frac{\lambda}{4} \varphi^4(x) \right], \qquad (2.50)$$

where $x = (\mathbf{x}, x_4)$ and $x_4 = t$. Note that in the imaginary-time formalism the symmetry between space and time is manifest, since the metric tensor is simply equal to the Kronecker $\delta_{\mu\nu}$. Consequently, we shall not distinguish between upper and lower indices $\mu, \nu, \dots$. One often speaks of the *Euclidean formalism*, since the space–time symmetries of the theory consist of Euclidean rotations, reflections and translations.

The partition function is given by

$$Z = \int D\varphi \, e^{S(\varphi)}, \qquad (2.51)$$

where the integral is over all functions periodic in the time direction, $\varphi(\mathbf{x}, t + \beta) = \varphi(\mathbf{x}, t)$, with $\beta = T^{-1}$ the inverse temperature.

The path integral Z will be given a precise definition with the lattice regularization, by a straightforward generalization of the example of quantum mechanics with one degree of freedom. Let x_μ be restricted to a four-dimensional hypercubic lattice,

$$x_\mu = m_\mu a, \quad m_\mu = 0, 1, \dots, N - 1, \qquad (2.52)$$

where a is the lattice distance. The size of the hypercubic box is $L = Na$ and its space–time volume is L^4. The notation

$$\sum_x \equiv a^4 \sum_{m_1=0}^{N-1} \cdots \sum_{m_4=0}^{N-1} \equiv a^4 \sum_m \qquad (2.53)$$

will be used in this book. For smooth functions $f(x)$ we have in the continuum limit

$$\sum_x f(x) \to \int_0^L d^4x \, f(x), \quad N \to \infty, \quad a = L/N \to 0, \quad L \text{ fixed.} \qquad (2.54)$$

We have put $x = 0$ at the edge of the box. If we want it in the middle of the box we can choose $m_\mu = -N/2 + 1, -N/2 + 2, \dots, N/2$. Below we shall choose such a labeling for Fourier modes and we shall assume N to be even in the following.

The scalar field on the lattice is assigned to the sites x, we write φ_x. The part of the action without derivatives is transcribed to the lattice as $\sum_x (\mu^2 \varphi_x^2/2 + \lambda \varphi_x^4/4)$.

Derivatives can be replaced by differences. We shall use the notation

$$\partial_\mu \varphi_x = \frac{1}{a}(\varphi_{x+a\hat{\mu}} - \varphi_x), \qquad (2.55)$$

$$\partial'_\mu \varphi_x = \frac{1}{a}(\varphi_x - \varphi_{x-a\hat{\mu}}), \qquad (2.56)$$

where $\hat{\mu}$ is a unit vector in the μ direction. For smooth functions $f(x)$,

$$\partial_\mu f(x), \ \partial'_\mu f(x) \to \frac{\partial}{\partial x_\mu} f(x), \quad a \to 0. \qquad (2.57)$$

It is convenient to use periodic boundary conditions (such that the lattice has no boundary), which are specified by

$$\varphi_{x+Na\hat{\mu}} = \varphi_x, \qquad (2.58)$$

and, for example,

$$\partial_4 \varphi_{\mathbf{x},(N-1)a} = \frac{1}{a}(\varphi_{\mathbf{x},0} - \varphi_{\mathbf{x},(N-1)a}). \qquad (2.59)$$

With periodic boundary conditions the derivative operators ∂_μ and ∂'_μ are related by 'partial summation' (the analog of partial integration)

$$\sum_x \varphi_{1x} \, \partial_\mu \varphi_{2x} = -\sum_x \partial'_\mu \varphi_{1x} \, \varphi_{2x}. \qquad (2.60)$$

In matrix notation,

$$\partial_\mu \varphi_x = (\partial_\mu)_{xy} \, \varphi_y, \qquad (2.61)$$

∂'_μ is minus the transpose of ∂_μ, $\partial'_\mu = -\partial_\mu^{\mathrm{T}}$:

$$(\partial_\mu)_{xy} = \frac{1}{a}(\delta_{x+a\hat{\mu},y} - \delta_{x,y}), \qquad (2.62)$$

$$(\partial'_\mu)_{xy} = \frac{1}{a}(\delta_{x,y} - \delta_{x-a\hat{\mu},y}) = -(\partial_\mu)_{yx} = -(\partial_\mu^{\mathrm{T}})_{xy}. \qquad (2.63)$$

After these preliminaries, the path integral will now be defined by

$$Z = \int D\varphi \, e^{S(\varphi)}, \qquad (2.64)$$

$$\int D\varphi = \prod_x \left(c \int_{-\infty}^{\infty} \right) d\varphi_x, \quad \prod_x \equiv \prod_{x \, m}, \qquad (2.65)$$

$$S(\varphi) = -\sum_x \left(\frac{1}{2} \partial_\mu \varphi_x \partial_\mu \varphi_x + \frac{\mu^2}{2} \varphi_x^2 + \frac{\lambda}{4} \varphi_x^4 \right), \qquad (2.66)$$

$$c = a/\sqrt{2\pi}. \qquad (2.67)$$

Note that $c\varphi$ is dimensionless. The dimension of φ follows from the requirement that the action S is dimensionless. In d space–time dimensions,

$$[\varphi] = a^{-(d-2)/2}, \quad c = a^{(d-2)/2}. \qquad (2.68)$$

The factor $1/\sqrt{2\pi}$ is an inessential convention, chosen such that there is no additional factor in the expression for the transfer operator (2.74) below, which would lead to an additional constant in the Hamiltonian (2.80).

The lattice action was chosen such that for smooth functions $f(x)$, $S(f) \to S_{\text{cont}}(f)$ in the *classical* continuum limit $a \to 0$. However, it is useful to keep in mind that typical field configurations φ_x contributing to the path integral are not smooth at all on the lattice scale. The previous sentence is meant in the following sense. The factor $Z^{-1} \exp S(\varphi)$ can be interpreted as a normalized probability distribution for an ensemble of field configurations φ_x. Drawing a typical field configuration φ from this ensemble, e.g. one generated by a computer with some Monte Carlo algorithm, one finds that it varies rather wildly from site to site on the lattice. This has the consequence that different discretizations (e.g. different discrete differentiation schemes) may lead to different answers for certain properties, although this should not be the case for physically observable properties. The discussion of continuum behavior in the *quantum* theory is a delicate matter, which involves concepts like renormalization, scaling and universality.

2.6 Transfer operator for the scalar field

The derivation of the transfer operator for the scalar field on the lattice follows the steps made earlier in the example with one degree of freedom. For later use we generalize to different lattice spacings for time and space, a_t and a, respectively. We use the notation $x_4 = t = na_t$, $\varphi_x = \varphi_{n,\mathbf{x}}$, with $n = 0, 1, \ldots, N-1$ and $\varphi_{N,\mathbf{x}} = \varphi_{0,\mathbf{x}}$. Then the action can be written as

$$S(\varphi) = -a_t \sum_n \sum_{\mathbf{x}} \frac{1}{2a_t^2} (\varphi_{n+1,\mathbf{x}} - \varphi_{n,\mathbf{x}})^2 - a_t \sum_n V(\varphi_n), \quad (2.69)$$

$$V(\varphi_n) = \sum_{\mathbf{x}} \left[\frac{1}{2} \sum_{j=1}^{3} \partial_j \varphi_{n,\mathbf{x}} \partial_j \varphi_{n,\mathbf{x}} + \frac{\mu^2}{2} \varphi_{n,\mathbf{x}}^2 + \frac{\lambda}{4} \varphi_{n,\mathbf{x}}^4 \right]. \qquad (2.70)$$

The transfer operator $\hat{T}$ is defined by its matrix elements

$$\langle \varphi_{n+1} | \hat{T} | \varphi_n \rangle = c^{N^3} \exp \left[-a_t \sum_{\mathbf{x}} \frac{1}{2a_t^2} (\varphi_{n+1,\mathbf{x}} - \varphi_{n,\mathbf{x}})^2 \right]$$

$$\times \exp \left[-a_t \frac{1}{2} (V(\varphi_{n+1}) + V(\varphi_n)) \right], \qquad (2.71)$$

such that

$$Z = \left(\prod_x \int d\varphi_x \right) \langle \varphi_N | \hat{T} | \varphi_{N-1} \rangle \cdots \langle \varphi_1 | \hat{T} | \varphi_0 \rangle \qquad (2.72)$$

$$= \operatorname{Tr} \hat{T}^N. \qquad (2.73)$$

The transfer operator $\hat{T}$ can be written in the form

$$\hat{T} = \exp \left[-a_t \frac{1}{2} V(\hat{\varphi}) \right] \exp \left[-a_t \frac{1}{2} \sum_{\mathbf{x}} \hat{\pi}_{\mathbf{x}}^2 \right] \exp \left[-a_t \frac{1}{2} V(\hat{\varphi}) \right], \qquad (2.74)$$

where $\hat{\pi}_{\mathbf{x}}$ is the canonical conjugate operator of $\hat{\varphi}_{\mathbf{x}}$, with the property

$$[\hat{\varphi}_{\mathbf{x}}, \hat{\pi}_{\mathbf{y}}] = i a^{-3} \delta_{\mathbf{x},\mathbf{y}}. \qquad (2.75)$$

To check (2.74) we take matrix elements between $|\varphi_n\rangle$ and $\langle \varphi_{n+1}|$ and compare with (2.71). Using

$$e^{-a_t \frac{1}{2} V(\hat{\varphi})} |\varphi_n\rangle = e^{-a_t \frac{1}{2} V(\varphi_n)} |\varphi_n\rangle, \qquad (2.76)$$

we see that (2.74) is correct provided that

$$\langle \varphi_{n+1} | e^{-a_t \frac{1}{2} \sum_{\mathbf{x}} \hat{\pi}_{\mathbf{x}}^2} |\varphi_n\rangle = c^{N^3} \exp \left[-a_t \sum_{\mathbf{x}} (\varphi_{n+1,\mathbf{x}} - \varphi_{n,\mathbf{x}})^2 / 2a_t^2 \right]. \qquad (2.77)$$

This relation is just a product over $\mathbf{x}$ of relations of the one-degree-of-freedom type

$$\langle q_1 | e^{-\hat{p}^2/2\xi} | q_2 \rangle = \sqrt{\frac{\xi}{2\pi}} e^{-\xi(q_1 - q_2)^2/2}, \qquad (2.78)$$

with the identification, for given $\mathbf{x}$, $q = a\varphi$, $\hat{p} = a^2 \hat{\pi} \to -i \partial/\partial q$, and $|\varphi\rangle = \sqrt{a} |q\rangle$ (such that $\langle \varphi' | \varphi \rangle = a \langle q' | q \rangle = a\delta(q' - q) = \delta(\varphi' - \varphi)$). It follows that

$$c = a \sqrt{\frac{\xi}{2\pi}}, \quad \xi = \frac{a}{a_t}. \qquad (2.79)$$

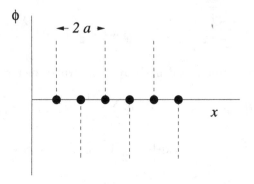

Fig. 2.2. Shortest wave length of a lattice field.

Making the formal continuous-time limit by letting $a_t \to 0$ and expanding $\hat{T} = 1 - a_t \hat{H} + \cdots$, we find a conventional-looking Hamiltonian[1] on a spatial lattice,

$$\hat{H} = \sum_x \left(\tfrac{1}{2}\hat{\pi}_{\mathbf{x}}^2 + \tfrac{1}{2}\partial_j\hat{\varphi}_{\mathbf{x}}\partial_j\hat{\varphi}_{\mathbf{x}} + \tfrac{1}{2}\mu^2\hat{\varphi}_{\mathbf{x}}^2 + \tfrac{1}{4}\lambda\hat{\varphi}_{\mathbf{x}}^4 \right) + O(a^2). \qquad (2.80)$$

2.7 Fourier transformation on the lattice

We record now some frequently used formulas involving the Fourier transform. The usual plane waves in a finite volume with periodic boundary conditions are given by

$$e^{ipx}, \quad p_\mu = n_\mu \frac{2\pi}{L}, \qquad (2.81)$$

where the n_μ are integers. We want to use these functions for (Fourier) transformations of variables. On the lattice the x_μ are restricted to $x_\mu = m_\mu a$, $m_\mu = 0, \ldots, N-1$, $L = Na$. There should not be more p_μ than x_μ; we take

$$n_\mu = -N/2 + 1, -N/2 + 2, \ldots, N/2. \qquad (2.82)$$

Indeed, the shortest wave length and largest wave vector are given by (cf. figure 2.2)

$$\lambda_{\min} = 2a, \quad p_{\max} = \frac{\pi}{a} = \frac{N}{2}\frac{2\pi}{L}. \qquad (2.83)$$

Apart from these intuitive arguments, the reason for (2.82) is the fact that (in d dimensions, $mn = m_\mu n_\mu$)

$$U_{mn} \equiv N^{-d/2} e^{i2\pi mn/N} \equiv N^{-d/2} \left(e^{ipx} \right)_{mn} \qquad (2.84)$$

is a unitary matrix,

$$U_{mn} U_{mn'}^* = \delta_{n,n'}. \qquad (2.85)$$

We check this for the one-dimensional case, $d = 1$:

$$U_{mn} U_{mn'}^* = \frac{1}{N} \sum_{m=0}^{N-1} r^m = \frac{1}{N} \frac{1 - r^N}{1 - r} = \bar\delta_{n,n'}, \qquad (2.86)$$

$$r \equiv e^{i2\pi(n-n')/N}, \qquad (2.87)$$

where

$$\bar\delta_{n,n'} \equiv 0, \quad n \neq n' \mod N \qquad (2.88)$$

$$= 1, \quad n = n' \mod N. \qquad (2.89)$$

We shall use this result in the form

$$\sum_x e^{-i(p-p')x} = \bar\delta_{p,p'} \equiv \prod_\mu \left(N|a_\mu| \bar\delta_{m_\mu,m'_\mu} \right), \qquad (2.90)$$

$$\sum_p e^{ip(x-x')} = \bar\delta_{x,x'} \equiv \prod_\mu \left(|a_\mu|^{-1} \bar\delta_{n_\mu,n'_\mu} \right), \qquad (2.91)$$

$$\sum_x \equiv \prod_\mu \left(|a_\mu| \sum_{m_\mu} \right), \qquad (2.92)$$

$$\sum_p \equiv \prod_\mu \left(\frac{1}{N|a_\mu|} \sum_{n_\mu} \right), \qquad (2.93)$$

where $|a_\mu|$ is the lattice spacing in the μ direction (unless stated otherwise, $|a_\mu| = a$). With this notation we can write the Fourier transformation of variables ('from position space to momentum space') and its inverse as

$$\tilde\varphi_p = \sum_x e^{-ipx} \varphi_x, \qquad (2.94)$$

$$\varphi_x = \sum_p e^{ipx} \tilde\varphi_p. \qquad (2.95)$$

For smooth functions $f(p)$ we have, in the infinite-volume limit $L = Na \to \infty$,

$$\sum_p f(p) = \frac{(\Delta p)^4}{(2\pi)^4} \sum_n f\left(\frac{2\pi n}{Na}\right) \tag{2.96}$$

$$\to \int_{-\pi/a}^{\pi/a} \frac{d^4 p}{(2\pi)^4} f(p), \quad N \to \infty, \quad a \text{ fixed}, \tag{2.97}$$

where $\Delta p = 2\pi/Na$.

2.8 Free scalar field

For $\lambda = 0$ we get the free scalar field action. For this case the path integral can be done easily. Assuming $\mu^2 \equiv m^2 > 0$, we write

$$S = -\sum_x \left(\tfrac{1}{2}\partial_\mu\varphi_x \partial_\mu\varphi_x + \tfrac{1}{2}m^2\varphi_x^2\right) \tag{2.98}$$

$$= \tfrac{1}{2}\sum_{xy} S_{xy}\,\varphi_x\varphi_y, \tag{2.99}$$

where

$$S_{xy} = -\sum_z \left[\sum_\mu (\bar\delta_{z+a\hat\mu,x} - \bar\delta_{z,x})(\bar\delta_{z+a\hat\mu,y} - \bar\delta_{z,y}) + m^2\bar\delta_{z,x}\bar\delta_{z,y}\right]. \tag{2.100}$$

It is useful to introduce an external source J_x, which can be chosen as we wish. The partition function with an external source is defined as

$$Z(J) = \int D\varphi \, \exp\left(S + \sum_x J_x\varphi_x\right). \tag{2.101}$$

The transformation of variables

$$\varphi_x \to \varphi_x + \sum_y G_{xy}J_y, \tag{2.102}$$

with G_{xy} minus the inverse of S_{xy},

$$S_{xy}G_{yz} = -\bar\delta_{x,z}, \tag{2.103}$$

brings $Z(J)$ into the form

$$Z(J) = Z(0) \, \exp\left(\tfrac{1}{2}G_{xy}J_xJ_y\right). \tag{2.104}$$

The integral $Z(0)$ is just a multiple Gaussian integral,

$$Z(0) = \int D\varphi \, \exp\left(-\tfrac{1}{2}G_{xy}^{-1}\varphi_x\varphi_y\right)$$

$$= \frac{1}{\sqrt{\det G^{-1}}} = \exp\left(\tfrac{1}{2}\ln\det G\right). \tag{2.105}$$

There is finite-temperature physics that can be extracted from the partition function $Z(0)$, but here we shall not pay attention to it.

The propagator G can be easily found in 'momentum space'. First we determine the Fourier transform of S_{xy}, using lattice units $a = 1$,

$$S_{p,-q} \equiv \sum_{xy} e^{-ipx+iqy} S_{xy} \tag{2.106}$$

$$= -\sum_z \left[\sum_\mu (e^{-ip\hat\mu} - 1)(e^{iq\hat\mu} - 1) + m^2\right] e^{-ipz+iqz}$$

$$= S_p \,\bar\delta_{p,q}, \tag{2.107}$$

$$-S_p = m^2 + \sum_\mu (2 - 2\cos p_\mu) \tag{2.108}$$

$$= m^2 + \sum_\mu 4\sin^2\left(\frac{p_\mu}{2}\right). \tag{2.109}$$

Since $S_{p,-q}$ is diagonal in momentum space, its inverse is given by

$$G_{p,-q} = G_p \,\bar\delta_{p,q}, \quad G_p = \frac{1}{m^2 + \sum_\mu (2 - 2\cos p_\mu)}. \tag{2.110}$$

From this we can restore the lattice distance by using dimensional analysis: $p \to ap$, $m \to am$ and $G_p \to a^2 G(p)$. This gives

$$G(p) = \frac{1}{m^2 + a^{-2}\sum_\mu (2 - 2\cos ap_\mu)}, \tag{2.111}$$

and in the continuum limit $a \to 0$,

$$G(p) = \frac{1}{m^2 + p^2 + O(a^2)}, \tag{2.112}$$

which is the usual covariant expression for the scalar field propagator. It is instructive to check that the corrections to the continuum form are already quite small for $ap_\mu < \tfrac{1}{2}$.

From the form (2.104) we calculate the correlation function of the free theory,

$$\langle \varphi_x \rangle = \left[\frac{\partial \ln Z}{\partial J_x} \right]_{J=0} = 0, \tag{2.113}$$

$$\langle \varphi_x \varphi_y \rangle = \left[\frac{\partial \ln Z}{\partial J_x \, \partial J_y} \right]_{J=0} = G_{xy}. \tag{2.114}$$

Hence, the propagator G is the correlation function of the system (cf. problem (v)).

We now calculate the time dependence of G_{xy}, assuming that the temporal extent of the lattice is infinite (zero temperature),

$$G(x - y) \equiv G_{xy} = \int_{-\pi}^{\pi} \frac{dp_4}{2\pi} \sum_{\mathbf{p}} e^{ip(x-y)} G(p), \tag{2.115}$$

$$G(\mathbf{x}, t) = \sum_{\mathbf{p}} e^{i\mathbf{p}\mathbf{x}} \int_{-\pi}^{\pi} \frac{dp_4}{2\pi} \frac{e^{ip_4 t}}{2b - 2\cos p_4}, \tag{2.116}$$

$$b = 1 + \frac{1}{2} \left(m^2 + \sum_{j=1}^{3} 4\sin^2 \frac{p_j}{2} \right), \tag{2.117}$$

where we reverted to lattice units. The integral over p_4 can be done by contour integration. We shall take $t > 0$. Note that, in lattice units, t is an integer and $b > 1$. With $z = \exp(ip_4)$ we have

$$I \equiv \int_{-\pi}^{\pi} \frac{dp_4}{2\pi} \frac{e^{ip_4 t}}{2b - 2\cos p_4} \tag{2.118}$$

$$= -\int \frac{dz}{2\pi i} \frac{z^t}{z^2 - 2bz + 1}, \tag{2.119}$$

where the integration is counter clockwise over the contour $|z| = 1$, see figure 2.3. The denominator has a pole at $z = z_-$ within the unit circle,

$$z^2 - 2bz + 1 = (z - z_+)(z - z_-),$$
$$z_\pm = b \pm \sqrt{b^2 - 1}, \quad z_+ z_- = 1, \quad z_+ > 1, \quad z_- < 1,$$
$$z_- = e^{-\omega}, \quad \cosh \omega = b, \quad \omega = \ln \left(b + \sqrt{b^2 - 1} \right). \tag{2.120}$$

The residue at $z = z_-$ gives

$$I = \frac{e^{-\omega t}}{2 \sinh \omega}, \tag{2.121}$$

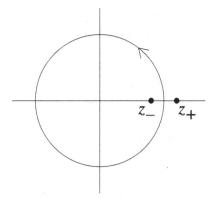

Fig. 2.3. Integration contour in the complex $z = e^{ip_4}$ plane. The crosses indicate the positions of the poles at $z = z_\pm$.

and it follows that (ω depends on $\mathbf{p}$)

$$G(\mathbf{x}, t) = \sum_{\mathbf{p}} \frac{e^{i\mathbf{px} - \omega t}}{2 \sinh \omega}. \tag{2.122}$$

Notice that the pole $z = z_-$ corresponds to a pole in the variable p_4 at $p_4 = i\omega$.

In the continuum limit ($m \to am$, $p_j \to ap_j$, $\omega \to a\omega$, $a \to 0$) we get the familiar Lorentz covariant expression,

$$\omega \to \sqrt{m^2 + \mathbf{p}^2}. \tag{2.123}$$

The form (2.122) is a sum of exponentials $\exp(-\omega t)$. For large t the exponential with the smallest ω, $\omega = m$, dominates,

$$G \propto e^{-mt}, \quad t \to \infty, \tag{2.124}$$

and we see that the correlation length of the system is $1/m$.

2.9 Particle interpretation

The free scalar field is just a collection of harmonic oscillators, which are coupled by the gradient term $\partial_j \varphi \partial_j \varphi$ in the action or Hamiltonian (2.80). We can diagonalize the transfer operator explicitly by taking similar steps to those for the harmonic oscillator. One then finds (cf. problem (iii)) creation and annihilation operators $\hat{a}_{\mathbf{p}}^\dagger$ and $\hat{a}_{\mathbf{p}}$, which are indexed by the Fourier label $\mathbf{p}$. The ground state $|0\rangle$ has the property $\hat{a}_{\mathbf{p}}|0\rangle = 0$ with energy $E_0 = \sum_{\mathbf{p}} \omega_{\mathbf{p}}/2$. The elementary excitations $|\mathbf{p}\rangle =$

$\hat{a}_{\mathbf{p}}^{\dagger}|0\rangle$ are interpreted as particles with momentum $\mathbf{p}$ and energy $\omega_{\mathbf{p}}$. This interpretation is guided by the fact that these states are eigenstates of the translation operators in space and time, namely, $\exp(\hat{H}t)$ (eigenvalue $\exp[(\omega_{\mathbf{p}} + E_0)t]$), and the spatial translation operator $\hat{U}_{\mathbf{x}}$ (eigenvalue $\exp(-i\mathbf{p}\mathbf{x})$, see problem (vii) for its definition).

In the continuum limit we recover the relativistic energy–momentum relation $\omega(\mathbf{p}) = \sqrt{m^2 + \mathbf{p}^2}$. The mass (rest energy) of the particles is evidently m. They have spin zero because they correspond to a scalar field under rotations (there are no further quantum numbers to characterize their state). They are *bosons* because the (basis) states are symmetric in interchange of labels: $|\mathbf{p}_1\mathbf{p}_2\rangle \equiv \hat{a}_{\mathbf{p}_1}^{\dagger}\hat{a}_{\mathbf{p}_2}^{\dagger}||0\rangle = \hat{a}_{\mathbf{p}_2}^{\dagger}\hat{a}_{\mathbf{p}_1}^{\dagger}||0\rangle = |\mathbf{p}_2\mathbf{p}_1\rangle$. The ground state is usually called 'the vacuum'.

For interacting fields the above creation and annihilation operators no longer commute with the Hamiltonian – they are said to create 'bare' particles. The 'dressed' particle states are the eigenstates of the Hamiltonian, but only the single-particle states have the simple free energy–momentum relation $\omega(\mathbf{p}) = \sqrt{m^2 + \mathbf{p}^2}$. Multi-particle states have in general interaction energy, unless the particles (i.e. their wavepackets) are far apart.

Using the spectral representation (cf. problem (viii))

$$\langle\varphi_x\varphi_y\rangle - \langle\varphi\rangle^2 = \sum_{\mathbf{p},\gamma\neq0} |\langle0|\hat{\varphi}_0|\mathbf{p},\gamma\rangle|^2 \exp[-\omega_{\mathbf{p},\gamma}|x_4 - y_4| + i\mathbf{p}(\mathbf{x} - \mathbf{y})],$$

$$\omega_{\mathbf{p},\gamma} = E_{\mathbf{p},\gamma} - E_0, \tag{2.125}$$

the particle properties can still be deduced from the correlation functions, e.g. by studying their behavior at large time differences, for which the states with lowest excitation energies (i.e. the particles) $\omega_{\mathbf{p}}$ dominate. Alternatively, one can diagonalize the transfer operator by variational methods.

These methods are very general and also apply to confining theories such as QCD. The quantum numbers of the particles excited by the fields out of the vacuum match those of the fields chosen in the correlation functions.

2.10 Back to real time

In (2.22) we analytically continued the lattice distance in the time direction a_t to imaginary values. If we want to go back to real time we have to keep track of a_t. For instance, the action (2.70) may be rewritten in lattice units as

$$S = -\xi \sum_x \tfrac{1}{2} \partial_4 \varphi_x \partial_4 \varphi_x$$

$$- \frac{1}{\xi} \sum_x \left[\frac{1}{2} \partial_j \varphi_x \partial_j \varphi_x + \frac{m^2}{2} \varphi_x^2 + \frac{\lambda}{4} \varphi^4 \right], \quad \xi = \frac{a}{a_t}, \quad (2.126)$$

which leads to the correlation function in momentum space

$$G_p = \frac{\xi}{m^2 + \sum_j (2 - 2\cos p_j) + \xi^2 (2 - 2\cos p_4)}. \quad (2.127)$$

We have to realize that the symbol a_t in the Euclidean notation was really $|a_t|$ (cf. below (2.24)) and that $|a_t| = i a_t = i |a_t| \exp(-i\varphi)$, $\varphi = \pi/2$, according to (2.22). Hence, restoring the φ dependence of ξ means

$$\xi \to |\xi| (-i e^{i\varphi}). \quad (2.128)$$

Rotating back to real time, we keep φ infinitesimally positive in order to avoid singularities in G_p, $\varphi: \pi/2 \to \epsilon$, $\epsilon > 0$ infinitesimal. This gives

$$G \to \frac{-i|\xi|}{m^2 + \sum_j (2 - 2\cos p_j) - |\xi|^2 (2 - 2\cos p_4) - i\epsilon}, \quad (2.129)$$

where we freely rescaled the infinitesimal ϵ by positive values, $[-i \exp(i\epsilon)]^2 = (-i + \epsilon)^2 = -1 - i\epsilon$; $(2 - 2\cos p_4)$ is also positive.

In the continuum limit $m \to am$, $p_j \to ap_j$, $p_4 \to |\xi|^{-1} a p_4$, $G_p \to a^{-2} \xi G(p)$, $a \to 0$ we obtain the Feynman propagator

$$G(p) \to \frac{-i}{m^2 + \mathbf{p}^2 - p_4^2 - i\epsilon} \equiv -i G_{\mathrm{M}}(p). \quad (2.130)$$

In continuum language the rotation to imaginary time is usually called a Wick rotation:

$$x^0 \to -ix_4, \quad p^0 \to -ip_4, \quad p_0 \to ip_4, \quad (2.131)$$

where $-i$ is meant to represent the rotation $\exp(-i\varphi)$, $\varphi: 0 \to \pi/2$ in the complex plane. For instance, one continues the Minkowski space propagator to the Euclidean-space correlation function

$$G_{\mathrm{M}}(x) = \int \frac{dp_0 \, d^3p}{(2\pi)^4} \frac{e^{ipx}}{m^2 + \mathbf{p}^2 - p_0^2 - i\epsilon} \quad (2.132)$$

$$\to i \int \frac{dp_4 \, d^3p}{(2\pi)^4} \frac{e^{ipx}}{m^2 + \mathbf{p}^2 + p_4^2} \quad (2.133)$$

$$= iG(x), \quad (2.134)$$

without encountering the singularities at $p_0 = \pm\sqrt{m^2 + \mathbf{p}^2} \mp i\epsilon$. Notice that $\exp(ipx)$ is invariant under the rotation: $\sum_{\mu=0}^{3} p_\mu x^\mu \to \sum_{\mu=1}^{4} p_\mu x_\mu$.

In (2.130) the timelike momentum is still denoted by p_4 instead of p_0, because the p_μ (and x_μ) in lattice units are just dummy indices denoting lattice points. The actual values of G in the scaling region $|x| \gg a$ are the same as in the continuum.

2.11 Problems

We use lattice units $a = 1$ unless indicated otherwise.

(i) *Restoration of rotation invariance*

Consider the free scalar field propagator in two dimensions

$$G_{xy} = \int \frac{d^2 p}{(2\pi)^2} \frac{e^{ip(x-y)}}{m^2 + 4 - 2\cos p_1 - 2\cos p_2}. \tag{2.135}$$

Let $x - y \to \infty$ along a lattice direction, or along the diagonal: $x - y = nt$, $t \to \infty$, $n = (1,0)$ or $n = (1,1)/\sqrt{2}$. The correlation length $\xi(n)$ in direction n is identified by $G \propto \exp(-t/\xi(n))$. Use the saddle-point technique to show that, along a lattice direction,

$$\xi^{-1} = \omega, \quad \cosh\omega = 1 + m^2/2, \tag{2.136}$$

whereas along the diagonal

$$\xi'^{-1} = \sqrt{2}\omega', \quad \cosh\omega' = 1 + m^2/4. \tag{2.137}$$

Discuss the cases $m \ll 1$ and $m \gg 1$. In particular show that in the first case

$$\xi'/\xi = 1 - m^2/48 + O(m^4). \tag{2.138}$$

In non-lattice units $m \to am$, and we see restoration of rotation invariance, $\xi'/\xi \to 1$ as $a \to 0$. Corrections are of order $a^2 m^2$.

(ii) *Real form of the Fourier transform*

Consider for simplicity one spatial dimension. Since φ_x is real, $\tilde{\varphi}_p^* = \tilde{\varphi}_{-p}$. Let $\tilde{\varphi}_p = \tilde{\varphi}_p' + i\tilde{\varphi}_p''$. The real and imaginary parts $\tilde{\varphi}_p'$ and $\tilde{\varphi}_p''$ satisfy $\tilde{\varphi}_p' = \tilde{\varphi}_{-p}'$ and $\tilde{\varphi}_p'' = -\tilde{\varphi}_{-p}''$. The $\tilde{\varphi}_p'$, $p \geq 0$, and $\tilde{\varphi}_{-p}''$, $p < 0$, may be considered independent variables equivalent to φ_x. Expressing φ_x in these variables gives the real form of the

Fourier transform, and the matrix O given by

$$O_{mn} = \frac{1}{\sqrt{N}}, \qquad\qquad n = 0,$$

$$= \sqrt{\frac{2}{N}} \cos\left(\frac{2\pi mn}{N}\right), \qquad n = 1, \ldots, \frac{N}{2} - 1,$$

$$= \frac{1}{\sqrt{N}}, \qquad\qquad n = \frac{N}{2},$$

$$= -\sqrt{\frac{2}{N}} \sin\left(\frac{2\pi mn}{N}\right), \qquad n = -\frac{N}{2} + 1, \ldots, -1,$$

$$\tag{2.139}$$

where $m = 0, \ldots, N-1$, is orthogonal: $O\,O^{\mathrm{T}} = \mathbb{1}$.

Similar considerations apply to canonical conjugate π_x and $\tilde{\pi}_p$. Verify that the operators $\hat{\tilde{\varphi}}_p$ and $\hat{\tilde{\pi}}_p$ satisfy the commutation relations

$$[\hat{\tilde{\varphi}}_p, \hat{\tilde{\pi}}_q^\dagger] = i\bar{\delta}_{p,q}, \quad [\hat{\tilde{\varphi}}_p, \hat{\tilde{\pi}}_q] = 0, \quad [\hat{\tilde{\varphi}}_p^\dagger, \hat{\tilde{\pi}}_q^\dagger] = 0, \quad [\hat{\tilde{\varphi}}_p^\dagger, \hat{\tilde{\pi}}_q] = i\bar{\delta}_{p,q},$$

$$\tag{2.140}$$

in addition to $[\hat{\tilde{\varphi}}_p, \hat{\tilde{\varphi}}_q] = [\hat{\tilde{\varphi}}_p, \hat{\tilde{\varphi}}_q^\dagger] = [\hat{\tilde{\pi}}_p, \hat{\tilde{\pi}}_q] = [\hat{\tilde{\pi}}_p, \hat{\tilde{\pi}}_q^\dagger] = 0.$

(iii) *Creation and annihilation operators*

For a free scalar field show that

$$\hat{T} = e^{-\sum_p m_p^2 |\hat{\tilde{\varphi}}_p|^2/4}\, e^{-\sum_p |\hat{\tilde{\pi}}_p|^2/2}\, e^{-\sum_p m_p^2 |\hat{\tilde{\varphi}}_p|^2/4},$$

$$m_p^2 = m^2 + 2(1 - \cos p), \tag{2.141}$$

where $|\hat{\tilde{\varphi}}_p|^2 = \hat{\tilde{\varphi}}_p^\dagger \hat{\tilde{\varphi}}_p$, etc. Hence, the transfer operator has the form $\hat{T} = \prod_p \hat{T}_p$.

Obtain the commutation relations of the creation $(\hat{a}_p^\dagger)$ and annihilation $(\hat{a}_p)$ operators defined by

$$\hat{a}_p = \sqrt{\frac{1}{2\sinh \omega_p}}\, [\sinh(\omega_p)\, \hat{\tilde{\varphi}}_p + i\hat{\tilde{\pi}}_p^\dagger]. \tag{2.142}$$

Using the results derived for the harmonic oscillator, show that the energy spectrum is given by

$$E = L \sum_p \left(N_p + \tfrac{1}{2}\right) \omega_p, \quad \cosh \omega_p = 1 + \tfrac{1}{2} m_p^2. \tag{2.143}$$

where N_p is the occupation number of the mode p (recall that in our notation $L \sum_p = \sum_n$, $p = 2\pi n/L$).

Verify that

$$\hat{\varphi}_x = \sum_p \sqrt{\frac{1}{2 \sinh \omega_p}} \left(e^{ipx} \hat{a}_p + e^{-ipx} \hat{a}_p^\dagger \right). \qquad (2.144)$$

(iv) *Ground-state wave functional*

For the free scalar field, write down the wave function for the ground state in the coordinate representation, $\Psi_0(\varphi) = \langle \varphi | 0 \rangle$.

(v) *Correlation functions*

We define expectation values

$$\langle \varphi_{x_1} \cdots \varphi_{x_n} \rangle = \frac{1}{Z(J)} \int D\varphi \, e^{S(\varphi) + J_x \varphi_x} \, \varphi_{x_1} \cdots \varphi_{x_n}, \qquad (2.145)$$

and correlation functions (connected expectation values)

$$G_{x_1 \cdots x_n} = \langle \varphi_{x_1} \cdots \varphi_{x_n} \rangle_{\text{conn}} = \frac{\partial}{\partial J_{x_1}} \cdots \frac{\partial}{\partial J_{x_1}} \ln Z(J). \qquad (2.146)$$

Verify that

$$G_x = \langle \varphi_x \rangle, \qquad (2.147)$$

$$G_{x_1 x_2} = \langle \varphi_{x_1} \varphi_{x_2} \rangle - \langle \varphi_{x_1} \rangle \langle \varphi_{x_2} \rangle. \qquad (2.148)$$

Give similar expressions for the three- and four-point functions $G_{x_1 x_2 x_3}$ and $G_{x_1 x_2 x_3 x_4}$. Note that $\langle \varphi_x \rangle$ may be non-zero in cases of spontaneous symmetry breaking even when $J_x = 0$.

(vi) *Ground-state expectation values of Heisenberg operators*

On an $L^3 \times \beta$ space–time lattice, verify that

$$\langle \varphi_x \varphi_y \rangle = \frac{\text{Tr} \, e^{-(\beta - x_4 + y_4) \hat{H}} \hat{\varphi}_{\mathbf{x}} \, e^{-(x_4 - y_4) \hat{H}} \hat{\varphi}_{\mathbf{y}}}{\text{Tr} \, e^{-\beta \hat{H}}}, \qquad (2.149)$$

where $x_4 > y_4$ and $J = 0$.

Let $|n\rangle$ be a complete set of energy eigenstates of the Hamiltonian,

$$\hat{H} |n\rangle = E_n |n\rangle. \qquad (2.150)$$

The ground state $|0\rangle$ has lowest energy, E_0. Show that for $\beta \to \infty$ (zero temperature)

$$\langle \varphi_x \varphi_y \rangle = \langle 0 | T \hat{\varphi}_x \hat{\varphi}_y | 0 \rangle, \qquad (2.151)$$

where T is the time ordering 'operator' and $\hat{\varphi}_x$ is the Heisenberg operator

$$\hat{\varphi}_{\mathbf{x}, x_4} = e^{x_4 \hat{H}} \hat{\varphi}_{\mathbf{x}, 0} e^{-x_4 \hat{H}}. \qquad (2.152)$$

(vii) *Translation operator*

The translation operator $\hat{U}_{\mathbf{x}}$ may be defined by

$$\hat{U}_{\mathbf{x}}|\varphi_{\mathbf{y}}\rangle = |\varphi_{\mathbf{x}-\mathbf{y}}\rangle, \tag{2.153}$$

with $|\varphi_{\mathbf{y}}\rangle$ the factor in the tensor product $|\varphi\rangle = \prod_{\mathbf{y}}|\varphi_{\mathbf{y}}\rangle$. This operator has the properties

$$\hat{U}_{\mathbf{x}}^{\dagger}\hat{\varphi}_{\mathbf{y}}\hat{U}_{\mathbf{x}} = \hat{\varphi}_{\mathbf{x}-\mathbf{y}}, \tag{2.154}$$

$$\hat{U}_{\mathbf{x}}^{\dagger}\hat{\pi}_{\mathbf{y}}\hat{U}_{\mathbf{x}} = \hat{\pi}_{\mathbf{x}-\mathbf{y}}, \tag{2.155}$$

such that the expectation value of e.g. $\hat{\varphi}_{\mathbf{x}}$ in an actively translated state $|\psi'\rangle \equiv \hat{U}_{\mathbf{z}}|\psi\rangle$ behaves in a way to be expected intuitively: $\langle\psi'|\hat{\varphi}_{\mathbf{x}}|\psi'\rangle = \langle\psi|\hat{\varphi}_{\mathbf{x}-\mathbf{z}}|\psi\rangle$.

Verify that for periodic boundary conditions the Hamiltonian is translation invariant,

$$\hat{U}_{\mathbf{x}}^{\dagger}\hat{H}\hat{U}_{\mathbf{x}} = \hat{H}. \tag{2.156}$$

(viii) *Spectral representation*

Let $|\mathbf{p}, \gamma\rangle$ be simultaneous eigenvectors of $\hat{H}$ and $\hat{U}_{\mathbf{x}}$ (γ is some label needed to specify the state in addition to $\mathbf{p}$),

$$\hat{H}|\mathbf{p}, \gamma\rangle = E_{\mathbf{p},\gamma}|\mathbf{p}, \gamma\rangle, \quad \hat{U}_{\mathbf{x}}|\mathbf{p}, \gamma\rangle = e^{-i\mathbf{p}\mathbf{x}}|\mathbf{p}, \gamma\rangle \tag{2.157}$$

Derive the spectral representation for zero temperature:

$$\langle\varphi_x\varphi_y\rangle - \langle\varphi\rangle^2 = \sum_{\mathbf{p},\gamma\neq 0} |\langle 0|\hat{\varphi}_0|\mathbf{p}, \gamma\rangle|^2$$
$$\times \exp[-\omega_{\mathbf{p},\gamma}|x_4 - y_4| + i\mathbf{p}(\mathbf{x} - \mathbf{y})],$$
$$\omega_{\mathbf{p},\gamma} = E_{\mathbf{p},\gamma} - E_0, \tag{2.158}$$

where $\gamma \neq 0$ indicates that the ground state is not included.

3

$O(n)$ models

In this chapter we study scalar field models with $O(n)$ symmetry described by the Euclidean action

$$S = -\int d^4x \left[\tfrac{1}{2}\partial_\mu\varphi^\alpha\partial_\mu\varphi^\alpha + \tfrac{1}{2}\mu^2\varphi^\alpha\varphi^\alpha + \tfrac{1}{4}\lambda(\varphi^\alpha\varphi^\alpha)^2\right], \qquad (3.1)$$

where $\varphi^\alpha(x) = \alpha = 0,\ldots,n-1$ is an n-vector in 'internal space'. The action is invariant under $O(n)$, the group of orthogonal transformations in n dimensions. For $n = 4$ this action describes the scalar Higgs sector of the Standard Model. It can also be used as an effective low-energy action for pions. Since the models are relatively simple they serve as a good arena for illustrating scaling and universality, concepts of fundamental importance in quantum field theory.

It turns out that scalar field models (in four dimensions) become 'trivial' in the sense that the interactions disappear very slowly when the lattice distance is taken to zero. The interpretation and implication of this interesting phenomenon will be also be discussed.

3.1 Goldstone bosons

We have seen in section 1.2 that the one-component classical scalar field (i.e. $n = 1$) can be in two different phases, depending on the sign of μ^2, namely a 'broken phase' in which the ground-state value $\varphi_g \neq 0$, and a 'symmetric phase' in which $\varphi_g = 0$. For $n > 1$ there are also two phases and we shall see that in the case of continuous internal symmetry the consequence of spontaneous symmetry breaking is the appearance of massless particles, called Goldstone bosons.†

---

† Actually, this is true in space–time dimensions $\geq$ 3. In one and two space–time dimensions spontaneous breaking of a continuous symmetry is not possible (Merwin–Wagner theorem, Coleman's theorem).

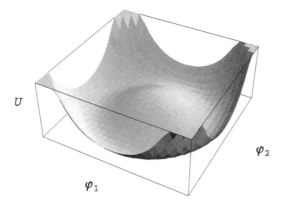

Fig. 3.1. Shape of U for $n = 2$ for $\mu^2 < 0$.

The potential

$$U = \tfrac{1}{2}\mu^2\varphi^2 + \tfrac{1}{4}\lambda(\varphi^2)^2 \qquad (3.2)$$

has a 'wine-bottle-bottom' shape, also called 'Mexican-hat' shape, if $\mu^2 < 0$ (figure 3.1). It is clear that for $\mu^2 > 0$ the ground state is unique ($\varphi_{\mathrm{g}} = 0$) but that for $\mu^2 < 0$ it is infinitely degenerate. The equation $\partial U/\partial\varphi^k = 0$ for the minima, $(\mu^2 + \lambda\varphi^2)\varphi^\alpha = 0$, has the solution

$$\varphi_{\mathrm{g}}^\alpha = v\delta_{\alpha,0}, \quad v^2 = -\mu^2/\lambda \quad (\mu^2 < 0), \qquad (3.3)$$

or any $O(n)$ rotation of this vector. To force the system into a definite ground state we add a symmetry-breaking term to the action (the same could be done in the one-component φ^4 model),

$$\Delta S = \int dx\, \epsilon\varphi^0(x), \quad \epsilon > 0. \qquad (3.4)$$

The constant ϵ has the dimension of (mass)^3. The equation for the stationary points now reads

$$(\mu^2 + \lambda\varphi^2)\varphi^\alpha = \epsilon\delta_{\alpha 0}. \qquad (3.5)$$

With the symmetry breaking (3.4) the ground state has $\varphi_{\mathrm{g}}^\alpha$ pointing in the $\alpha = 0$ direction,

$$\varphi_{\mathrm{g}}^\alpha = v\delta_{\alpha 0}, \quad (\mu^2 + \lambda v^2)v = \epsilon. \qquad (3.6)$$

Consider now small fluctuations about φ_{g}. The equations of motion (field equations) read

$$(-\partial^2 + \mu^2 + \lambda\varphi^2)\varphi^\alpha = \epsilon\delta_{\alpha 0}, \quad \partial^2 \equiv \nabla^2 - \partial_t^2. \qquad (3.7)$$

Linearizing around $\varphi = \varphi_{\mathrm{g}}$, writing

$$\varphi^0 = v + \sigma, \quad \varphi^k = \pi_k, \quad k = 1, \ldots, n-1, \qquad (3.8)$$

we find

$$(-\partial^2 + m_\sigma^2)\sigma = 0, \quad (-\partial^2 + m_\pi^2)\pi_k = 0, \qquad (3.9)$$

with

$$m_\sigma^2 = \mu^2 + 3\lambda v^2 = 2\lambda v^2 + \epsilon/v, \qquad (3.10)$$

$$m_\pi^2 = \mu^2 + \lambda v^2 = \epsilon/v. \qquad (3.11)$$

For $\mu^2 > 0$, $v = 0$ and $m_\sigma^2 = m_\pi^2 = \mu^2$, whereas for $\mu^2 < 0$, $v > 0$ and the σ particle is heavier than the π particles. For $\epsilon \to 0$ the π particles become *massless*,

$$m_\pi^2 \approx \epsilon/v_0 \to 0, \quad v_0 = v_{|\epsilon=0}. \qquad (3.12)$$

The simple effective $O(n)$ model reproduces the important features of Goldstone's theorem: spontaneous symmetry breaking of a continuous symmetry leads to massless particles, the Goldstone bosons. For small explicit symmetry breaking the Goldstone bosons get a squared mass proportional to the strength of the breaking. The massless modes correspond to oscillations along the vacuum valley of the 'Mexican hat'.

As mentioned earlier, the $O(4)$ model is a reasonable model for the effective low-energy interactions of pions amongst themselves. The particles $\pi^\pm$ and π^0 are described by the fields $\pi_k(x)$. The σ field (after which the model is named the σ model) corresponds to the very broad σ resonance around 900 MeV. The model loses its validity at such energies, for example the ρ mesons with mass 770 MeV are completely neglected.

3.2 $O(n)$ models as spin models

We continue in the quantum theory. The lattice regularized action will be taken as

$$S = -\sum_x \left[\tfrac{1}{2}\partial_\mu \varphi_x^\alpha \partial_\mu \varphi_x^\alpha + \tfrac{1}{2}m_0^2 \varphi_x^\alpha \varphi_x^\alpha + \tfrac{1}{4}\lambda_0 (\varphi_x^\alpha \varphi_x^\alpha)^2 \right]. \qquad (3.13)$$

We have changed the notation for the parameters: $\mu^2 \to m_0^2$, $\lambda \to \lambda_0$. The subscript 0 indicates that these are 'bare' or 'unrenormalized' parameters that differ from the physical 'dressed' or 'renormalized' values which are measured in experiments.

We shall mostly use lattice units, $a = 1$. Using $\partial_\mu \varphi_x^\alpha = \varphi_{x+\hat\mu}^\alpha - \varphi_x^\alpha$, the action can be rewritten in the form

$$S = \sum_{x\mu} \varphi_x^\alpha \varphi_{x+\hat\mu}^\alpha - \sum_x \left[\tfrac{1}{2}(2d + m_0^2)\varphi^2 + \tfrac{1}{4}\lambda_0(\varphi^2)^2 \right], \qquad (3.14)$$

where d is the number of space–time dimensions. Another standard choice of parameters is obtained by writing

$$\varphi^\alpha = \sqrt{2\kappa}\,\phi^\alpha, \quad m_0^2 = \frac{1 - 2\lambda}{\kappa} - 2d, \quad \lambda_0 = \frac{\lambda}{\kappa^2}, \qquad (3.15)$$

which brings S into the form

$$S = 2\kappa \sum_{x\mu} \phi_x^\alpha \phi_{x+\hat\mu}^\alpha - \sum_x \left[\phi_x^\alpha \phi_x^\alpha + \lambda(\phi_x^\alpha \phi_x^\alpha - 1)^2 \right]. \qquad (3.16)$$

The partition function is given by

$$Z = \left(\prod_{x\alpha} \int_{-\infty}^{\infty} d\phi_x^\alpha \right) \exp S \equiv \int D\mu(\phi) \exp\left(2\kappa \sum_{x\mu} \phi_x \phi_{x+\hat\mu} \right), \qquad (3.17)$$

where we have introduced an integration measure $D\mu(\phi)$, which is the product of probability measures $d\mu(\phi)$ for a single site,

$$D\mu(\phi) = \prod_x d\mu(\phi_x), \quad d\mu(\phi) = d^n\phi \, \exp[-\phi^2 - \lambda(\phi^2 - 1)^2]. \qquad (3.18)$$

Note that λ has to be positive in order that the integrations $\int d\mu(\phi)$ make sense.

The second form in (3.17) shows Z as the partition function of a generalized Ising model, a typical model studied in statistical physics. For $\lambda \to \infty$ the distribution $d\mu(\phi)$ peaks at $\phi^2 = 1$,

$$\frac{\int d\mu(\phi)\, f(\phi)}{\int d\mu(\phi)} \to \frac{\int d\Omega_n\, f(\phi)}{\int d\Omega_n}, \qquad (3.19)$$

where $\int d\Omega_n$ is the integral over the unit sphere S^n in n dimensions. In particular, for $n = 1$,

$$\frac{\int d\mu(\phi)\, f(\phi)}{\int d\mu(\phi)} \to \frac{1}{2}[f(1) + f(-1)]. \qquad (3.20)$$

Hence, for $n = 1$ and $\lambda \to \infty$ we get precisely the Ising model in d dimensions. For $n = 3$, $d = 3$ the model is called the Heisenberg model for a ferromagnet. The $O(n)$ models on the lattice are therefore also called (generalized) spin models.

3.3 Phase diagram and critical line

The spin model aspect makes it plausible that the models can be in a broken (ferromagnetic) or in a symmetric (paramagnetic) phase, such that in the thermodynamic limit and for zero temperature

$$\langle \phi_x^\alpha \rangle \equiv v^\alpha \neq 0, \quad \kappa > \kappa_c(\lambda), \tag{3.21}$$

$$= 0, \quad \kappa < \kappa_c(\lambda). \tag{3.22}$$

Here $\kappa_c(\lambda)$ is the boundary line between the two phases in the λ–κ plane.

We can give a mean-field estimate of κ_c as follows. Consider a site x. The probability for ϕ_x^α is proportional to $d\mu(\phi_x) \exp[2\kappa\phi_x^\alpha \sum_\mu (\phi_{x+\hat\mu}^\alpha + \phi_{x-\hat\mu}^\alpha)]$. Assume that we may approximate ϕ^α at the $2d$ neighbors of x by their average value, $\sum_\mu (\phi_{x+\hat\mu}^\alpha + \phi_{x-\hat\mu}^\alpha) \to 2dv^\alpha$. Then the average value of ϕ_x^α can be written as

$$\langle \phi_x^\alpha \rangle = \frac{\int d\mu(\phi)\, \phi^\alpha \exp(4\kappa d\phi^\beta v^\beta)}{\int d\mu(\phi) \exp(4\kappa d\phi^\beta v^\beta)}. \tag{3.23}$$

By consistency we should have $\langle \phi_x^\alpha \rangle = v^\alpha$, or

$$v^\alpha = \frac{1}{z(J)} \frac{\partial}{\partial J_\alpha} z(J)_{|J=4\kappa dv}, \tag{3.24}$$

$$z(J) = \int d\mu(\phi) \exp(J_\alpha \phi^\alpha). \tag{3.25}$$

The integral $z(J)$ can be calculated analytically in various limits, numerically otherwise. The basics are already illustrated by the Ising case $n = 1$, $\lambda = \infty$,

$$z(J) = z(0) \cosh(J), \tag{3.26}$$

$$v = \tanh(4\kappa dv), \quad n = 1, \lambda = \infty. \tag{3.27}$$

The equation for v can be analyzed graphically, see figure 3.2. As $\kappa \searrow \kappa_c$, evidently $v \to 0$. Then we can expand

$$v = \tanh(4\kappa dv) = 4\kappa dv - \tfrac{1}{3}(4\kappa dv)^3 + \cdots, \tag{3.28}$$

$$\kappa_c = \frac{1}{4d}, \tag{3.29}$$

$$v^2 \propto (\kappa - \kappa_c), \quad \kappa \searrow \kappa_c. \tag{3.30}$$

Analysis for arbitrary n and λ leads to similar conclusions,

$$z(J) = z(0)\langle 1 + \phi^\alpha J_\alpha + \tfrac{1}{2}\phi^\alpha \phi^\beta J_\alpha J_\beta + \cdots \rangle_1 \tag{3.31}$$

$$= z(0)\left[1 + \frac{1}{2}\frac{\langle \phi^2 \rangle_1}{n} J_\alpha J_\alpha + \cdots\right], \tag{3.32}$$

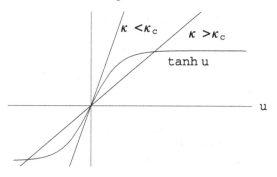

Fig. 3.2. Mean-field equation $u/4d\kappa = \tanh u$, $u = 4d\kappa v$, for $n = 1$, $\lambda = \infty$.

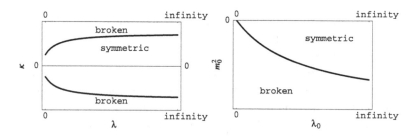

Fig. 3.3. Critical lines in the λ–κ plane and the m_0^2–λ_0 plane (qualitative).

where we used the notation

$$\langle F \rangle_1 = \frac{\int d\mu(\phi)\, F(\phi)}{\int d\mu(\phi)}, \tag{3.33}$$

and

$$\langle \phi^\alpha \phi^\beta \rangle_1 = \delta_{\alpha\beta} \frac{\langle \phi^2 \rangle_1}{n}, \tag{3.34}$$

for the one-site averages. So we find

$$\kappa_c(\lambda) = \frac{n}{4d\langle \phi^2 \rangle_1} = \frac{n}{4d}, \quad \lambda = \infty, \tag{3.35}$$

$$= \frac{1}{2d}, \quad \lambda = 0. \tag{3.36}$$

The behavior $v^2 \propto (\kappa - \kappa_c)$ is typical for a second-order phase transition in the mean-field approximation. The line $\kappa = \kappa_c(\lambda)$ is a critical line in parameter space where a second-order phase transition takes place. Note that in general m_0^2 is negative at the phase boundary (cf. (3.15) and figure 3.3). The critical exponent β in

$$v \propto (\kappa - \kappa_c)^\beta \tag{3.37}$$

differs in general from the mean-field value $\beta = \frac{1}{2}$. This is the subject of the theory of critical phenomena, and indeed, that theory is crucial for quantum fields. In four dimensions, however, it turns out that there are only small corrections to mean-field behavior.

We have restricted ourselves here to the region $\kappa > 0$. For $\kappa < 0$ the story more or less repeats itself, we then get an antiferromagnetic phase for $\kappa < -\kappa_c(\lambda)$. The region with negative κ can be mapped onto the region of positive κ by the transformation $\phi_x^\alpha \to (-1)^{x_1+x_2+\cdots+x_d}\phi_x^\alpha$.

It is important that the phase transition is of second order rather than, for example, of first order. In a second-order transition the correlation length diverges as a critical point is approached. The correlation length ξ can then be interpreted as the physical length scale and, when physical quantities are expressed in terms of ξ, the details on the scale of the lattice distance become irrelevant. The correlation length is defined in terms of the long-distance behavior of the correlation function,

$$G_{xy}^{\alpha\beta} \equiv \langle \phi_x^\alpha \phi_y^\beta \rangle - \langle \phi_x^\alpha \rangle \langle \phi_y^\beta \rangle \tag{3.38}$$

$$\propto |x - y|^{2-d-\eta} e^{-|x-y|/\xi}, \quad |x - y| \to \infty. \tag{3.39}$$

Here ξ may in principle depend on the direction we take $|x-y|$ to infinity, but the point is that it becomes independent of that direction (a lattice detail) as $\xi \to \infty$. In the symmetric phase ξ is independent of α and β. The exponent η is another critical exponent.

The correlation length is the inverse mass gap, the Compton wave length of the lightest particle, in lattice units,

$$\xi = 1/am. \tag{3.40}$$

This can be understood from the spectral representation

$$G_{xy}^{\alpha\beta} = \sum_{\mathbf{p},\gamma \neq 0} \langle 0 | \phi_\mathbf{0}^\alpha | \mathbf{p}\gamma \rangle \langle \mathbf{p}\gamma | \phi_\mathbf{0}^\beta | 0 \rangle e^{i\mathbf{p}(\mathbf{x}-\mathbf{y})-\omega_{\mathbf{p}\gamma}|x_4-y_4|}, \tag{3.41}$$

where $|0\rangle$ is the ground state (vacuum), $|\mathbf{p}\gamma\rangle$ are states with total momentum $\mathbf{p}$, distinguished by other quantum numbers γ, and $\omega_{\mathbf{p}\gamma} = E_{\mathbf{p},\gamma} - E_0$ is the difference in energy from the ground state. This representation is obtained by writing the path integral in terms of the transfer operator and its eigenstates in the limit of zero temperature, using translation invariance (cf. problem (viii) in chapter 2). Expression (3.41) is a sum of exponentials $\exp(-\omega t)$, $t = |x_4 - y_4|$. For large t the exponential with smallest ω dominates, $G \propto \exp(-\omega_{min} t)$, hence $\xi = 1/\omega_{min}$, with $\omega_{min} = m$ the minimum energy or mass gap.

In the broken phase we expect Goldstone bosons (section 3.1). If these are made sufficiently heavy by adding an explicit symmetry-breaking term $\sum_x \epsilon \varphi_x^n$ to the action (cf. equation (3.11)), we can expect two mass gaps: m_σ for the components of $G^{\alpha\beta}$ parallel to v^α and m_π for the components perpendicular to v^α. When the explicit symmetry breaking is diminished, $2m_\pi$ becomes less than m_σ and the σ particle becomes unstable, $\sigma \to 2\pi$. Then the large-distance behavior for the σ correlation function is controlled by $2m_\pi$ rather than by the mass m_σ of the unstable particle. Since m_π is expected to be zero in absence of explicit symmetry breaking, the transverse correlation length will be infinite in this case (for infinite volume).

The region near the phase boundary line where $\xi \gg 1$ is called the scaling region. In this region, at large distances $|x - y|$, the correlation function G_{xy} is expected to become a universal scaling function (independent of lattice details, with $1/m$ as the only relevant length scale rather than a).

3.4 Weak-coupling expansion

Expansion of the path-integral expectation value

$$\langle F(\varphi) \rangle = \frac{1}{Z} \int D\varphi \, e^{S(\varphi)} F(\varphi), \qquad (3.42)$$

$$S(\varphi) = -\sum_x \left[\frac{1}{2}\partial_\mu \varphi^\alpha \partial_\mu \varphi^\alpha + \frac{1}{2}m_0^2 \varphi^2 + \frac{1}{4}\lambda_0(\varphi^2)^2 \right], \qquad (3.43)$$

in powers of λ_0 leads to Feynman diagrams in terms of the free propagator and vertex functions. For simplicity we shall deal with the symmetric phase, which starts out with $m_0^2 > 0$ in the weak-coupling expansion. The free propagator is given by

$$^0G_{xy}^{\alpha\beta} = \delta_{\alpha\beta} \sum_p e^{ip(x-y)} \frac{1}{m_0^2 + \sum_\mu(2 - 2\cos p_\mu)}, \qquad (3.44)$$

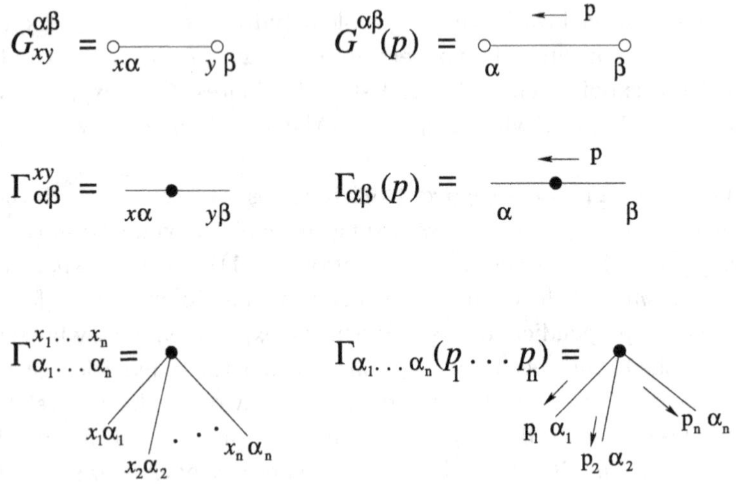

Fig. 3.4. Diagrams for ^0G, $^0\Gamma_{(2)}$ and $^0\Gamma_{(n)}$. Notice the convention of attaching a small circle at the end of external lines that represent propagators; without this $\circ$ the external line does not represent a propagator.

which is minus the inverse of the free second-order vertex function $\delta_{\alpha\beta}S_{xy}$ (recall (2.99) and (2.108)), which we shall denote here by $^0\Gamma_{(2)}$. In momentum space

$$^0\Gamma_{\alpha\beta}(p) = -\delta_{\alpha\beta}\left[m_0^2 + \sum_\mu (2 - 2\cos p_\mu)\right]. \qquad (3.45)$$

The bare (i.e. lowest-order) vertex functions $^0\Gamma_{(n)}$ are defined by the expansion of the action S around the classical minimum $\varphi_x^\alpha = v^\alpha$,

$$S(\varphi) = \sum_n \frac{1}{n!} {}^0\Gamma^{x_1\cdots x_n}_{\alpha_1\cdots\alpha_n}(\varphi_{x_1}^{\alpha_1} - v^{\alpha_1})\cdots(\varphi_{x_n}^{\alpha_n} - v^{\alpha_n}). \qquad (3.46)$$

Since they correspond to a translationally invariant theory, their Fourier transform contains a $\bar\delta$ function expressing momentum conservation modulo 2π (cf. (2.90)),

$$\sum_{x_1\cdots x_n} e^{-ip_1x_1\cdots -ip_nx_n}\,{}^0\Gamma^{x_1\cdots x_n}_{\alpha_1\cdots\alpha_n} = {}^0\Gamma_{\alpha_1\cdots\alpha_n}(p_1\cdots p_n)\,\bar\delta_{p_1+\cdots+p_n,0}. \qquad (3.47)$$

In the symmetric phase ($v^\alpha = 0$) there is only one interaction vertex function, the four-point function

$$
{}^0T^{wxyz}_{\alpha\beta\gamma\delta} = -2\lambda_0(\delta_{\alpha\beta}\delta_{\gamma\delta} + \delta_{\alpha\gamma}\delta_{\beta\delta} + \delta_{\alpha\delta}\delta_{\beta\gamma})\delta_{wx}\delta_{wy}\delta_{wz},
$$

$$
{}^0T_{\alpha_1\cdots\alpha_n}(p_1\cdots p_n) = -2\lambda_0\, s_{\alpha\beta\gamma\delta}, \tag{3.48}
$$

$$
s_{\alpha\beta\gamma\delta} \equiv \delta_{\alpha\beta}\delta_{\gamma\delta} + \delta_{\alpha\gamma}\delta_{\beta\delta} + \delta_{\alpha\delta}\delta_{\beta\gamma}. \tag{3.49}
$$

The free propagators and vertex functions are illustrated in figure 3.4.

It can be shown that disconnected subdiagrams without external lines ('vacuum bubbles') cancel out between the numerator and the denominator in the above expectation values. The expectation values can be rewritten in terms of vertex functions, which are simpler to study because they have fewer diagrams in a given order in λ_0. The two- and four-point functions can be expressed as

$$
\langle\varphi^{\alpha_1}_{x_1}\varphi^{\alpha_2}_{x_2}\rangle = G^{\alpha_1\alpha_2}_{x_1 x_2} \equiv G^{12}, \tag{3.50}
$$

$$
\langle\varphi^{\alpha_1}_{x_1}\varphi^{\alpha_2}_{x_2}\varphi^{\alpha_3}_{x_3}\varphi^{\alpha_4}_{x_4}\rangle = G^{12}G^{34} + G^{13}G^{24} + G^{14}G^{23} + G^{1234}, \tag{3.51}
$$

and the vertex functions $\Gamma_{(2)}$ and $\Gamma_{(4)}$ can be identified by writing

$$
G^{12} = -\Gamma^{-1}_{12}, \tag{3.52}
$$

$$
G^{1234} = G^{11'}G^{22'}G^{33'}G^{44'}\Gamma_{1'2'3'4'}, \tag{3.53}
$$

where as usual repeated indices are summed. Notice that Γ_{123} is zero in the symmetric phase.

To one-loop order Γ_{12} and Γ_{1234} are given by the connected diagrams in figure 3.5,

$$
\Gamma_{12} = {}^0T_{12} + \tfrac{1}{2}{}^0T_{1234}\,{}^0G^{34}, \tag{3.54}
$$

$$
\Gamma_{1234} = {}^0T_{1234} + \tfrac{1}{2}{}^0T_{1256}\,{}^0G^{55'}\,{}^0G^{66'}\,{}^0T_{5'6'34}
$$

$$
+ \text{two permutations.} \tag{3.55}
$$

In momentum space, we have conservation of momentum modulo 2π at each vertex. This may be replaced by ordinary momentum conservation because all functions in momentum space have period 2π anyway. We find for the two-point vertex function

$$
\Gamma_{\alpha_1\alpha_2}(p) = -(m_0^2 + \hat{p}^2)\delta_{\alpha_1\alpha_2}
$$

$$
+ \frac{1}{2}(-2\lambda_0)s_{\alpha_1\alpha_2\alpha_3\alpha_4}\,\delta_{\alpha_3\alpha_4}\sum_l \frac{1}{m_0^2 + \hat{l}^2} \tag{3.56}
$$

$$
\equiv -\delta_{\alpha_1\alpha_2}[\,m_0^2 + \hat{p}^2 + \lambda_0(n+2)I(m_0)\,], \tag{3.57}
$$

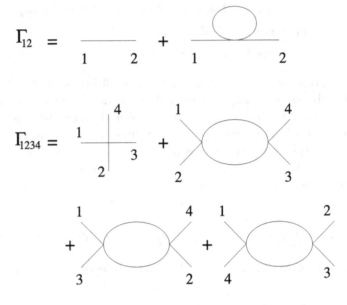

Fig. 3.5. Diagrams for Γ_{12} and Γ_{1234} to one-loop order.

and for the four-point vertex function

$$\Gamma_{\alpha_1\alpha_2\alpha_3\alpha_4}(p_1p_2p_3p_4) = -2\lambda_0 s_{\alpha_1\alpha_2\alpha_3\alpha_4}$$
$$+ \frac{1}{2}(2\lambda_0)^2 s_{\alpha_1\alpha_2\alpha_5\alpha_6} s_{\alpha_3\alpha_4\alpha_5\alpha_6} \sum_l \frac{1}{m_0^2 + \hat{l}^2}$$
$$\times \frac{1}{m_0^2 + 2\sum_\mu(1 - \cos(l + p_1 + p_2)_\mu)}$$
$$+ \text{ two permutations} \tag{3.58}$$
$$\equiv -2\lambda_0 s_{\alpha_1\alpha_2\alpha_3\alpha_4} + 2\lambda_0^2 t_{\alpha_1\alpha_2\alpha_3\alpha_4} J(m_0, p_1 + p_2)$$
$$+ \text{ two permutations}.$$

Here

$$\hat{l}^2 = 2\sum_\mu(1 - \cos l_\mu), \tag{3.59}$$

and similarly for $\hat{p}^2$, and (using the condensed notation $\delta_{12} = \delta_{\alpha_1\alpha_2}$ etc.)

$$s_{1234} = \delta_{12}\delta_{34} + \delta_{13}\delta_{24} + \delta_{14}\delta_{23}, \tag{3.60}$$
$$t_{1234} = s_{1256}s_{3456} = \delta_{12}\delta_{34}(n+4) + 2\delta_{13}\delta_{24} + 2\delta_{14}\delta_{23}. \tag{3.61}$$

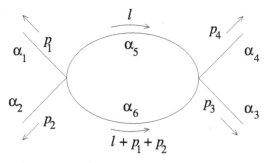

Fig. 3.6. Momentum flow.

The functions I and J are given by

$$I(m_0) = \int_{-\pi}^{\pi} \frac{d^4 l}{(2\pi)^4} \frac{1}{m_0^2 + \hat{l}^2}, \tag{3.62}$$

$$J(m_0, p) = \int_{-\pi}^{\pi} \frac{d^4 l}{(2\pi)^4} \frac{1}{\{m_0^2 + \hat{l}^2\}\{m_0^2 + 2\sum_{\mu}(1 - \cos(l + p)_{\mu})\}}.$$

We assumed an infinite lattice, $\sum_l \to \int d^4 l / (2\pi)^4$. The momentum flow in the second term in (3.58) is illustrated in figure 3.6

We are interested in the scaling forms of I and J. Let us therefore restore the lattice distance a. The functions I and J have dimensions a^{-2} and a^0, respectively. We are interested in $a^{-2} I(am_0)$ and $J(am_0, ap)$, for $a \to 0$. This suggests expanding in powers of a and keeping only terms nonvanishing as $a \to 0$. For I we need terms of order a^0 and a^2, for J only terms of order a^0. Consider first I. A straightforward expansion $1/(a^2 m_0^2 + \hat{l}^2) = \sum_n (-a^2 m_0^2)^n / (\hat{l}^2)^{n+1}$ leads to divergences in the loop integrals at the origin $l = 0$. There are various ways to deal with this situation. Here we shall give just one. Intuitively we know that the region near the origin in momentum space corresponds to continuum physics. Let us split the integration region into a ball round the origin with radius δ and the rest, with $a \ll \delta$. The radius δ is sent to zero, such that, for the integrand in the region $|l| < \delta$, we may use the continuum form l^2 for $\hat{l}^2$. Then

$$I = I_{|l| < \delta} + I_{|l| > \delta}, \tag{3.63}$$

$$I_{|l| < \delta}(am_0) = \int_{|l| < \delta} \frac{d^4 l}{(2\pi)^4} \frac{1}{a^2 m_0^2 + l^2} = \frac{2\pi^2}{(2\pi)^4} \int_0^{\delta} l^3 \, dl \frac{1}{a^2 m_0^2 + l^2}$$

$$= \frac{1}{16\pi^2}\left[\delta^2 - a^2 m_0^2 \ln\left(\frac{a^2 m_0^2 + \delta^2}{a^2 m_0^2}\right)\right]$$

$$= \frac{1}{16\pi^2}\left[-a^2 m_0^2 \ln \delta^2 + a^2 m_0^2 \ln(a^2 m_0^2)\right] + O(a^4, \delta^2).$$

$$(3.64)$$

With symbols like $O(a^2)$ we shall mean terms proportional to a^2 or $a^2 \ln a^2$. Note that expressing also l in physical units, $l \to al$, would bring $a^{-2} I_{|l|<\delta}$ into continuum form with a spherical cutoff δ/a. The integral $I_{|l|>\delta}$ can be expanded in a^2 without encountering $\ln(a^2 m_0^2)$ terms,

$$I_{|l|>\delta}(am_0) = I_{|l|>\delta}(0) + I'_{|l|>\delta}(0)a^2 m_0^2 + O(a^4)$$

$$= I(0) + I'_{|l|>\delta}(0)a^2 m_0^2 + O(a^4, \delta^2). \qquad (3.65)$$

where $I' \equiv \partial I/\partial(a^2 m_0^2)$. Instead, we encounter $\ln \delta^2$ terms in $I'_{|l|>\delta}(0)$. However, these cancel out against the $\ln \delta^2$ term in (3.64) because the complete integral is independent of δ. So we get

$$I(am_0) = C_0 - C_2 a^2 m_0^2 + \frac{1}{16\pi^2}a^2 m_0^2 \ln(a^2 m_0^2), \qquad (3.66)$$

$$C_0 = I(0) = 0.154933\ldots \qquad (3.67)$$

$$C_2 = \lim_{\delta \to 0}\left[\int_{-\pi, |l|>\delta}^{\pi} \frac{d^4 l}{(2\pi)^4} \frac{1}{(\hat{l}^2)^2} + \frac{1}{16\pi^2}\ln\delta^2\right] \qquad (3.68)$$

$$= 0.0303457\ldots. \qquad (3.69)$$

The function J can be evaluated in similar fashion. We need $J(am_0, ap)$ for $a \to 0$. For $a = 0$ the integral for J is logarithmically divergent at the origin. To deal with this we use the same procedure,

$$J = J_{|l|<\delta} + J_{|l|>\delta}, \qquad (3.70)$$

$$J_{|l|<\delta} = \int_{|l|<\delta} \frac{d^4 l}{(2\pi)^4} \frac{1}{[a^2 m_0^2 + l^2][a^2 m_0^2 + (l+ap)^2]}, \qquad (3.71)$$

$$J_{|l|>\delta} = \int_{\pi, |l|>\delta}^{-\pi} \frac{d^4 l}{(2\pi)^4} \frac{1}{(\hat{l}^2)^2} + O(a^2) \qquad (3.72)$$

($J_{|l|>\delta}$ can be expanded in powers of a, the term linear in a vanishes). With the help of the identity

$$\frac{1}{[a^2 m_0^2 + l^2][a^2 m_0^2 + (l+ap)^2]}$$

$$= \int_0^1 dx \frac{1}{\{x[a^2 m_0^2 + l^2] + (1-x)[a^2 m_0^2 + (l+ap)^2]\}^2} \qquad (3.73)$$

and the transformation of variable $l' = l + (1 - x)ap$ we get for the inner-region integral

$$J_{|l|<\delta} = \int_0^1 dx \int_D \frac{d^4 l'}{(2\pi)^4} \frac{1}{[a^2 m_0^2 + l'^2 + x(1-x)a^2 p^2]^2}. \tag{3.74}$$

Here the domain of integration D is obtained from the ball with radius δ by shifting it over $(1-x)ap$. Replacing D by the original ball with radius δ leads to an error of order a, which may be neglected. (The difference between the two integration regions has a volume $O(ap\delta^3)$, the integrand is $O(\delta^{-4})$.) Then

$$
\begin{aligned}
J_{|l|<\delta} &= \int_0^1 dx \, \frac{2\pi^2}{(2\pi)^4} \int_0^\delta l^3 \, dl \, \frac{1}{[a^2 m_0^2 + l^2 + x(1-x)a^2 p^2]^2} \\
&= \frac{1}{16\pi^2} \int_0^1 dx \left[\ln(a^2 \Delta + \delta^2) - \ln(a^2 \Delta) - \frac{\delta^2}{a^2 \Delta + \delta^2} \right] \\
&= \frac{1}{16\pi^2} \left[\ln \delta^2 - \int_0^1 dx \, \ln(a^2 \Delta) - 1 \right] + O(a^2), \tag{3.75}
\end{aligned}
$$

$$\Delta \equiv m^2 + x(1-x)p^2. \tag{3.76}$$

Combining the term $\ln \delta^2 / 16\pi^2$ with $J_{|l|>\delta}$ as in (3.68) we get

$$J(am_0, ap) = -\frac{1}{16\pi^2} \int_0^1 dx \, \ln[\, a^2(m_0^2 + x(1-x)p^2)] + C_2 - \frac{1}{16\pi^2} + O(a^2). \tag{3.77}$$

(We expect errors $O(a^2)$, i.e. not $O(a)$: a will appear together with the external momentum as ap_μ or as $a^2 m_0^2$, and there will not be odd powers of p_μ because of cubic symmetry, including reflections.)

Summarizing, we have obtained the following continuum forms for the vertex functions (in physical units):

$$
\begin{aligned}
\Gamma_{\alpha\beta}(p) &= -\delta_{\alpha\beta} \left\{ m_0^2 + p^2 + \lambda_0(n+2) \left[\frac{C_0}{a^2} - C_2 m_0^2 \right. \right. \\
&\qquad \left. \left. + \frac{1}{16\pi^2} m_0^2 \ln(a^2 m_0^2) \right] \right\}, \tag{3.78}
\end{aligned}
$$

$$
\begin{aligned}
\Gamma_{\alpha_1\alpha_2\alpha_3\alpha_4}(p_1 p_2 p_3 p_4) &= -2\lambda_0 s_{\alpha_1\alpha_2\alpha_3\alpha_4} \\
&\quad + 2\lambda_0^2 t_{\alpha_1\alpha_2\alpha_3\alpha_4} \left\{ C_2 - \frac{1}{16\pi^2} - \frac{1}{16\pi^2} \right. \\
&\qquad \left. \times \int_0^1 dx \, \ln[a^2(m_0^2 + x(1-x)(p_1 + p_2)^2)] \right\} \\
&\quad + \text{two permutations}, \tag{3.79}
\end{aligned}
$$

We see that $\Gamma_{(2)}$ and $\Gamma_{(4)}$ are, respectively, quadratically and logarithmically divergent as $a \to 0$.

3.5 Renormalization

Perturbative renormalization theory tells us that, when we rescale the correlation functions $G^{(n)}$ by a suitable factor $Z^{-n/2}$ and express them in terms of a suitable renormalized mass parameter m_R and renormalized coupling constant λ_R, the result is finite as $a \to 0$. The renormalized $G_R^{(n)} = Z^{-n/2}G^{(n)}$ are the correlation functions of renormalized fields $\varphi_R = Z^{-1/2}\varphi$. From (3.53) we see that the renormalized vertex functions are then given by

$$\Gamma_{(n)}^R = Z^{n/2}\Gamma_{(n)}. \tag{3.80}$$

The wave function renormalization constant Z and the renormalized mass parameter m_R may be defined by the first two terms of the expansion

$$\Gamma_{\alpha\beta}(p) = -Z^{-1}(m_R^2 + p^2 + O(p^4))\delta_{\alpha\beta}. \tag{3.81}$$

Since the one-loop diagram for $\Gamma_{(2)}$ is momentum independent, the order λ contribution to Z vanishes in the $O(n)$ model,

$$Z = 1 + O(\lambda^2). \tag{3.82}$$

For m_R we find from (3.78)

$$m_R^2 = m_0^2 + \lambda_0(n+2)\left[C_0 a^{-2} - C_2 m_0^2 + \frac{1}{16\pi^2}m_0^2\ln(a^2 m_0^2)\right]. \tag{3.83}$$

A renormalized coupling constant λ_R may be defined in terms of $\Gamma_{(4)}$ at zero momentum, by writing

$$\Gamma_{\alpha_1\alpha_2\alpha_3\alpha_4}^R(0,0,0,0) = -2\lambda_R\, s_{\alpha_1\alpha_2\alpha_3\alpha_4}. \tag{3.84}$$

From the result (3.79) for the four-point function, using (3.82) and

$$t_{1234} + t_{1324} + t_{1423} = (n+8)s_{1234}, \tag{3.85}$$

we find

$$\lambda_R = \lambda_0 + \lambda_0^2 \frac{n+8}{16\pi^2}[\ln(a^2 m_0^2) + c], \tag{3.86}$$

$$c = -\frac{16\pi^2}{n+8}\left(C_2 - \frac{1}{16\pi^2}\right). \tag{3.87}$$

To express the correlation functions in terms of m_R and λ_R we consider λ_R as an expansion parameter and invert (3.83), (3.86),

$$m_0^2 = m_R^2 - \lambda_R(n+2)\left[C_0 a^{-2} - C_2 m_R^2 + \frac{1}{16\pi^2} m_R^2 \ln(a^2 m_R^2)\right]$$
$$+ O(\lambda_R^2), \tag{3.88}$$

$$\lambda_0 = \lambda_R - \lambda_R^2 \frac{n+8}{16\pi^2}[\ln(a^2 m_R^2) + c] + O(\lambda_R^3). \tag{3.89}$$

Inserting these relations into (3.78), (3.79) gives the renormalized vertex functions

$$\Gamma_{\alpha\beta}^R(p) = -\delta_{\alpha\beta}(m_R^2 + p^2) + O(\lambda_R^2), \tag{3.90}$$

$$\Gamma_{\alpha_1\alpha_2\alpha_3\alpha_4}^R(p_1 p_2 p_3 p_4) = -2\lambda_R s_{\alpha_1\alpha_2\alpha_3\alpha_4} - 2\lambda_R^2 t_{\alpha_1\alpha_2\alpha_3\alpha_4}$$
$$\times \frac{1}{16\pi^2} \int_0^1 dx \ln\left(\frac{m_R^2 + x(1-x)(p_1+p_2)^2}{m_R^2}\right)$$
$$+ \text{two permutations} + O(\lambda_R^3), \tag{3.91}$$

which are indeed independent of the lattice spacing a. Notice that the constants C_0, C_1 and C_2 are absent: all reference to the lattice has disappeared from the renormalized vertex functions.

To this order the mass m of the particles is equal to m_R. The mass m is given by the value of $-p^2$ where $\Gamma_{(2)}$ is zero and $G^{(2)}$ has a pole. In higher orders the mass m will be different from the renormalized mass parameter m_R: $m = m_R(1 + O(\lambda_R^2))$.

The $O(n)$ tensor structure in (3.84) is the general form of $\Gamma_{(4)}$ at a symmetry point where $(p_1 + p_2)^2 = (p_1 + p_3)^2 = (p_1 + p_4)^2 \equiv \mu^2$. We can therefore also define a 'running renormalized coupling' $\bar{\lambda}(\mu)$ at momentum scale μ by

$$\Gamma_{\alpha_1\alpha_2\alpha_3\alpha_4}^R(p_1 p_2 p_3 p_4) = -2\bar{\lambda}(\mu)s_{\alpha_1\alpha_2\alpha_3\alpha_4}, \quad \text{symmetry point } \mu, \tag{3.92}$$

which gives

$$\bar{\lambda}(\mu) = \lambda_0 + \lambda_0^2 \frac{n+8}{16\pi^2}\left\{\int_0^1 dx \ln[a^2 m_0^2 + x(1-x)a^2\mu^2] + c\right\}. \tag{3.93}$$

Expressing the running coupling in terms of λ_R and m_R leads to

$$\bar{\lambda}(\mu) = \lambda_R + \lambda_R^2 \frac{n+8}{16\pi^2}\int_0^1 dx \ln[1 + x(1-x)\mu^2/m_R^2] + O(\lambda_R^3)$$

$$= \lambda_R, \qquad \mu = 0, \tag{3.94}$$

$$\approx \lambda_R + \lambda_R^2 \frac{n+8}{16\pi^2}[\ln(\mu^2/m_R^2) - 2], \quad \mu^2 \gg m_R^2. \tag{3.95}$$

The running coupling indicates the strength of the interactions at momentum scale μ. Expressing the vertex function (3.91) in terms of this running coupling shows that, at large momenta, terms of the type $\lambda_R^2 \ln[(p_1 + p_2)^2/m_R^2]$ are replaced by $\bar{\lambda}^2(\mu) \ln[(p_1 + p_2)^2/\mu^2]$. So, on choosing μ^2 equal to values of $(p_i + p_j)^2$ that typically occur in a given situation, the logarithms are generically not large and the strength of the four-point vertex on this momentum scale is expressed by $\bar{\lambda}(\mu)$.

3.6 Renormalization-group beta functions

The renormalized quantities do not depend explicitly on the lattice distance, all dependence on a is absorbed by the relations between m_0, λ_0 and m_R, λ_R. Thus it seems that we can take the continuum limit $a \to 0$ in the renormalized quantities. Changing a while keeping m_R and λ_R fixed implies that m_0 and λ_0 must be chosen to depend on a, as given by (3.88) and (3.89). We see that $a^2 m_0^2$ decreases and becomes negative as a decreases, even in the symmetric phase. This we found earlier in the mean-field approximation. However, the bare λ_0 increases as a decreases and beyond a certain value we can no longer trust perturbation theory in λ_0. Neither can we trust (3.89) if a becomes too small, since then the coefficient of λ_R^2 blows up.

Let us look at the problem in another way. Consider what happens to λ_R as we approach the phase boundary at fixed λ_0. In (3.86) we may replace to this order m_0 by m_R,

$$\lambda_R = \lambda_0 + \lambda_0^2 \frac{n+8}{16\pi^2} [\ln(a^2 m_R^2) + c] + O(\lambda_0^3). \tag{3.96}$$

We see that λ_R decreases as a decreases, but when the logarithm becomes too large the perturbative relation breaks down. We can extract more information by differentiating with respect to a and writing the result in terms of λ_R,

$$\beta_R(\lambda_R) = \left[a \frac{\partial \lambda_R}{\partial a} \right]_{\lambda_0} = \left[a m_R \frac{\partial \lambda_R}{\partial a m_R} \right]_{\lambda_0}$$
$$= \beta_1 \lambda_0^2 + O(\lambda_0^3) \tag{3.97}$$
$$= \beta_1 \lambda_R^2 + \beta_2 \lambda_R^3 + \cdots, \tag{3.98}$$
$$\beta_1 = \frac{n+8}{8\pi^2}. \tag{3.99}$$

The function $\beta_R(\lambda_R)$ is one of the renormalization-group functions introduced by Callan and by Symanzik. For a clear derivation of the

Callan–Symanzik equations in our context see [20]. They are dimensionless functions which may be expressed in terms of renormalized vertex functions and are given by renormalized perturbation theory as a series $\sum_k \beta_k \lambda_R^k$. This means that the higher-order terms of the form $\lambda_0^k [\ln(am_R)]^l$ can be rearranged in terms of powers of λ_R with coefficients that do not depend any more on $\ln(am_R)$. This is the justification for rewriting (3.97) in terms of λ_R.

Integration of $\partial \lambda_R / \partial t = -\beta_1 \lambda_R^2$ gives

$$\lambda_R = \frac{\lambda_1}{1 + \lambda_1 \beta_1 t}, \quad t \equiv -[\ln(am_R) + c/2], \qquad (3.100)$$

where λ_1 is an integration constant, $\lambda_1 = \lambda_0 + O(\lambda_0^2)$. As $a \to 0$, $t \to \infty$ and we see that λ_R approaches zero. The approximation of using only the lowest-order approximation to the beta function is therefore self-consistent.

Let us try the beta-function trick on λ_0 to see whether we can determine how it depends on a if we keep λ_R fixed. From (3.89) we find

$$\left[a \frac{\partial \lambda_0}{\partial a} \right]_{\lambda_R} \equiv -\beta_0(\lambda_0) = -\beta_1 \lambda_0^2 + \cdots . \qquad (3.101)$$

Note the change of sign compared with (3.98). Integrating this equation gives

$$\lambda_0 = \frac{\lambda_2}{1 - \lambda_2 \beta_1 t}, \qquad (3.102)$$

where $\lambda_2 = \lambda_R + O(\lambda_R^2)$. We see that λ_0 blows up at the 'Landau pole' $t = 1/\lambda_2 \beta_1$, but, of course, before reaching this value the first-order approximation to $\beta_0(\lambda_0)$ breaks down.

Consider next the beta function $\bar{\beta}(\bar{\lambda})$ for the running coupling $\bar{\lambda}(\mu)$ on momentum scale μ. From (3.95) we see that, for large $\mu \gg m_R$,

$$\left[\mu \frac{\partial \bar{\lambda}(\mu)}{\partial \mu} \right]_{\lambda_R, m_R} \equiv \bar{\beta}(\bar{\lambda}) = \beta_1 \bar{\lambda}^2 + \cdots, \quad \mu \gg m_R, \qquad (3.103)$$

again with the same universal coefficient for the first-order term in its expansion. The solution is similar to that for λ_0,

$$\bar{\lambda} = \frac{\lambda_3}{1 - \lambda_3 \beta_1 \ln(\mu/m_R)}. \qquad (3.104)$$

The effective coupling $\bar{\lambda}$ increases with momentum scale μ. To see if it can become arbitrarily large we need to go beyond the weak-coupling expansion.

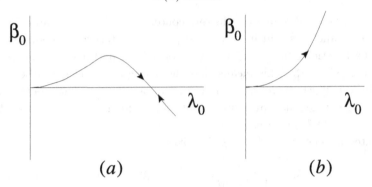

Fig. 3.7. Two possible shapes of $\beta_0(\lambda_0)$. The arrows denote the flow of λ_0 for increasing $t = -\ln(am_R) +$ constant.

We end this section by speculating about different shapes of the beta function β_0 for the bare coupling constant. Two typical possibilities are shown in figure 3.7. In case (a) there is a fixed point λ^* that attracts the flow of λ_0 for increasing 'time' t. Near λ^* we can linearize

$$\frac{\partial \lambda_0}{\partial t} = -A(\lambda_0 - \lambda^*), \tag{3.105}$$

$$\lambda_0 - \lambda^* = C\exp(-At), \quad t \to \infty, \tag{3.106}$$

where C is an integration constant. The large-t behavior can be rewritten in the form

$$\xi = \frac{1}{am_R} \propto (\lambda^* - \lambda_0)^{-\nu}, \quad \nu = 1/A, \tag{3.107}$$

which shows that the critical exponent ν is determined by the slope of the beta function at the fixed point. Since t can go to infinity without a problem, a continuum *limit* $a \to 0$ is possible for case (a).

In case (b) the beta function does not have a zero, apart from the origin $\lambda_0 = 0$. Supposing a behavior

$$\frac{\partial \lambda_0}{\partial t} = A\lambda_0^\alpha, \quad \lambda_0 \to \infty, \alpha > 0, A > 0, \tag{3.108}$$

leads to the asymptotic solution

$$\lambda_0^{-(\alpha-1)} = -A(\alpha - 1)(t - t_1), \tag{3.109}$$

where we assumed $\alpha > 1$. In this case λ_0 becomes infinite in a finite 'time' $t = t_1$. Since $\lambda_0 = \infty$ is the largest value λ_0 can take, t cannot go beyond t_1, a cannot go to zero and a continuum *limit* is not possible.

A similar discussion can be given for the running coupling $\bar{\lambda}$. Cases (a) and (b) also illustrate possible behaviors of the running coupling for large momentum scales μ. In case (a) the running coupling approaches λ^* as $\mu \to \infty$, whereas in case (b) $\bar{\lambda}$ goes to infinity on some large but finite momentum scale μ_1.

A fixed point like λ^* is called ultraviolet stable as it attracts the running coupling when $\mu \to \infty$, while the fixed point at the origin is called infrared stable as it attracts the running coupling for $\mu \to 0$. Case (a) is like the situation in three Euclidean dimensions (with a reflection about the horizontal axis), whereas we shall see in the following that in four dimensions the situation is like case (b).

The main conclusion in this section is that $\lambda_R \to 0$ as we approach the phase boundary at fixed sufficiently small λ_0. To see whether we can avoid a noninteracting theory in the continuum *limit*, we need to be able to investigate large λ_0. This can be done with the hopping expansion and with numerical simulations.

3.7 Hopping expansion

Consider the partition function in the form

$$Z = \int D\mu(\phi) \prod_{x\mu} \exp(2\kappa \phi_x^\alpha \phi_{x+\hat{\mu}}^\alpha), \qquad (3.110)$$

where $D\mu(\phi) = \prod_x d\mu(\phi_x)$ is the product of one-site measures defined in (3.18). Expansion in κ (hopping expansion) leads to products of one-site integrals of the form

$$\int D\mu(\phi) \equiv Z_0 = \left(\int d\mu(\phi)\right)^{\#\,\text{sites}}, \qquad (3.111)$$

$$\int D\mu(\phi)\,\phi_x^\alpha \phi_y^\beta = \delta_{xy} Z_0 \langle \phi^\alpha \phi^\beta \rangle_1, \qquad (3.112)$$

$$\int D\mu(\phi)\,\phi_x^\alpha \phi_y^\beta \phi_z^\gamma = 0, \qquad (3.113)$$

$$\int D\mu(\phi)\,\phi_x^\alpha \phi_x^\beta \phi_x^\gamma \phi_x^\delta = Z_0 \langle \phi^\alpha \phi^\beta \phi^\gamma \phi^\delta \rangle_1, \qquad (3.114)$$

etc., where $\#$ sites is the total number of lattice sites. Odd powers of ϕ vanish in the one-site average

$$\langle F \rangle_1 = \int d\mu(\phi)\,F(\phi) \Big/ \int d\mu(\phi). \qquad (3.115)$$

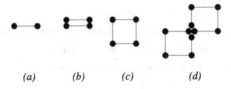

Fig. 3.8. Diagrams in the expansion of $\exp(2\kappa \sum_{x\mu} \phi_x^\alpha \phi_{x+\hat\mu}^\alpha)$.

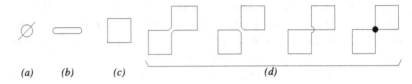

(a)   *(b)*   *(c)*   *(d)*

Fig. 3.9. The diagrams of figure 3.8 after integration over ϕ. The fat dot denotes the four-point vertex γ_4.

Before integration over ϕ, each term in the expansion can be represented by a dimer diagram 'on the lattice', as illustrated in figure 3.8. The dots indicate the fields ϕ. The integration over ϕ leads to diagrams as shown in figure 3.9.

The one-site integrals can be treated as a mini field theory, with propagators $g^{\alpha\beta}$ and vertex functions $\gamma_{\alpha_1\cdots\alpha_4}$, $\gamma_{\alpha_1\cdots\alpha_6}$, $\ldots$. For instance, $\gamma_{\alpha\beta\gamma\delta}$ can be defined by

$$\langle \phi^\alpha \phi^\beta \rangle_1 = g^{\alpha\beta}, \tag{3.116}$$

$$\langle \phi^\alpha \phi^\beta \phi^\gamma \phi^\delta \rangle_1 = g^{\alpha\beta} g^{\gamma\delta} + g^{\alpha\gamma} g^{\beta\delta} + g^{\alpha\delta} g^{\beta\gamma} + g^{\alpha\beta\gamma\delta}, \tag{3.117}$$

$$g^{\alpha\beta\gamma\delta} = g^{\alpha\alpha'} g^{\beta\beta'} g^{\gamma\gamma'} g^{\delta\delta'} \gamma_{\alpha'\beta'\gamma'\delta'}, \tag{3.118}$$

analogously to (3.53). By $O(n)$ symmetry we have

$$g^{\alpha\beta} = \delta_{\alpha\beta}\, g, \quad g = \frac{\langle \phi^2 \rangle_1}{n}, \tag{3.119}$$

$$g^{\alpha\beta\gamma\delta} = s_{\alpha\beta\delta\gamma} \frac{\langle (\phi^2)^2 \rangle_1}{n^2 + 2n}, \tag{3.120}$$

$$\gamma_{\alpha\beta\gamma\delta} = s_{\alpha\beta\delta\gamma}\, \gamma_4, \quad \gamma_4 = \frac{n^3}{n+2} \frac{\langle (\phi^2)^2 \rangle_1}{\langle \phi^2 \rangle_1^4} - \frac{n^2}{\langle \phi^2 \rangle_1^2}, \tag{3.121}$$

where $s_{\alpha\beta\delta\gamma} = \delta_{\alpha\beta}\delta_{\gamma\delta} + \cdots$ has been defined in (3.49). For small λ, $\gamma_4 \propto \lambda$, whereas for $\lambda \to \infty$, $\gamma_4 \to -2n^4/(n+2)$.

As usual, one expects that disconnected diagrams cancel out in expressions for the vertex functions, and that the two-point function, $G_{xy}^{\alpha\beta} = \langle \phi_x^\alpha \phi_y^\beta \rangle$, can be expressed as a sum of connected diagrams. It

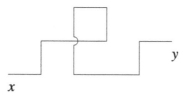

Fig. 3.10. Random-walk contribution to the propagator.

is instructive to make an approximation for the two-point function in which the vertex functions $\gamma_{(4)}$, $\gamma_{(6)}$, ..., are neglected at first. This leads to the random-walk approximation

$$G_{xy}^{\alpha\beta} = \delta_{\alpha\beta} \sum_{L=0}^{\infty} (2\kappa)^{L} g^{L+1} (H^{L})_{xy} + \text{'interactions'}, \qquad (3.122)$$

illustrated in figure 3.10. Here 'interactions' denote the neglected contributions proportional to $\gamma_{(4)}$, $\gamma_{(6)}$, ..., and we introduced the hopping matrix

$$H_{uv} = \sum_{\mu} (\delta_{u+\hat{\mu}-v,0} + \delta_{v+\hat{\mu}-u,0}). \qquad (3.123)$$

Applying this matrix e.g. to the vector $\delta_{v,x}$ gives a non-zero answer only for u's that are nearest neighbors of x, i.e. all sites that can be reached from x in 'one step'. Applying H once more corresponds to making one more step in all possible directions, etc. In this way a random walk is built up by successive application of H. Each link in the expansion contributes a factor 2κ, and each site a factor g. In momentum space we get

$$G_{xy}^{\alpha\beta} = \delta_{\alpha\beta} \int_{-\pi}^{\pi} \frac{d^4 p}{(2\pi)^4} e^{ip(x-y)} g \sum_{L} (2\kappa g)^{L} H(p)^{L} \qquad (3.124)$$

$$= \delta_{\alpha\beta} \int_{-\pi}^{\pi} \frac{d^4 p}{(2\pi)^4} e^{ip(x-y)} \frac{g}{1 - 2\kappa g H(p)}, \qquad (3.125)$$

where

$$H(p) = \sum_{x} e^{-ipx} H_{x,0} = \sum_{\mu} 2 \cos p_{\mu}. \qquad (3.126)$$

In the random-walk approximation the two-point correlation function has the free-field form. For small momenta we identify the mass param-

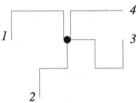

Fig. 3.11. Four random walks correlated by the one site γ_4.

eter m_{R} and the wavefunction-renormalization constant Z_ϕ,

$$G^{\alpha\beta}(p) = \delta_{\alpha\beta}\frac{Z_\phi}{m_{\mathrm{R}}^2 + p^2 + O(p^4)}, \tag{3.127}$$

$$Z_\phi = (2\kappa)^{-1}, \quad m_{\mathrm{R}}^2 = (2g\kappa)^{-1} - 2d. \tag{3.128}$$

This Z_ϕ corresponds to $Z_\varphi = 1$ (cf. (3.15)). When $m_{\mathrm{R}} \to 0$ we enter the scaling region. In the random-walk approximation this occurs at $\kappa = \kappa_{\mathrm{c}} = 1/4gd$, which is the mean-field value (3.36). This is not so surprising as the mean-field approximation is good for $d \to \infty$, when also the random-walk approximation is expected to be good, because the chance of self-intersections in the walk, where $\gamma_{(4)}$, $\gamma_{(6)}$, ... come into play, goes to zero. Notice that κ_{c} is also the radius of convergence of the expansion (3.124).

Within the random-walk approximation we have the estimate for the renormalized coupling (cf. (3.80)) as illustrated in figure 3.11,

$$-2\lambda_{\mathrm{R}} = Z^2\gamma_4 = \frac{\gamma_4}{4\kappa_{\mathrm{c}}^2}, \tag{3.129}$$

$$\lambda_{\mathrm{R}} \to \lambda_0, \qquad \lambda \to 0, \tag{3.130}$$

$$\to \left(\frac{2d}{n}\right)^2 \frac{n^2}{n+2} = \frac{32}{3}, \quad \lambda \to \infty, d = 4, n = 4. \tag{3.131}$$

This indicates already that λ_{R} is not infinite at $\lambda = \infty$.

The partition function and expectation values can be expressed as a systematic expansion in κ. This is called the hopping expansion because the random-walk picture suggests propagation of particles by 'hopping' from one site to the next. By the analogy of κ with the inverse temperature in the Ising model, the expansion is known in statistical physics as the high-temperature expansion, or, with increasing sophistication, the linked-cluster expansion. Using computers to help with the algebra, the expansion can be carried out to high orders (see e.g. [22] and references therein).

A good property of the hopping expansion is that it has a non-zero radius of convergence, for any fixed $\lambda \in (0, \infty)$. This is in contrast to the weak-coupling expansion, which is an asymptotic expansion (as is typical for saddle-point expansions) with zero radius of convergence (see for example [13]). An expansion $f(x) = \sum_{k=0}^{\infty} f_k x^k$ is called asymptotic if

$$\left| f(x) - \sum_{k=0}^{N} f_k x^k \right| = O(x^{N+1}). \tag{3.132}$$

For fixed finite N the sum gives an accurate approximation to $f(x)$, for sufficiently small x. The expansion need not converge as $N \to \infty$ and for a given x there is an optimum N beyond which the approximation becomes worse.

3.8 Lüscher–Weisz solution

Using the hopping expansion in combination with the Callan–Symanzik renormalization-group equations, Lüscher and Weisz showed how the $O(n)$ models in four dimensions can be solved to a good approximation [20, 21, 22, 23]. The coefficients of the hopping series were calculated to 14th order and the Callan–Symanzik beta functions were used to three-loop order. The cases $n = 1$ [20, 21] and $n = 4$ [23] were worked out in detail. The interested reader is urged to study these lucid papers which contain a lot of information on field theory. We shall review the highlights for the $O(4)$ model.

The critical $\kappa_c(\lambda)$ is estimated from the radius of convergence of the hopping expansion to be monotonically increasing from $\kappa_c = \frac{1}{8}$ at $\lambda = 0$ to $\kappa_c = 0.304\,11(6)$ at $\lambda = \infty$. An important aspect of the results is the carefully estimated errors in various quantities. For simplicity, we shall not quote the errors anymore in the following. Along the line $\kappa = 0.98\,\kappa_c$ in the κ–λ plane the hopping expansion still converges well, with the mass parameter m_R decreasing from 0.40 to 0.28 and the renormalized coupling λ_R increasing from 0 to 3.2 as λ increases from 0 to ∞. At a slightly smaller $\kappa < \kappa_c$ such that $m_R = 0.5$, $\lambda_R = 4.3$ for $\lambda = \infty$.

Remarkably, $\lambda_R = 3.2$ may be considered as relatively weak coupling. Let us rewrite the beta function

$$m_R \frac{\partial \lambda_R}{\partial m_R} = \beta_R(\lambda_R) = \beta_1 \lambda_R^2 + \beta_2 \lambda_R^3 + \cdots, \tag{3.133}$$

in terms of a natural variable $\tilde{\lambda} \equiv \beta_1 \lambda_R$,

$$m_R \frac{\partial \tilde{\lambda}}{\partial m_R} = \tilde{\lambda}^2 + \frac{\beta_2}{\beta_1^2} \tilde{\lambda}^3 + \cdots. \tag{3.134}$$

The results

$$\beta_1 = \frac{n+8}{8\pi^2}, \quad \beta_2 = -\frac{9n+42}{(8\pi^2)^2}, \tag{3.135}$$

give $\beta_2/\beta_1^2 \approx -0.54$. Then $\lambda_R = 3.2$ means $\tilde{\lambda} \approx 0.41$ and the two-loop term in (3.134) is only about 20% of the one-loop term. This indicates that renormalized perturbation theory may be applicable for these couplings. The next (three-loop) term in the series is again positive and Lüscher and Weisz reason that the true beta function in this coupling region may be between the two- and three-loop values.

A basic assumption made in order to proceed is that renormalized perturbation theory is valid for sufficiently small λ_R, even if the bare λ is infinite. This may seem daring if one thinks of deriving renormalized perturbation theory from the bare weak-coupling expansion. However, it appears natural from the point of view of Wilson's renormalization theory in terms of an effective action with an effective cutoff, or from the point of view of effective actions, or Schwinger's Source Theory, which uses unitarity to obtain higher-order approximations in an expansion in a physical coupling parameter (e.g. λ_R).

Using the beta function calculated in renormalized perturbation theory, Lüscher and Weisz integrate the Callan–Symanzik equations toward the critical point $m_R = 0$. (The variable κ is traded for m_R.) As we have seen in (3.100) this leads to the conclusion that the renormalized coupling vanishes at the phase boundary, which is thus established even for bare coupling $\lambda = \infty$ (!).

The integration is done numerically, using (3.133). Sufficiently deep in the scaling region we may integrate by expansion,

$$\frac{\partial \lambda_R}{\partial \ln m_R} = \beta_R(\lambda_R), \tag{3.136}$$

$$\ln m_R = \int^{\lambda_R} \frac{dx}{\beta_R(x)}, \tag{3.137}$$

$$= \int^{\lambda_R} dx \left[\frac{1}{\beta_1 x^2} - \frac{\beta_2}{\beta_1^2 x} + O(1) \right], \tag{3.138}$$

$$= -\frac{1}{\beta_1 \lambda_R} - \frac{\beta_2}{\beta_1^2} \ln(\beta_1 \lambda_R) + \ln C_1 + O(\lambda_R). \tag{3.139}$$

Here C_1 is an integration constant, which becomes dependent on the bare λ once the solution is matched to the hopping expansion. (Part of the integration constant is written as $-(\beta_2/\beta_1^2)\ln\beta_1$.) Notice that knowledge of β_2 is needed in order to be able to define $C_1(\lambda)$ as $\lambda_R \to 0$. Eq. (3.139) can also be written as

$$m_R = C_1(\beta_1\lambda_R)^{-\beta_2/\beta_1^2}e^{-1/\beta_1\lambda_R}[1+O(\lambda_R)], \qquad (3.140)$$

which shows that m_R depends non-analytically on λ_R for $\lambda_R \to 0$. Lüscher and Weisz show that similarly

$$Z = C_2[1+O(\lambda_R)], \qquad (3.141)$$

$$\kappa_c - \kappa = C_3 m_R^2(\lambda_R)^{\delta_1/\beta_1}[1+O(\lambda_R)], \qquad (3.142)$$

where δ_1 is a Callan–Symanzik coefficient similar to the β's.

From these equations follow the scalings laws, $\tau = 1 - \kappa/\kappa_c \to 0$:

$$m_R \to C_4\tau^{1/2}|\ln\tau|^{\delta_1/2\beta_1}, \qquad (3.143)$$

$$\lambda_R \to \frac{2}{\beta_1}|\ln\tau|^{-1}, \qquad (3.144)$$

$$Z \to C_2. \qquad (3.145)$$

We recognize that the behavior (3.144) follows from (3.100). Note that (3.143) shows that the correlation-length critical exponent ν has almost the mean-field value $\nu = \frac{1}{2}$: it is modified only by a power of $\ln\tau$.

In the scaling limit all information about the renormalized coupling coming from the hopping expansion is contained in $C_1(\lambda)$, which increases monotonically with decreasing λ. For small bare coupling C_1 can be calculated with the weak-coupling expansion. In fact, inserting the expansion (3.86) for λ_R into (3.140) and expanding in λ_0 leads to

$$\ln C_1(\lambda) = \frac{1}{\beta_1\lambda_0} + \frac{\beta_2}{\beta_1^2}\ln(\beta_1\lambda_0) - \frac{c}{2} + O(\lambda_0). \qquad (3.146)$$

For infinite bare coupling Lüscher and Weisz find $C_1(\infty) = \exp(1.5)$. The fact that $C_1(\lambda)$ decreases as λ increases corresponds to the intuition that for given m_R, the renormalized coupling increases with λ. Conversely, for given λ_R, the smallest lattice spacing (smallest m_R) is obtained with the largest λ, i.e. $\lambda = \infty$.

The hopping expansion holds in the region of the phase diagram connected to the line $\kappa = 0$, i.e. the symmetric phase. Lüscher and Weisz extended these results into the physically relevant broken phase, where relations similar to (3.140)–(3.145) were obtained with coefficients

C' (the Callan–Symanzik coefficients are the same in both phases). They considered the critical theory at $\kappa = \kappa_c$ and used perturbation theory in $\kappa - \kappa_c$, or equivalently m_R^2, to connect the symmetric and broken phases. This is done again using renormalized perturbation theory with the results

$$C_1'(\lambda) = 1.435\, C_1(\lambda), \quad C_{2,3}'(\lambda) = C_{2,3}(\lambda). \tag{3.147}$$

Another definition was chosen for the renormalized coupling in the broken phase, which is very convenient:

$$\lambda_R = \frac{m_R^2}{2v_R^2}, \quad v_R \equiv Z_\pi^{-1/2} v = Z_\pi^{-1/2}\langle\phi\rangle, \tag{3.148}$$

where Z_π is the wave-function renormalization constant of the Goldstone bosons. This choice is identical in form to the classical relation between the coupling, mass and vacuum expectation value (cf. (3.10)). The renormalized coupling in the broken phase cannot be defined at zero momentum, as in the symmetric phase, because the massless Goldstone bosons would lead to infrared divergences (in absence of explicit symmetry breaking). Using Z_π in the definition of v_R allows the identification of v_R with the pion decay constant f_π in the application of the $O(4)$ model to low-energy pion physics, or with the electroweak scale of 246 GeV in the application to the Standard Model.

The renormalization-group equations were numerically integrated again in the broken phase, this time for increasing m_R, until the renormalized λ_R became too large and the perturbative beta function could no longer be trusted. We mention here the result $\lambda_R < 3.5$ for $m_R < 0.5$, at $\lambda = \infty$. Hence, also in the broken phase the renormalized coupling is relatively small even at the edge of the scaling region, taken somewhat arbitrarily to be at $m_R = 0.5$, and the renormalized coupling goes to zero in the continuum limit $m_R \to 0$.

Figure 3.12 shows lines of constant renormalized coupling with varying κ/κ_c for the case $n = 1$ [21]. For a given λ_R we can go deeper into the scaling region, i.e. approach the critical line $\kappa/\kappa_c = 1$ by increasing the bare coupling λ. This behavior was also found in the weak-coupling expansion, but the results there became invalid as λ grew too big. Here we see that the behavior continues for large λ and that the line $\lambda = \infty$ is reached before reaching the critical line. The critical line can be approached arbitrarily closely only for arbitrarily small renormalized coupling. It follows that the beta function of the model has to correspond to case (b) in figure 3.7.

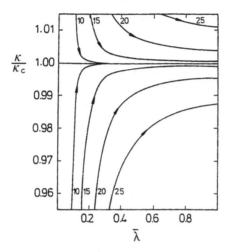

Fig. 3.12. Lines of constant renormalized coupling for the case $n = 1$ determined by Lüscher and Weisz. The lines are labeled by the value of $g_R \equiv 6\lambda_R$. The bare coupling λ increases from 0 to ∞ as the LW parameter $\bar{\lambda}$ goes from 0 to 1. From [21].

For the $O(4)$ model, the figure corresponding to 3.12 is similar, except that the values of λ_R at a given am_R are smaller [23]. The first beta-function coefficient increases with n, so one expects the renormalization effects to be larger than for $n = 1$.

Let us recall here another well-known criterion for a coupling being small or large: the unitarity bound. This is the value of the renormalized coupling at which the lowest-order approximation to the elastic scattering amplitude T becomes larger than a bound deduced from the unitarity of the scattering matrix S. In a partial wave state of definite angular momentum (e.g. the s-wave) the scattering matrix is finite dimensional, its eigenvalues are phase factors $S = \exp(i2\delta)$, with δ the standard phase shifts. Since the lowest-order (Born) approximation is real and $T = (S - 1)/i = 2\exp(i\delta)\sin\delta$ has a real part $\in (-1, 1)$, one requires the Born approximation for $|T|$ to be smaller than 1. This gives an upper bound on λ_R: the unitarity bound. The maximum values of the renormalized coupling at $m_R = 0.5$ turn out to be smaller than the unitarity bound (in the broken phase the maximum λ_R is only about two thirds of this bound).

Summarizing, the results show that the $O(n)$ models (in particular the cases $n = 1$ and 4) in four dimensions are 'trivial': the renormalized coupling vanishes in the continuum limit. Since we want of course an interacting model we cannot take the lattice distance to zero. The model is to be interpreted as an effective model that is valid at momenta much smaller than the cutoff π (π/a in physical units). For not too large renormalized coupling the cutoff can be huge and lattice artifacts very small. At the scale of the cutoff the model loses its validity, and in realistic applications new physical input is needed. Where this happens depends on the circumstances. The relevance of these results for the Standard Model will be discussed later.

3.9 Numerical simulation

With numerical simulations we get non-perturbative results albeit on finite lattices. Simulations provide furthermore a valuable kind of insight into the properties of the systems, which is complementary to expansions in some parameter.

The lattice is usually taken of the form $N^3 \times N_t$, with $N = 4, 6, 8, \ldots$, and N_t of the same order. For simplicity we shall assume that $N_t = N$ in the following. For the $O(4)$ model sizes 10^4–16^4 are already very useful. Expectation values

$$\langle O \rangle = \frac{1}{Z} \int D\phi \, \exp[S(\phi)] \, O(\phi) \tag{3.149}$$

are evaluated by producing a set of field configurations $\{\phi_x^\alpha\}_j$, $j = 1, \ldots, K$, which is distributed according to the weight factor $\exp S(\phi)$, giving the approximate result

$$\langle O \rangle \approx \overline{O} \equiv \frac{1}{K} \sum_{j=1}^{K} O(\phi_j), \tag{3.150}$$

with a statistical error $\propto 1/\sqrt{K}$. The ensemble is generated with a stochastic process, e.g. using a Metropolis or a Langevin algorithm. We shall give only a brief description of the Monte Carlo methods and the analysis of the results. Monte Carlo methods are described in more detail in [4, 6, 10].

For example, a Langevin simulation produces a sequence $\phi_{x,n}^\alpha$, $n = 1, 2, \ldots$, by the rule

$$\phi_{x,n+1}^\alpha = \phi_{x,n}^\alpha + \delta \frac{\partial S(\phi_n)}{\partial \phi_{x,n}^\alpha} + \sqrt{2\delta} \, \eta_{x,n}^\alpha, \tag{3.151}$$

where $\eta_{x,n}^\alpha$ are Gaussian pseudo-random numbers with unit variance and zero mean,

$$\langle \eta_{x,n}^\alpha \rangle = 0, \quad \langle \eta_{x,n}^\alpha \eta_{x',n'}^{\alpha'} \rangle = \delta_{\alpha\alpha'} \delta_{xx'} \delta_{nn'}, \tag{3.152}$$

and δ is a step size related to the Langevin time t by $t = \delta n$. It can be shown that as $t \to \infty$, the ϕ's become distributed according to the desired $\exp S(\phi)$, up to terms of order δ (cf. problem (viii)). Using a small δ such as 0.01, the system reaches equilibrium after some time, in units related to the mass gap of the model, and configurations ϕ_j may be recorded every $\Delta t = 1$, say. The finite δ produces a systematic error, which can be reduced by taking δ sufficiently small, or by using several δ's and extrapolating to $\delta = 0$. The configurations j and $j+1$ are usually correlated, such that the true statistical error is larger than the naive standard deviation

$$\sqrt{\frac{1}{K} \sum_{j=1}^K \left(O(\phi_j) - \overline{O} \right)^2} \tag{3.153}$$

but there are methods to take care of this.

The Metropolis algorithm is often preferred over the Langevin one, since it does not suffer from systematic step-size errors $\propto \delta$ and it is often more efficient. Research into efficient algorithms is fascinating and requires good insight into the nature of the system under investigation. New algorithms are being reported every year in the 'Lattice proceedings'.

Since the lattices are finite, we have to take into account systematic errors due to scaling ($O(a)$) violations and finite-size (L) effects ($L = Na$). It is important to determine these systematic errors and check that they accord with theoretical scaling and finite-size formulas. We can then attempt to extrapolate to infinite volume and zero lattice spacing.

Typical observables O for the $O(n)$ models are the average 'magnetization' $\bar{\phi}^\alpha = \sum_x \phi_x^\alpha / N^4$, the average 'energy' $-S/N^4$, which reduces to the average 'link' $\sum_{x\mu} \phi_x^\alpha \phi_{x+\hat{\mu}}^\alpha / 4N^4$ in the limit $\lambda \to \infty$, and products like $\phi_x^\alpha \phi_y^\beta$ giving correlation functions upon averaging. The free energy $F = -\ln Z$ itself cannot be obtained directly by Monte Carlo methods, but may be reconstructed, e.g. by integrating $\partial F / \partial \kappa = -2\langle \sum_{x\mu} \phi_x^\alpha \phi_{x+\hat{\mu}}^\alpha \rangle$.

The correlation function $G_{xy}^{\alpha\beta} = \langle \phi_x^\alpha \phi_y^\beta \rangle - \langle \phi_x^\alpha \rangle \langle \phi_y^\beta \rangle$ is used to determine the masses of particles. With periodic boundary conditions it depends only on the difference $x - y$. For example, in the symmetric phase the spectral representation can be written as

$$\sum_{\mathbf{x}} e^{-i\mathbf{px}} G^{\alpha\alpha}_{\mathbf{x},t;0,0} = \sum_{\gamma} |\langle 0|\phi^{\alpha}|\mathbf{p},\gamma\rangle|^2 \left[e^{-\omega_{\mathbf{p},\gamma}t} + e^{-\omega_{\mathbf{p},\gamma}(N_t-t)} \right], \quad (3.154)$$

where finite-temperature (finite N_t) corrections of the form $\propto$ $\langle \mathbf{p}'\gamma'|\phi^{\alpha}_x|\mathbf{p}\gamma\rangle$ have been neglected. Choosing zero momentum $\mathbf{p}$, one may fit the propagator data for large t and $N_t - t$ to

$$R \cosh\left[m\left(t - \frac{N_t}{2} \right) \right], \quad R = |\langle 0|\phi^{\alpha}|0\alpha\rangle|^2 \exp\left(-m\frac{N_t}{2} \right), \quad (3.155)$$

where $m = \omega_{\min}$ is the mass of the particle with the quantum numbers of ϕ^{α}. It is assumed that the contributions of the next energy levels ω' with the same quantum numbers (such as three particle intermediate states), which have relative size $\exp[-(\omega'-m)t]$, can be neglected for sufficiently large times. Alternatively, one can try to determine the renormalized mass and wave-function renormalization constant in momentum space from eq. (3.81), but this does not give the particle mass directly. Only when the particle is weakly coupled is $m_R \approx m$. The higher-order correlation functions (such as the four-point functions) require in general much better statistics than do the propagators.

For illustration we show first some early results in the symmetric phase. Figure 3.13 shows the particle mass and the renormalized coupling $g_R = 6\lambda_R$ as functions of the spatial size N in a simulation at infinite bare coupling [24]. We see that the interactions cause the finite-volume mass to increase over the infinite-volume value (the linear extent in physical units, Lm, changes by roughly a factor of two). The results for the coupling constant roughly agree within the errors with those obtained by Lüscher and Weisz using the hopping expansion. Figure 3.14 shows a result [25] for the dressed propagator (correlation function) analyzed in momentum space. The fact that the propagator resembles so closely a free propagator, apart from renormalization, is an indication that the effective interactions are not very strong, despite the large bare coupling.

The broken phase is physically more interesting. Although there is rigorously no phase transition in a finite volume, the difference between the symmetric- and broken-phase regions in parameter space is clear in the simulations. The phase boundary is somewhat smeared out by finite-volume effects. In the broken phase of the $O(n)$ model for $n > 1$, there is a preferred direction, along $\langle \phi^{\alpha} \rangle = v^{\alpha} \neq 0$, and one considers the longitudinal and transverse modes $G_{\sigma} = v^{-2} v^{\alpha} v^{\beta} G^{\alpha\beta}$ and $G_{\pi} = (\delta_{\alpha\beta} - v^{-2} v^{\alpha} v^{\beta}) G^{\alpha\beta}/(n-1)$. The latter correspond to the Gold-

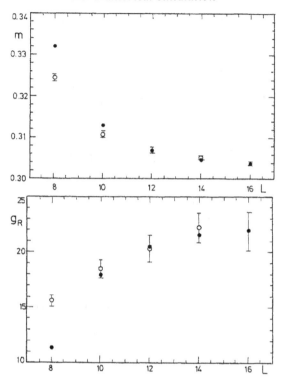

Fig. 3.13. Finite-size dependence of m and $g_R = 6\lambda_R$ in a simulation in the symmetric phase ($L = N$, $\lambda = \infty$). The full circles correspond to a finite-size dependence expected from renormalized perturbation theory. From [24].

stone bosons. The σ particle can decay into the π particles, which leads to complications in the analysis of the numerical data. The Goldstone bosons lead to strong finite-size effects. Finite-size effects depend on the range of the interactions, the correlation length, which is infinite for the Goldstone bosons. However, the finite size also gives a non-zero mass to the Goldstone bosons. These effects have to be taken into account in the analysis of the simulation results. The theoretical analysis is based on effective actions, using 'chiral perturbation theory' or 'renormalized perturbation theory'.

Consider the magnetization observable $\bar{\phi}^\alpha = \sum_x \phi_x^\alpha / N^4$. An impression of its typical distribution is illustrated in figure 3.15. The difference between the symmetric and broken phase is clear, yet the figure suggests correctly that the angular average leads to $\langle \bar{\phi}^\alpha \rangle = 0$ also in the broken-phase region. In a finite volume there is no spontaneous

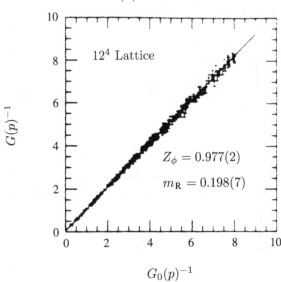

Fig. 3.14. Dressed propagator in momentum space plotted as a function of $\sum_\mu 4\sin^2(p_\mu/2)$, at $m_0^2 = -24.6$, $\lambda_0 = 100$. From [25].

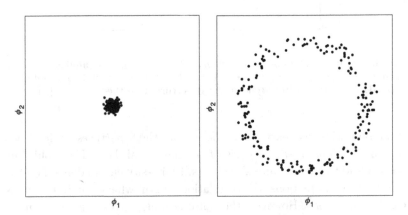

Fig. 3.15. Qualitative illustration of the probability distribution of $\bar{\phi}^\alpha$ at finite volume for $n = 2$ in the symmetric phase (left) and the broken phase (right).

symmetry breaking. To formulate a precise definition of v^α, we introduce an explicit symmetry-breaking term into the action, which 'pulls' the spins along a direction, say 0,

$$\Delta S = \epsilon \sum_x \phi_x^0, \tag{3.156}$$

and define

$$v^\alpha = \lim_{\epsilon \to 0} \lim_{L \to \infty} \langle \phi_x^\alpha \rangle, \tag{3.157}$$

where the order of the limits is crucial. To understand this somewhat better, one introduces the constrained effective potential $V_L(\bar{\phi})$, which is obtained by integrating over all field configurations with the constraint $\bar{\phi}^\alpha = \sum_x \phi_x^\alpha / L^4$,

$$\exp(-L^4 V_L(\bar{\phi})) = \int D\phi \, \exp[S(\phi)] \, \delta\left(\bar{\phi}^\alpha - \sum_x \phi_x^\alpha / L^4\right), \tag{3.158}$$

such that

$$Z = \int d^n \bar{\phi} \, \exp\left[-L^4 V_L(\bar{\phi})\right], \tag{3.159}$$

and

$$\langle \bar{\phi}^\alpha \rangle = \frac{\int d^n \bar{\phi} \, \exp\left[-L^4 V_L(\bar{\phi})\right] \bar{\phi}^\alpha}{\int d^n \bar{\phi} \, \exp\left[-L^4 V_L(\bar{\phi})\right]}. \tag{3.160}$$

The fact that the effective potential comes with a factor L^4 is easily understood from the lowest-order approximation in $\lambda \to 0$, which is obtained by simply inserting the constant $\bar{\phi}^\alpha$ for ϕ_x^α in the classical action,

$$S(\bar{\phi}) = -L^4 V_L(\bar{\phi}) = -N^4[(1 - 8\kappa)\bar{\phi}^2 + \lambda(\bar{\phi}^2 - 1)^2 - \epsilon\bar{\phi}^0], \tag{3.161}$$

where we used the form (3.16) of the action in lattice units. In this classical approximation the constraint effective potential is independent of L. The exact constraint effective potential is only weakly dependent on L, for sufficiently large L, and as L increases the integrals in (3.160) are accurately given by the saddle-point approximation, i.e. by the sum over the minima of $V_L(\bar{\phi})$. In absence of the ϵ term there is a continuum of saddle points and $\langle \bar{\phi}^\alpha \rangle = 0$ even in the broken phase. A unique saddle point is obtained, however, for non-zero ϵ. If we let ϵ go to zero after the infinite-volume limit, $\langle \bar{\phi}^\alpha \rangle$ remains non-zero. For more information on the constraint effective potential, see e.g. [26].

This technique of introducing explicit symmetry breaking is used in simulations [27] as shown in figure 3.16. A simpler estimate of the infinite-volume value v of the magnetization is obtained with the 'rotation method', in which the magnetization of each individual configuration is rotated to a standard direction before averaging. The resulting $\langle |\bar{\phi}| \rangle$ can be fitted to a form $v + \text{constant} \times N^{-2}$.

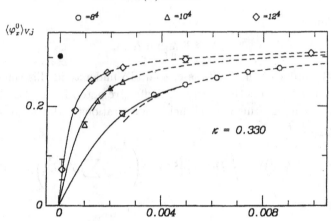

Fig. 3.16. Plots of $\langle \bar{\phi}^4 \rangle$ as a function of $j = \epsilon$ in the $O(4)$ model for various lattice sizes. The data are fitted to the theoretical behavior (curves) and extrapolated to infinite-volume and $\epsilon = 0$, giving the full circle in the upper left-hand corner. From [27].

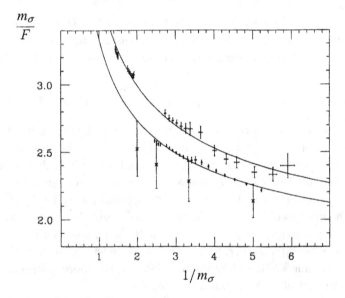

Fig. 3.17. Results on $m_\sigma / F = m_\sigma / v_R$ as a function of the correlation length $1/m_\sigma$, for the 'standard (usual) action' (lower data) and a 'Symanzik-improved action' (upper data). The crosses are results of Lüscher and Weisz obtained with the hopping expansion. The bare coupling $\lambda = \infty$. The curves are interpolations based on renormalized perturbation theory. From [28].

As a last example we show in figure 3.17 results on the renormalized coupling $\sqrt{2\lambda_R} = m_R/v_R$ [28]. Data are shown for the action considered here (the 'standard action') and for a 'Symanzik-improved action'. We see that the data for the standard action agree with the results obtained with the hopping expansion in the previous section, within errors. The Symanzik-improved action has next-to-nearest-neighbor couplings such that the $O(a^2)$ errors are eliminated in the classical continuum limit. It is not clear *a priori* that this leads to better scaling in the quantum theory, because the scalar field configurations that contribute to the path integral are not smooth on the lattice scale, but it is interesting that the different regularization leads to somewhat larger renormalized couplings for a given correlation length.

In conclusion, the numerical simulations have led to accurate results which fully support the theoretical understanding that the $O(n)$ models are 'trivial'.

3.10 Real-space renormalization group and universality

One of the cornerstones of quantum field theory is universality: the physical properties emerging in the scaling region are to a large extent independent of the details of formulating the theory on the scale of the cutoff. The physics of the $O(n)$ models is expected to be independent of the lattice shape, the addition of next-nearest-neighbor couplings, next-next-nearest-neighbor couplings, ..., or higher-order terms $(\phi^2)^k$, $k = 3, 4, \ldots$ (of course, in its Ising limit or non-linear sigma limit where $\phi^2 = 1$ such higher-order terms no longer play a role). More precisely, the physical outcome of the models falls into universality classes, depending on the symmetries of the system and the dimensionality of space–time. Our understanding of universality comes from the renormalization group *à la* Wilson [29, 30] ('block spinning', see e.g. [11]), and from the weak-coupling expansion. We shall sketch the ideas using the one-component scalar field as an example, starting with the block spinning approach used in the theory of critical phenomena.

In the real-space renormalization-group method one imagines integrating out the degrees of freedom with wave lengths of order of the lattice distance and expressing the result in terms of an effective action for the remaining variables. On iterating this procedure one obtains the effective action describing the theory at physical ($\gg a$) distance scales.

Let $\bar{\phi}_{\bar{x}}$ be the average of ϕ_x over a region of linear size s around $\bar{x}$,

$$\bar{\phi}_{\bar{x}} = \sum_x B_{\bar{x},x}\phi_x. \tag{3.162}$$

The averaging function $B(\bar{x}, x)$ is concentrated near sites $\bar{x}$ on a coarser lattice that are a distance s apart in units of the original lattice. We could simply take values $\bar{x}_\mu = 2x_\mu$ with $B_{x,\bar{x}} = z\sum_\mu \delta_{\bar{x}\pm\hat{\mu},x}$ ('blocking'), or a smoother Gaussian average $B = z\sum_x \exp(-(x-\bar{x})^2/2s)$, with suitable normalization factors z. The effective action $\bar{S}$ is defined by

$$e^{\bar{S}(\bar{\phi})} = \int D\phi\, e^{S(\phi)} \prod_{\bar{x}} \delta\left(\bar{\phi}_{\bar{x}} - \sum_x B_{\bar{x},x}\phi_x\right), \tag{3.163}$$

and it satisfies

$$\int D\bar{\phi}\, e^{\bar{S}(\bar{\phi})} = \int D\phi\, e^{S(\phi)}. \tag{3.164}$$

After a few iterations the effective action has many types of terms, so one is led to consider general actions of the form

$$S(\phi) = \sum_\alpha K_\alpha O_\alpha(\phi). \tag{3.165}$$

Here O_α denotes terms of the schematic form $(\partial_\mu\phi\partial_\mu\phi)^k$, $(\phi^2)^k$, $\ldots$ ($k = 1, 2, \ldots$). The new effective action can then again be written in the form

$$\bar{S}(\bar{\phi}) = \sum_\alpha \bar{K}_\alpha O_\alpha(\bar{\phi}). \tag{3.166}$$

The scale factor z in the definition of the averaging function B is chosen such that the coefficient of $\partial_\mu\bar{\phi}\partial_\mu\bar{\phi}$ is equal to $\frac{1}{2}$, in lattice units of the coarse $\bar{x}$ lattice, in order that the new coefficients $\bar{K}_\alpha$ do not run away after many iterations. Because of the locality of the averaging function one expects the action $\bar{S}$ to be local too, i.e. the range of the couplings in $\bar{S}$ is effectively finite, and one expects the dependence of the coefficients $\bar{K}_\alpha$ on K_α to be analytic. One iteration thus constitutes a renormalization-group transformation

$$\bar{K}_\alpha = T_\alpha(K). \tag{3.167}$$

We can still calculate correlation functions and quantities of physical interest with the new fields $\bar{\phi}$. For these the highest-momentum contributions are suppressed by the averaging, as can be seen by expressing them in terms of the original fields ϕ, but contributions from physical momenta which are low compared to the cutoff are unaffected. In particular the correlation length in units of the original lattice distance is unchanged. However, in units of the $\bar{x}$ lattice distance the correlation length is smaller by a factor $1/s$. Each iteration the correlation length is shortened by a factor $1/s$ and when it is of order one we imagine stopping the iterations. We can then still extract the physics on the momentum scales of order of the mass scale. If we want to discuss scales ten times higher, we can stop iterating when the correlation length is still of order ten.

In the infinite dimensional space of coupling constants K_α there is a hypersurface where the correlation length is infinite, the critical surface. We want to start our iterations very close to the critical surface because we want a large correlation length on the original lattice, which means that we are able to do many iterations before the correlation length is of order unity. If there is a fixed point K^*,

$$T_\alpha(K^*) = K_\alpha^*, \tag{3.168}$$

then we can perform many iterations near such a point without changing the K_α very much. At such a fixed point the correlation length does not change, so it is either zero or infinite. We are of course particularly interested in fixed points in the critical surface. Linearizing about such a critical fixed point (on the critical surface),

$$\bar{K}_\alpha - K_\alpha^* = M_{\alpha\beta}(K_\beta - K_\beta^*), \quad M_{\alpha\beta} = [\partial T_\alpha/\partial K_\beta]_{K=K^*}, \tag{3.169}$$

it follows that the eigenvalues λ_i of $M_{\alpha\beta}$ determine the attractive ($\lambda_i < 1$) or repulsive ($\lambda_i > 1$) directions of the 'flow'. These directions are given by the corresponding eigenvectors e_i^α, which determine the combinations $e_i^\alpha O_\alpha$.

One expects only a few repulsive eigenvalues, called 'relevant', while most of them are attractive and called 'irrelevant'. Eigenvalues $\lambda_i = 1$ are called marginal. Further away from the fixed point the attractive and repulsive directions will smoothly deform into attractive and repulsive curves. The marginal directions will also turn into either attractive or repulsive curves.

Let us start the iteration somewhere on the critical surface. Then the flow stays on the surface. Suppose that the flow on the surface is attracted to a critical fixed point K^*. Next let us start very close to the critical surface. Then the flow will at first still be attracted to K^*, but, since with each iteration the correlation length decreases by a factor $1/s$, the flow moves away from the critical surface and eventually turns away from the fixed point. Hence the critical fixed point has at least one relevant direction away from the critical surface.

Suppose there is only one such relevant direction (and its opposite on the other side of the critical surface). Then, after many iterations the flow just follows the flow-line through this relevant direction. The physics is then completely determined by the flow-line through the relevant direction: the physical trajectory (also called the renormalized trajectory). To the relevant direction there corresponds the only free parameter we end up with: the ratio cutoff/mass, Λ/m (where $\Lambda = \pi/a$). This ratio is determined by the initial distance to the critical surface, or equivalently, by the number of iterations and the final distance to the critical surface where we stop the iterations. However, the mass just sets the dimensional scale of the theory and there is no physical free parameter at all under these circumstances. All the physical properties (e.g. the renormalized vertex functions and the renormalized coupling λ_R) are fixed by the properties of the physical trajectory. On the other hand, each additional relevant direction offers the possibility of an additional free physical parameter, which may be tuned by choosing appropriate initial conditions.

Many years of investigation have led to the picture that there is only one type of critical fixed point in the $O(n)$ symmetric models, which has only one relevant direction corresponding to the mass as described above, and one marginal but attractive direction corresponding to the renormalized coupling. This means that eventually the renormalized coupling will vanish after an infinite number of iterations (triviality). This is the reason that the fixed point is called 'Gaussian', for the corresponding effective action is quadratic. However, because the renormalized coupling is marginal and therefore changes very slowly near the critical point, it can still be substantially different from zero even after very many iterations (very large Λ/m ratios). With a given number of iterations we can imagine maximizing the renormalized coupling over all possible initial actions parameterized by K_α, giving an upper bound on the renormalized coupling. Within its upper bound the renormalized coupling is then still a free parameter in the models. The situation is illustrated in figure 3.18.

For the massless theory the correlation length is infinite, so we start on the critical surface. The flow is attracted to K^*, which determines the physics outcome. The marginally attractive direction corresponds in the massless case to the running renormalized coupling $\bar{\lambda}(\mu)$ at some physical momentum scale μ. Each iteration the maximum momentum scale is lowered by a factor $1/s$ and, after many iterations, the ratio (maximum momentum scale)/cutoff is very small. We stop the iteration

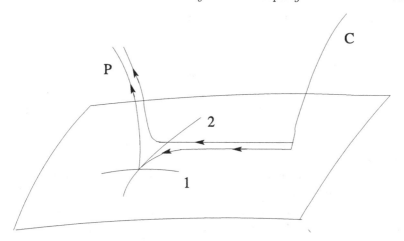

Fig. 3.18. Renormalization group-flow in ϕ^4 theory. The line C represents the 'canonical surface' of actions of the standard form $S = 2\kappa \sum_{x\mu} \phi_x \phi_{x+\hat{\mu}} - \sum_x [\phi_x^2 + \lambda(\phi_x^2 - 1)^2]$. The line P represents the physical trajectory. Direction 1 is an irrelevant direction, direction 2 represents the marginal direction corresponding to the renormalized coupling. Shown are two flows starting from a point in C, one near the critical surface and on this surface.

when the maximum momentum scale is of order μ. For a given number of iterations the running coupling can still vary within its upper bound. As the number of iterations goes to infinity, μ has to go to zero and $\bar{\lambda}(\mu) \to 0$ because the flow along the marginal direction is attracted to zero coupling. So the massless theory can be defined by taking the number of iterations ($\propto \Lambda/\mu$) large but finite, and $\bar{\lambda}(\mu) \to 0$ as $\mu \to 0$.

The critical fixed points of the real-space renormalization-group transformation give a very attractive explanation of universality.

3.11 Universality at weak coupling

To formulate a general action at weak coupling we start with the form (3.16) and first make a scale transformation $\phi = \phi'/\sqrt{\lambda}$, which brings the action into the form

$$S(\phi') = \frac{1}{\lambda} \sum_x \left[2\kappa \sum_\mu \phi'_x \phi'_{x+\hat{\mu}} - \phi_x'^2 - (\phi_x'^2 - 1)^2 \right]. \qquad (3.170)$$

We see that λ appears as a natural expansion parameter for a saddle-point expansion, while the other coefficients in the action are of order one in lattice units. A natural generalization is given by

$$S(\phi') = \frac{1}{\lambda} \sum_x \left[2\kappa \sum_\mu \phi'_x \phi'_{x+\hat{\mu}} + \kappa' \sum_{\mu<\nu} \phi'_x \phi'_{x+\hat{\mu}+\hat{\nu}} \right. \tag{3.171}$$

$$\left. + \kappa'' \sum_\mu \phi'_x \phi'_{x+2\hat{\mu}} + \kappa''' \sum_\mu \phi'^2_x \phi'^2_{x+\hat{\mu}} + \cdots - \sum_{k=1}^{\infty} v_{2k} \phi'^{2k}_x \right],$$

which still has the symmetry $\phi \to -\phi$. The coefficients in this expression are supposed to be of order 1.

The parameter λ enters in the same place as Planck's constant $\hbar$ when we introduced the path-integral quantization method, before we set it equal to 1. It can be shown that the expansion in powers of $\hbar$ corresponds to an expansion in the number of loops in Feynman diagrams. For this reason the weak-coupling expansion is called the semiclassical expansion.

For convenience in the following we shall use the original continuum-motivated parameterization (3.13) with the field $\varphi = \phi/\sqrt{2\kappa}$ and rewrite (3.171) in the form

$$S = -\frac{1}{\lambda_0} \sum_x \left[\frac{1}{2} \partial_\mu \varphi'_x \partial_\mu \varphi'_x + z \partial_\mu \varphi'^2_x \partial_\mu \varphi'^2_x + \cdots + \sum_k u_{2k} \varphi'^{2k}_x \right]$$

$$= -\sum_x \left[\frac{1}{2} \partial_\mu \varphi_x \partial_\mu \varphi_x + \lambda_0 z \partial_\mu \varphi^2_x \partial_\mu \varphi^2_x + \cdots + \sum_k \lambda_0^{k-1} u_{2k} \varphi^{2k}_x \right],$$
$$\tag{3.172}$$

where $\varphi' = \sqrt{\lambda_0}\varphi$. Here again the coefficients $z, \ldots,$ and u_{2k} are supposed to be dimensionless numbers of order unity, with the exception of u_2 which becomes $m_{0c}^2 = O(\lambda_0)$ at the phase boundary (this is special to the continuum parameterization). It is instructive to rewrite the generic action (3.172) in physical units,

$$S = -\sum_x \left(\frac{1}{2} \partial_\mu \varphi_x \partial_\mu \varphi_x + a^2 \lambda_0 z \partial_\mu \varphi^2_x \partial_\mu \varphi^2_x + \cdots \right.$$

$$\left. + \sum_k a^{2k-4} \lambda_0^{k-1} u_{2k} \varphi^{2k}_x \right), \tag{3.173}$$

where now $\partial_\mu \varphi_x = (\varphi_{x+a_\mu} - \varphi_x)/a$ and $\sum_x$ contains a factor a^4. The higher-dimensional operators are accompanied by powers of the lattice distance a such that the action is dimensionless.

In the classical continuum limit $a \to 0$ we end up with just the φ^4 theory, with u_2 chosen such that $m^2 = 2u_2 a^{-2}$ remains finite. In other

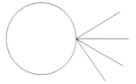

Fig. 3.19. Contribution of the bare six-point vertex to $\Gamma_{(4)}$.

words, the bare two- and four-point vertex functions take their usual continuum limits and the higher-order bare vertex functions vanish. In non-trivial orders of the semiclassical expansion, the powers of a in the bare vertex functions can be compensated by the divergences in the loop diagrams. For example, consider the effect of the term $-\sum_x \lambda_0^2 u_6 a^2 \varphi_x^6$ on the four-point vertex at one-loop order as given by the diagram in figure 3.19. The bare vertex function in momentum space is $-6! \, \lambda_0^2 u_6 a^2$ and the contribution to $\Gamma_{(4)}$ is given by

$$
-\tfrac{1}{2} 6! \, \lambda_0^2 u_6 a^2 \int_{-\pi/a}^{\pi/a} \frac{d^4 l}{(2\pi)^4} \frac{1}{m_0^2 + a^{-2} \sum_\mu (2 - 2\cos a l_\mu)} = -\tfrac{1}{2} 6! \lambda_0^2 u_6 C_0,
$$

(3.174)

in the limit $a \to 0$ (the constant C_0 was defined in (3.67)).

By looking at more examples one may convince oneself that the higher-order bare vertex functions just lead to new expressions for the vertex functions in terms of the coefficients in the action, which have, however, the same momentum dependence as before. All lattice artifacts end up in constants like C_0, and in the relation between λ_R and m_R^2 to λ_0 and m_0^2, such that the renormalized vertex functions, once expressed in terms of the renormalized coupling λ_R and renormalized mass parameter m_R, are universal, order by order in perturbation theory.

There is one aspect worth mentioning: the effect of the lattice symmetries. Consider the two-point vertex function in one-loop order, which has the form $\Gamma_{(2)}(p) = a^{-2} f(ap, am_0)$ on dimensional grounds. For $a \to 0$ this takes the form $a^{-2}(\tau a^2 m_0^2 + \tau_{\mu\nu} a^2 p_\mu p_\nu) +$ logarithms. We have seen in section 3.4 how the logarithms emerge from the integration over the loop variable near the origin in momentum space where the lattice expressions take their classical continuum form: the terms containing logarithms are covariant under continuous rotations. What about the polynomial $\tau_{\mu\nu} p_\mu p_\nu$? Its coefficient $\tau_{\mu\nu}$ depends on lattice details, the loop integrations over the cosines near the edge of the Brillouin zone in momentum space. Here the lattice symmetries come to help. The polynomial has to be invariant under the cubic rotations

$$R^{(\rho\sigma)}: \ x_\rho \to x_\sigma, \ x_\sigma \to -x_\rho, \ x_{\mu \neq \rho,\sigma} \to x_\mu \qquad (3.175)$$

and axis reversals

$$I^{(\rho)}: \ x_\rho \to -x_\rho, \ x_{\mu \neq \rho} \to x_\mu. \qquad (3.176)$$

There is only one such polynomial: $p^2 = p_1^2 + \cdots + p_4^2$. So the lattice symmetries and dimensional effects are important in order to get covariant renormalized vertex functions. Dimensional analysis showed that the above polynomial is at most of second order and even in $p_\mu \to -p_\mu$ because of axis-reversal symmetry. Note that there is more than one fourth-order polynomial $\tau_{\kappa\lambda\mu\nu} p_\kappa p_\lambda p_\mu p_\nu$ that is invariant under the lattice symmetries. Such polynomials go together with dimensional couplings, such as cutoff effects $\propto a^2$. The polynomials are called *contact terms*, because they correspond in position space to Dirac delta functions and derivatives thereof.

If we destroy the space–time symmetry of the lattice, e.g. by having different couplings in the space and time directions, then we may have to *tune* the couplings in the action to regain covariance in the scaling region.

3.12 Triviality and the Standard Model

Arguments that scalar field models are trivial in the sense that they become non-interacting when the regularization is removed were first given by Wilson, using his formulation of the renormalization group [29, 30]. The arguments imply that triviality should hold within a universality class of bare actions, e.g. next-to-nearest-neighbor couplings, ... In the previous sections we reviewed some calculations and numerical simulations leading to accurate determination of the renormalized coupling in the $O(4)$ model in the broken phase. The $O(4)$ model may be identified with the scalar Higgs sector of the Standard Model, and we shall now review the implications and applications of triviality.

First we review how the $O(4)$ model is embedded in the Standard Model. The action for the Higgs field is given by

$$S_{\mathrm{H}} = -\int d^4x \left[(D_\mu \varphi)^\dagger D_\mu \varphi + m_0^2 \varphi^\dagger \varphi + \lambda_0 (\varphi^\dagger \varphi)^2 \right], \quad (3.177)$$

$$D_\mu \varphi = \left(\partial_\mu - ig_1 \frac{1}{2} B_\mu - ig_2 W_\mu^k \frac{\tau_k}{2} \right) \varphi, \qquad (3.178)$$

where τ_k are the Pauli matrices, $\varphi = (\varphi_u, \varphi_d)^{\mathrm{T}}$ is the complex Higgs doublet and B_μ, and W_μ^k are the $U(1) \times SU(2)$ electroweak gauge fields. Setting the gauge couplings to zero, the action becomes equivalent to the $O(4)$ model,

$$S_{|g=0} = -\int d^4x \, [\partial_\mu \varphi^\dagger \partial_\mu \varphi + m_0^2 \varphi^\dagger \varphi + \lambda_0 (\varphi^\dagger \varphi)^2] \tag{3.179}$$

$$= -\int d^4x \left[\frac{1}{2} \partial_\mu \varphi^\alpha \partial_\mu \varphi^\alpha + \frac{m_0^2}{2} \varphi^\alpha \varphi^\alpha + \frac{\lambda_0}{4} (\varphi^\alpha \varphi^\alpha)^2 \right],$$

$$\varphi_u = \frac{1}{\sqrt{2}}(\varphi^2 + i\varphi^1), \quad \varphi_d = \frac{1}{\sqrt{2}}(\varphi^0 - i\varphi^3). \tag{3.180}$$

The Higgs field enters also in Yukawa couplings with the fermions. In terms of a matrix field ϕ defined by

$$\phi \equiv \sqrt{2} \begin{pmatrix} \varphi_d^* & \varphi_u \\ -\varphi_u^* & \varphi_d \end{pmatrix}, \tag{3.181}$$

$$= \varphi^0 + i\varphi^k \tau_k = \rho V, \quad V \in SU(2), \quad \rho > 0, \tag{3.182}$$

the Yukawa couplings to the quarks can be expressed as

$$S_Y = -\int d^4x \, \bar{\psi}_{cg} (P_{\mathrm{R}} \phi y_g + y_g P_{\mathrm{L}} \phi^\dagger) \psi_{cg}. \tag{3.183}$$

Here $P_{\mathrm{L,R}} = (1 \mp \gamma_5)/2$ are the projectors on the left- and right-handed fermion fields and the summation is over the QCD colors c and generations g. The Yukawa couplings y are diagonal in $SU(2)$ doublet space, $y = y_u(1+\tau_3)/2 + y_d(1-\tau_3)/2$. The Yukawa couplings to the leptons are similar (in the massless neutrino limit the right-handed neutrino fields decouple).

If we insert the vacuum expectation value of the scalar field

$$\varphi = \frac{1}{\sqrt{2}} \begin{pmatrix} 0 \\ v \end{pmatrix}, \tag{3.184}$$

$$\phi = v\mathbb{1}, \tag{3.185}$$

in the action, we find the masses of the vector bosons W and Z and the photon A, from the terms quadratic in the gauge fields, and the masses of the fermions from the Yukawa couplings. Choosing renormalization conditions such that the 'tree-graph' relations remain valid after renormalization, we have

$$m_W^2 = \tfrac{1}{4} g_{2\mathrm{R}}^2 v_{\mathrm{R}}^2, \quad m_Z^2 = \tfrac{1}{4}(g_{1\mathrm{R}}^2 + g_{2\mathrm{R}}^2) v_{\mathrm{R}}^2, \quad m_A = 0, \tag{3.186}$$

$$m_f = y_{\mathrm{R}f} v_{\mathrm{R}}, \quad m_{\mathrm{H}}^2 = 2\lambda_{\mathrm{R}} v_{\mathrm{R}}^2, \tag{3.187}$$

where f denotes the fermion. From experiment we know

$$v_R = 246 \text{ GeV}, \quad g_{1R} = 0.34, \quad g_{2R} = 0.64. \tag{3.188}$$

The electroweak gauge couplings are effectively quite small (recall the typical factors of g^2/π^2 that occur in perturbative expansions). The Yukawa couplings are generally much smaller, as follows from (3.187) and the fact that the fermion masses are generally small (< 5 GeV) on the electroweak scale. Even the much heavier top quark, which has a mass of about 175 GeV has a Yukawa coupling $y_t \approx 0.71$, which is not very large either. The (running) QCD gauge coupling of the strong interactions is also relatively small on the electroweak scale of 100 GeV: $g_{3R} \approx 1.2$.

To discuss the implications of triviality of the $O(4)$ model, let us assume for the moment that all the gauge and Yukawa couplings can be treated as perturbations on scales v_R or higher. Furthermore, assume that the Higgs mass is non-zero (we shall comment on these assumptions below). It then follows from the triviality of the $O(4)$ model that the Standard Model itself must be 'trivial'. Because a non-zero Higgs mass implies $\lambda_R \neq 0$, the triviality leads to the conclusion that the regularization cannot be removed completely. Consequently the model must lose its validity on the regularization scale. New physical input is required on this momentum or equivalent distance scale.

It would obviously be very interesting if we could predict at which scale new physics has to come into play. To some extent this can be done as follows. If the Higgs mass is not too large such that $\lambda_R = m_H^2/2v_R$ is in the perturbative domain, we can use eq. (3.140) to calculate the cutoff $\Lambda = \pi/a$ in the lattice regularization,

$$\Lambda = m_H C(\beta_1 \lambda_R)^{\beta_2/\beta_1^2} \exp(1/\beta_1 \lambda_R)[1 + O(\lambda_R)], \tag{3.189}$$

where $C = \pi/C_1'(\lambda_0)$. The constant C_1 is minimal, hence Λ maximal, for infinite bare coupling λ_0. We shall assume this in the following, with $C_1'(\infty) = 6.4$ (the value obtained by Lüscher and Weisz). As an example, $m_H = 100$ GeV gives $\lambda_R = 0.083$ and $\Lambda = 7 \times 10^{36}$ GeV. This value for Λ is far beyond the Planck mass $O(10^{19})$ GeV for which gravity cannot be neglected. Certainly new physics comes into play at the Planck scale, so effectively the regulator scale for $m_H = 100$ GeV may be considered to be irrelevantly high. On the other hand, when m_H increases, Λ decreases. When λ_R becomes too large eq. (3.189) can no longer be trusted, but we still have non-perturbative results for λ_R and the corresponding Λ/m_H anyhow. For example, for $m_H = 615$ GeV ($m_H/v_R = 2.5$) figure

3.17 shows that $1/am_H \approx 3$; hence $\Lambda \approx 3\pi m_H = 5800$ GeV, for the standard action. For the Symanzik-improved action this would be $\Lambda \approx 8300$ GeV.

So we can compute a cutoff scale Λ from knowledge of the Higgs mass, but this Λ is clearly regularization dependent (the dependence of C in (3.189) on λ_0 also indicates a regularization dependence, cf. c in eq. (3.146)). For values of m_H deep in the scaling regime Λ is very sensitive to changes in m_H, but at the edge of the scaling region, e.g. for $\Lambda/m_H \approx 6$, this dependence is greatly reduced.

This supports the idea of establishing an upper bound on the Higgs mass: given a criterion for allowed scaling violations, there is an upper bound on m_H [31]. For example, requiring $\Lambda/m_H > 2\pi$ $(am_H < \frac{1}{2})$, we get an upper bound on m_H from results like figure 3.17. This should then be maximized over all possible regularizations. Figure 3.17 shows that the standard and Symanzik-improved actions give the bounds $m_H/v_R \lesssim 2.7$ and 3, respectively. A way to search through regularization space systematically has been advocated especially by Neuberger [32]. To order $1/\Lambda^2$ all possible regularizations (including ones formulated in the continuum) can be represented by a three-parameter action on the F_4 lattice, which has more rotational symmetry than does the hypercubic lattice. It is believed that the results of this program will not lead to drastic changes in the above result $m_H/v_R \lesssim 3$.

The Pauli–Villars regularization in the continuum appears to give much larger Λ's than the lattice [33]. The problem with relating various regularization schemes lies in the fact that it is not immediately clear what the physical implications of a requirement like $\Lambda/m_H > 2\pi$ are. One may correlate Λ to regularization artifacts (mimicking 'new physics') in physical quantities, such as the scattering amplitude for the Goldstone bosons. Requiring, in a given regularization, that such an amplitude differs by less than 5%, say, from the value obtained in renormalized perturbation theory, would determine Λ and the corresponding upper bound on m_H in that regularization. The significance of such criteria is unclear, however.

At this point it is useful to recall one example in which nature (QCD) introduces 'new physics'. The $O(4)$ model may also be interpreted as giving an effective description of the three pions, which are Goldstone bosons with masses around 140 MeV due to explicit symmetry breaking. The expectation value v_R is equal to the pion decay constant, $v_R = f_\pi = 93$ MeV. The analog of the Higgs particle is the very broad σ resonance around 900 MeV. The low-energy pion physics is approximately described by the $O(4)$ model. However, since $m_\sigma/v_R \approx 10$ is far above the upper bounds found above, the cutoff needed in this application of the $O(4)$ model is very low, probably even below m_σ, and the model is not expected to describe physics at the σ scale. Indeed, there is 'new physics' in the form of the well-known ρ resonance with a mass of 770 MeV and width of about 150 MeV.

Let us now discuss the assumptions of neglecting the effect of the gauge and Yukawa couplings. The gauge couplings $g_{2,3}$ are asymptotically free and their effective size is even smaller on the scale of the cutoff. So it seems reasonable that their inclusion will not cause large deviations from the above results. The $U(1)$ coupling g_1 is not asymptotically free and its effective strength grows with the momentum scale. However, its size on the Planck scale is still small. If we accept not putting the cutoff beyond the Planck scale anyway, then also the gauge coupling g_1 may be expected to have little influence. The same can be said about the Yukawa couplings, which are also not asymptotically free (possibly with the exception of the top-quark coupling).

These expectations have been studied in some detail. An important result based on $O(4)$ Ward identities is that relations like $m_W^2 = g_2^2 v_R^2/4$ are still valid to first order in g_2^2 on treating the Higgs self-coupling non-perturbatively [31, 34]. This may be seen as justifying a definition of the g_R such that (3.186) is exact.

Of course, it is desirable to verify the above expectations non-perturbatively. A lattice formulation of the Standard Model is difficult because of problems with fermions on a lattice (cf. section 8.4). However, lattice formulations of gauge-Higgs systems and to a certain extent Yukawa models are possible and have been much studied over the years. The lattice formulation of gauge-Higgs systems has interesting aspects having to do with the gauge-invariant formulation of the Higgs phenomenon, presentation and discussion of which here would lead too far [35].

It turns out that the Yukawa couplings are also 'trivial' and that the maximum renormalized coupling is also relatively weak, see for example [36]. Numerical simulations have set upper bounds on the masses of possible hitherto undiscovered generations of heavy fermions (including heavy neutrinos), as well as the influence of such generations on the Higgs-mass bound.

Finally, the assumption made above, namely that $m_H \neq 0$, is justified by theoretical arguments for a *lower bound* on m_H, which are based on the effect that the fermions and gauge bosons induce on the Higgs self-couplings (for reviews, see [37, 38]).

3.13 Problems

(i) *Six-point vertex*

Determine the Feynman diagrams for the six-point vertex function in the φ^4 theory in the one-loop approximation. For one of these diagrams, write down the corresponding mathematical expression in lattice units $(a = 1)$ and in physical units $(a \neq 1)$. Show that it converges in the limit $a \to 0$, to the expression one would write down directly in the continuum.

(ii) *Renormalized coupling for mass zero*

In the massless $O(n)$ model λ_R is ill defined. In this case $\bar{\lambda}(\mu)$ is still a good renormalized coupling. Give the renormalized four-point vertex function $\Gamma^R_{\alpha_1 \cdots \alpha_4}(p_1 \cdots p_4)$ in terms of $\bar{\lambda}(\mu)$.

(iii) *Critical κ and m_0 at weak coupling*

What are the critical values of the bare mass m^2_{0c} (in lattice units) and the hopping parameter κ_c to first order in λ_0 in the weak-coupling expansion?

(iv) *Minimal subtraction*

To obtain renormalized vertex functions in the weak-coupling expansion, wavefunction, mass, and coupling-constant renormalizations are needed. Here we concentrate on the latter. We substitute the bare λ_0 for a series in terms of a renormalized λ (not to be confused with the λ in the lattice parameterization (3.15)),

$$\lambda_0 = \lambda Z_\lambda(\lambda, \ln a\mu),$$

$$Z_\lambda(\lambda, \ln a\mu) = 1 + \sum_{n=1}^{\infty} \sum_{k=0}^{n} Z_{nk} \lambda^n (\ln a\mu)^k$$

$$= \sum_{k=0}^{\infty} Z_k(\lambda)(\ln a\mu)^k. \tag{3.190}$$

Terms vanishing as $a \to 0$ have been neglected, order by order in perturbation theory. From the point of view of obtaining finite renormalized vertex functions we can be quite liberal and allow any choice of the coefficients Z_{nk} leading to a series in λ for physical quantities for which the a dependence cancels out.

The renormalized λ depends on a physical scale μ but not on a, it is a 'running coupling', whereas λ_0 is supposed to depend on a but not on μ. Introducing a reference mass μ_1, we write

$$\lambda_0(t) = \lambda(s) Z_\lambda(\lambda(s), s - t), \quad t = -\ln(a\mu_1), \quad s = \ln(\mu/\mu_1). \tag{3.191}$$

From $0 = d\lambda_0/ds$ we find

$$\left(\frac{\partial}{\partial s} + \beta \frac{\partial}{\partial \lambda}\right) \lambda Z_\lambda(\lambda, s - t) = 0, \qquad (3.192)$$

where

$$\beta = \frac{\partial \lambda}{\partial s}. \qquad (3.193)$$

Using the above expansion for Z_λ in terms of powers $(\ln a\mu)^k = (t - s)^k$, show that

$$(k + 1)Z_k(\lambda) + \beta \frac{\partial}{\partial \lambda}(\lambda Z_k(\lambda)) = 0, \quad k = 0, 1, 2, \ldots, \qquad (3.194)$$

and hence that the β function is given by

$$\beta(\lambda) = -\frac{Z_1(\lambda)}{\partial(\lambda Z_0(\lambda))/\partial \lambda}. \qquad (3.195)$$

In a minimal subtraction scheme one does not 'subtract' more than is necessary to cancel out the $\ln(a\mu)$'s, and one chooses $Z_0(\lambda) \equiv 1$. Notice that there is a whole class of minimal subtraction schemes: we may replace $\ln(a\mu)$ by $\ln(a\mu) + c$, with c some numerical constant, since such a c is equivalent to a redefinition of μ. It follows that the beta function in a minimal subtraction scheme can be read off from the coefficients of the terms involving only a single power of $\ln a\mu$:

$$\beta(\lambda) = -Z_1(\lambda). \qquad (3.196)$$

Show that in minimal subtraction the beta function $\beta_0(\lambda_0)$ for λ_0 is identical to $\beta(\lambda_0)$.

Assuming that the beta function is given, solve eq. (3.192) with the boundary condition $Z_\lambda(\lambda, 0) = Z_0(\lambda) = 1$.

(v) *Mass for small κ*

The hopping result (3.128) shows that the mass parameter m_R is infinite for $\kappa = 0$. For small κ we see from (3.122) that $G_{xy} \propto (2g\kappa)^{L_{xy}} = \exp(-m_{xy}|x - y|)$, where L_{xy} is the minimal number of steps between x and y. We can identify a mass $m_{xy} = -\ln(2g\kappa)\,(L_{xy}/|x - y|)$. For small κ, compare m_{xy} for x, y along a lattice direction and along a lattice diagonal with the results of problem (i) in chapter 2. Compare also with equations (2.117), (2.120) and (2.122), for the case that x and y are along a timelike direction in the lattice.

(vi) *An example of a divergent expansion*

Instructive examples of convergent and divergent expansions, in κ and λ, are given by

$$z(\kappa, \lambda) = \int_{-\infty}^{\infty} d\phi \exp(-\kappa\phi^2 - \lambda\phi^4) \qquad (3.197)$$

$$= \frac{\lambda^{-1/4}}{2} \sum_{k=0}^{\infty} \frac{\Gamma(k/2 + 1/4)}{k!} (-\kappa\lambda^{-1/2})^k \qquad (3.198)$$

$$= \kappa^{-1/2} \sum_{k=0}^{\infty} \frac{\Gamma(2k + 1/2)}{k!} (-\lambda\kappa^{-2})^k. \qquad (3.199)$$

Verify.

(vii) *A dimension-six four-point vertex*

Show that the dimension-six term $-\sum_x a^2 \lambda_0 z \partial_\mu \varphi_x^2 \partial_\mu \varphi_x^2$ in the general action (3.173) corresponds to the vertex function

$$^0\Gamma(p_1 \cdots p_4) = -8a^2 \lambda_0 z(-i)^2 K_\mu^*(p_1 + p_2) K_\mu^*(p_3 + p_4)$$

$$+ \text{two permutations}, \qquad (3.200)$$

$$K_\mu(p) = \frac{1}{ia}(e^{iap_\mu} - 1). \qquad (3.201)$$

In the classical continuum limit this vertex vanishes but in one-loop order it contributes to the two-point function $\Gamma(p)$ (cf. figure 3.5). Show that this contribution is given by

$$+4\lambda_0 z \left[2a^{-2} + p^2(C_0 - \tfrac{1}{8}) + O(a^2)\right], \qquad (3.202)$$

where C_0 is given in (3.67) and we used $(2\pi)^{-4} \int_{-\pi}^{\pi} d^4l \, \hat{l}_\mu^2/\hat{l}^2 = \tfrac{1}{4}$, independent of $\mu = 1, \ldots, 4$.

(viii) *Langevin equation and Fokker–Planck Hamiltonian*

Consider a probability distribution $P(\phi)$ for the field ϕ_x. One Langevin time step changes ϕ into ϕ' according to

$$\phi'_x = \phi_x + \sqrt{2\delta}\,\eta_x + \delta\,\frac{\partial S(\phi)}{\partial\phi_x}. \qquad (3.203)$$

This corresponds to $P(\phi) \rightarrow P'(\phi)$. The new $P'(\phi)$ may be determined as follows. Let $O(\phi)$ be an arbitrary observable, with average value $\int D\phi\, P(\phi)O(\phi)$. After a Langevin time step the new average value is $\int D\phi\, P(\phi)\langle O(\phi'(\phi))\rangle_\eta$, where $\langle \cdots \rangle_\eta$ denotes the average over the Gaussian random numbers η_x. By definition this new average value is equal to $\int D\phi\, P'(\phi)O(\phi)$, i.e.

$$\int D\phi\, P'(\phi)O(\phi) = \int D\phi\, P(\phi)\langle O(\phi'(\phi))\rangle_\eta. \qquad (3.204)$$

By expansion in $\sqrt{\delta}$, show that

$$\langle O(\phi')\rangle_\eta = O(\phi) + \delta \sum_x \left[\frac{\partial O(\phi)}{\partial \phi_x} \frac{\partial S(\phi)}{\partial \phi_x} + \frac{\partial^2 O(\phi)}{\partial \phi_x \, \partial \phi_x} \right] + O(\delta^2),$$
(3.205)

and consequently that

$$\frac{P' - P}{\delta} = \sum_x \frac{\partial}{\partial \phi_x} \left[\frac{\partial}{\partial \phi_x} - \frac{\partial S}{\partial \phi_x} \right] P \qquad (3.206)$$

$$\equiv -H_{\mathrm{FP}}\, P. \qquad (3.207)$$

The partial differential operator in ϕ-space, H_{FP}, is called the Fokker–Planck Hamiltonian. Using

$$\frac{\partial}{\partial \phi_x} e^{S/2} = e^{S/2} \left(\frac{\partial}{\partial \phi_x} + \frac{1}{2} \frac{\partial S}{\partial \phi_x} \right), \qquad (3.208)$$

show that $\tilde{P}$ defined by $P = e^{S/2}\tilde{P}$ satisfies

$$\frac{\tilde{P}' - \tilde{P}}{\delta} = -\tilde{H}\tilde{P} + O(\delta), \qquad (3.209)$$

$$\tilde{H} = \sum_x \left(-\frac{\partial}{\partial \phi_x} - \frac{1}{2} \frac{\partial S}{\partial \phi_x} \right)\left(\frac{\partial}{\partial \phi_x} - \frac{1}{2} \frac{\partial S}{\partial \phi_x} \right).$$

Show that $\tilde{H}$ is a Hermitian positive semidefinite operator, which has one eigenvalue equal to zero with eigenvector $\exp(S/2)$. Give arguments showing that, as $\delta \to 0$ and the number n of iterations goes to infinity, with $t = n\delta \to \infty$, P will tend to the desired distribution $\exp S$.

4

Gauge field on the lattice

Gauge invariance is formulated in position space (as opposed to momentum space), which makes the lattice very well suited as a regulator for gauge theories. In this chapter we shall first review the classical QED and QCD actions, then put these theories on the lattice and define gauge-invariant path integrals. Subsequently a natural quantum-mechanical Hilbert-space interpretation will be given. Gauge-invariant couplings to external sources will be shown to correspond to Wilson loops.

4.1 QED action

The QED action for electrons is given by

$$S = \int dx\, \mathcal{L}(x), \tag{4.1}$$

$$\mathcal{L}(x) = -\tfrac{1}{4} F_{\mu\nu}(x) F^{\mu\nu}(x) - \bar{\psi}(x)\gamma^\mu \left[\partial_\mu + ieA_\mu(x)\right]\psi(x) - m\bar{\psi}(x)\psi(x), \tag{4.2}$$

where ψ is the electron field, A_μ the photon field and

$$F_{\mu\nu}(x) = \partial_\mu A_\nu(x) - \partial_\nu A_\mu(x) \tag{4.3}$$

is the electromagnetic field-strength tensor. The γ^μ are Dirac matrices (cf. appendix D) acting on the ψ's, e is the elementary charge ($e > 0$) and m is the electron mass. It can be useful to absorb the coupling constant e in the vector potential:

$$A_\mu \to \frac{1}{e} A_\mu. \tag{4.4}$$

83

Then $\mathcal{L}$ takes the form •

$$\mathcal{L} = -\frac{1}{4e^2} F_{\mu\nu} F^{\mu\nu} - \bar{\psi}\gamma^\mu(\partial_\mu - iqA_\mu)\psi - m\bar{\psi}\psi, \qquad (4.5)$$

$$q = -1, \quad \text{for the electron}, \qquad (4.6)$$

in which the function of e as a coupling constant (characterizing the strength of the interaction) is separated from the charge q, which characterizes the behavior under gauge transformations.

The action is invariant under gauge transformations,

$$\psi'(x) = e^{i\omega(x)q}\psi(x), \quad \bar{\psi}'(x) = e^{-i\omega(x)q}\bar{\psi}(x), \qquad (4.7)$$

$$A'_\mu(x) = A_\mu(x) + \partial_\mu\omega(x), \qquad (4.8)$$

$$S(\bar{\psi}, \psi, A) = S(\bar{\psi}', \psi', A'), \qquad (4.9)$$

where $\omega(x)$ is real. The phase factors

$$\Omega(x) = e^{i\omega(x)} \qquad (4.10)$$

form a group, for each x: the gauge group $U(1)$. We may rewrite (4.7) and (4.8) entirely in terms of $\Omega(x)$,

$$\psi'(x) = \Omega(x)^q\psi(x), \quad \bar{\psi}'(x) = \bar{\psi}(x)\Omega^*(x)^q, \qquad (4.11)$$

$$A'_\mu(x) = A_\mu(x) + i\Omega(x)\partial_\mu\Omega^*(x). \qquad (4.12)$$

The covariant derivative

$$D_\mu \equiv \partial_\mu - iA_\mu q \qquad (4.13)$$

has the property that $D_\mu\psi(x)$ transforms just like $\psi(x)$ under gauge transformations:

$$D'_\mu\psi'(x) \equiv [\partial_\mu - iqA'_\mu(x)]\psi'(x) = \Omega(x)^q D_\mu\psi(x), \qquad (4.14)$$

such that $\bar{\psi}\gamma^\mu D_\mu\psi = \bar{\psi}'\gamma^\mu D'_\mu\psi'$.

Above we have interpreted the gauge transformations as belonging to the group $U(1)$. We may also interpret (4.10) as a unitary representation of the group (under addition) of real numbers, R:

$$\omega(x) \to e^{i\omega(x)q}, \quad \omega(x) \in R. \qquad (4.15)$$

Another unitary representation of R could be

$$\omega \to e^{i\omega T} \tag{4.16}$$

where T is a real number. For the group $U(1)$, however, the mapping

$$\Omega = e^{i\omega} \to D(\Omega) = e^{i\omega T} \tag{4.17}$$

is a representation only if T is an integer. If T is not an integer, $D(\Omega) = \Omega^T$ is not single valued as a function of Ω. Even if we restrict e.g. $\omega \in [-\pi, \pi]$ to make $\Omega \to e^{i\omega T}$ unique, the product rule would be violated for some Ω's. (For example, for $T = \frac{1}{2}$, $\Omega_1 \Omega_2 = \Omega_3$ with $\omega_{1,2} = 0.9\,\pi$, $\omega_3 = 1.8\,\pi - 2\pi = -0.2\,\pi$ would result in $e^{i\omega_1 T}\, e^{i\omega_2 T} = e^{i0.9\,\pi} \neq e^{i\omega_3 T} = e^{-i0.1\,\pi}$.)

If the gauge group were necessarily $U(1)$, charge would have to be quantized. Suppose there are fields ψ_r transforming with the representations $T = q_r =$ integer:

$$\psi_r'(x) = D^r(x)\psi_r(x), \quad D^r(x) = \Omega(x)^{q_r}. \tag{4.18}$$

Then we have to use a corresponding covariant derivative D_μ^r,

$$D_\mu^r = \partial_\mu - iq_r A_\mu(x), \tag{4.19}$$

such that the action density

$$\mathcal{L} = -\frac{1}{4e^2} F_{\mu\nu} F^{\mu\nu} - \sum_r \bar{\psi}_r \gamma^\mu D_\mu^r \psi_r - \sum_r m_r \bar{\psi}_r \psi_r \tag{4.20}$$

is $U(1)$-gauge invariant. It follows that the charges eq_r are a multiple of the fundamental unit e, which is called charge quantization. If the gauge group were R, there would be no need for charge quantization. In Grand Unified Theories the gauge group of electromagnetism is a $U(1)$ group which is embedded as a subgroup in the Grand Unified gauge group. This could provide the explanation for the quantization of charge observed in nature.

4.2 QCD action

The QCD action has the form

$$S = -\int d^4x \left(\frac{1}{4g^2} G_{\mu\nu}^k G^{k\mu\nu} + \bar{\psi}\gamma^\mu D_\mu \psi + \bar{\psi} m\psi \right). \tag{4.21}$$

The gauge group is $SU(3)$, the group of unitary 3×3 matrices with determinant equal to 1. An element of $SU(3)$ can be written as

$$\Omega = \exp(i\omega^k t_k), \qquad (4.22)$$

where the t_k, $k = 1, \ldots, 8$, are a complete set of Hermitian traceless 3×3 matrices. Then $\Omega^{-1} = \Omega^\dagger$ and

$$\operatorname{Tr} t_k = 0 \Rightarrow \det \Omega = \exp(\operatorname{Tr} \ln \Omega) = 1. \qquad (4.23)$$

The t_k are the generators of the group in the defining representation. A standard choice for these matrices is patterned after the $SU(2)$ spin matrices $\frac{1}{2}\sigma_k$ in terms of the Pauli matrices

$$\sigma_1 = \begin{pmatrix} 0 & 1 \\ 1 & 0 \end{pmatrix}, \quad \sigma_2 = \begin{pmatrix} 0 & -i \\ i & 0 \end{pmatrix}, \quad \sigma_3 = \begin{pmatrix} 1 & 0 \\ 0 & -1 \end{pmatrix}, \qquad (4.24)$$

namely,

$$t_k = \tfrac{1}{2}\lambda_k, \qquad (4.25)$$

with

$$\lambda_k = \begin{pmatrix} \sigma_k & & 0 \\ & & 0 \\ 0 & 0 & 0 \end{pmatrix}, \quad k = 1,2,3, \quad \lambda_8 = \frac{1}{\sqrt{3}} \begin{pmatrix} 1 & 0 & 0 \\ 0 & 1 & 0 \\ 0 & 0 & -2 \end{pmatrix},$$

$$\lambda_4 = \begin{pmatrix} 0 & 0 & 1 \\ 0 & 0 & 0 \\ 1 & 0 & 0 \end{pmatrix}, \quad \lambda_5 = \begin{pmatrix} 0 & 0 & -i \\ 0 & 0 & 0 \\ i & 0 & 0 \end{pmatrix},$$

$$\lambda_6 = \begin{pmatrix} 0 & 0 & 0 \\ 0 & 0 & 1 \\ 0 & 1 & 0 \end{pmatrix}, \quad \lambda_7 = \begin{pmatrix} 0 & 0 & 0 \\ 0 & 0 & -i \\ 0 & i & 0 \end{pmatrix}. \qquad (4.26)$$

These λ's are the well-known Gell-Mann matrices. They have the properties

$$\operatorname{Tr}(t_k t_l) = \tfrac{1}{2}\delta_{kl}, \qquad (4.27)$$

$$[t_k, t_l] = i f_{klm} t_m, \qquad (4.28)$$

where the f_{klm} are the real structure constants of $SU(3)$, totally anti-symmetric in k, l and m (cf. appendix A.1).

The quark fields ψ and $\bar{\psi}$ are in the defining representation of $SU(3)$. They carry three discrete indices,

$$\psi^{\alpha a f}: \begin{array}{ll} \alpha = 1, \ldots, 4, & \text{Dirac index;} \\ a = 1, 2, 3 & \text{(or red, white, blue), color index;} \\ f = 1, \ldots, n_{\text{f}} & \text{(or } u, d, s, c, b, t\text{),} \quad \text{flavor index.} \end{array} \quad (4.29)$$

The gauge transformations and the covariant derivative D_μ act on a, the Dirac gamma matrices act on α and the diagonal mass matrix m acts on f,

$$m = \text{diag}(m_u, m_d, m_s, \ldots). \quad (4.30)$$

We shall now explain the covariant derivative D_μ and the gauge-field term $G^{\mu\nu} G_{\mu\nu}$. It is instructive to assume for the moment that the ψ fields transform under some arbitrary unitary irreducible representation of the color group $SU(3)$, not necessarily the defining representation. For notational convenience we shall still denote the matrices by Ω; they can be written as

$$\Omega = e^{i\omega_k T_k}, \quad (4.31)$$

with T_k the generators in the chosen representation, which satisfy

$$[T_k, T_l] = i f_{klm} T_m, \quad (4.32)$$

$$\text{Tr}\,(T_k T_l) = \rho\, \delta_{kl}. \quad (4.33)$$

For the defining representation $\rho = \frac{1}{2}$, for the adjoint representation $\rho = 3$ (an expression for ρ is given in (A.47) in appendix A.2).

We assume that D_μ has a form similar to (4.13). However, here it is a matrix,

$$D_{\mu ab} = \delta_{ab} \partial_\mu - i G_\mu(x)_{ab}, \quad \text{or} \quad D_\mu = \partial_\mu - i G_\mu. \quad (4.34)$$

The matrix gauge field $G_\mu(x)$ should transform such that, under the gauge transformation

$$\psi'(x) = \Omega(x)\psi(x), \quad \bar{\psi}'(x) = \bar{\psi}(x)\Omega^\dagger(x), \quad (4.35)$$

$D_\mu \psi$ transforms just like ψ,

$$D'_\mu \psi'(x) = \Omega(x) D_\mu \psi(x) = \Omega(x) D_\mu \left[\Omega^\dagger(x)\psi'(x)\right]. \quad (4.36)$$

Treating ∂_μ as an operator gives the requirement

$$D'_\mu = \Omega D_\mu \Omega^\dagger, \quad (4.37)$$

or more explicitly

$$\partial_\mu - iG'_\mu = \Omega(\partial_\mu - iG_\mu)\Omega^\dagger$$
$$= \Omega\partial_\mu\Omega^\dagger + \partial_\mu - i\Omega G_\mu\Omega^\dagger. \tag{4.38}$$

It follows that G_μ has to transform as

$$G'_\mu = \Omega G_\mu\Omega^\dagger + i\Omega\partial_\mu\Omega^\dagger. \tag{4.39}$$

Note that this reduces to (4.12) for an Abelian group.

What is the general form of $G_\mu(x)$? How to parameterize this matrix field? Suppose $G_\mu = 0$. Then $G'_\mu = i\Omega\partial_\mu\Omega^\dagger$, so the parameterization of G_μ must at least incorporate the general form of $i\Omega\partial_\mu\Omega^\dagger$. We shall now show that the latter can be written as a linear superposition of the generators T_m. We write

$$i\Omega\partial_\mu\Omega^\dagger = i\Omega\,\frac{\partial}{\partial\omega^k}\Omega^\dagger\partial_\mu\omega^k = S_k\partial_\mu\omega^k, \tag{4.40}$$

where

$$S_k(\omega) = i\Omega(\omega)\frac{\partial}{\partial\omega^k}\Omega^\dagger(\omega). \tag{4.41}$$

Let $\omega + \varepsilon$ be a small deviation of ω and consider

$$\Omega(\omega)\Omega^\dagger(\omega + \varepsilon) = \Omega(\omega)\left[\Omega^\dagger(\omega) + \varepsilon^k\frac{\partial}{\partial\omega^k}\Omega^\dagger(\omega) + O(\varepsilon^2)\right]$$
$$= 1 - i\varepsilon^k S_k(\omega) + O(\varepsilon^2). \tag{4.42}$$

The left-hand side of this equation is only a small deviation of the unit matrix, so it is possible to write it as $1 - i\varphi_m(\omega,\varepsilon)T_m$, where φ_m is of order ε, so $\varphi_m(\omega,\varepsilon) = S_{km}(\omega)\varepsilon^k$. Hence, $S_k(\omega)$ can be written as

$$S_k(\omega) = S_{km}(\omega)T_m. \tag{4.43}$$

It will be shown in appendix A.2 (eq. (A.43)) that the coefficients $S_{km}(\omega)$ are independent of the representation chosen for Ω. It follows from (4.41) and (4.43) that the Ansatz

$$G_\mu(x) = G^m_\mu(x)T_m \tag{4.44}$$

incorporates the general form of $i\Omega\partial_\mu\Omega^\dagger$. Furthermore, since the generators T_m transform under the adjoint representation of the group (cf. appendix A.2)

$$\Omega T_m\Omega^\dagger = R^{-1}_{mn}T_n, \tag{4.45}$$

the form (4.44) is preserved under the gauge transformation (4.39):

$$\Omega G_\mu^m T_m \Omega^\dagger + i\Omega\partial_\mu\Omega^\dagger = \left(G_\mu^m R_{mn}^{-1} + S_{kn}\partial_\mu\omega^k\right) T_n,$$

$$G_\mu^{\prime n} = R_{nm}G_\mu^m + S_{kn}\partial_\mu\omega^k, \tag{4.46}$$

where we used $R^{-1} = R^{\mathrm{T}}$. The R_{nm}, S_{nm} and ω^k are real, so we may take (4.44) as the parameterization of $G_\mu(x)$ with real fields $G_\mu^m(x)$. Note that the transformation law for $G_\mu^m(x)$ depends only on the group (its adjoint representation), not on the particular representation chosen for Ω.

The gauge-field G_μ transforms inhomogeneously under the gauge group. The field tensor $G_{\mu\nu}$ is constructed out of G_μ such that it transforms homogeneously,

$$G'_{\mu\nu} = \Omega G_{\mu\nu}\Omega^\dagger. \tag{4.47}$$

Analogously to the electrodynamic case (4.3), $G_{\mu\nu}$ can be written as

$$G_{\mu\nu} = D_\mu G_\nu - D_\nu G_\mu = \partial_\mu G_\nu - \partial_\nu G_\mu - i[G_\mu, G_\nu]. \tag{4.48}$$

Using operator notation, this can also be written as

$$G_{\mu\nu} = [D_\mu, D_\nu]. \tag{4.49}$$

Indeed, using (4.37) we have

$$[D'_\mu, D'_\nu] = D'_\mu D'_\nu - (\mu \leftrightarrow \nu) = \Omega D_\mu D_\nu \Omega^\dagger - (\mu \leftrightarrow \nu) = \Omega [D_\mu, D_\nu]\Omega^\dagger, \tag{4.50}$$

which verifies the transformation rule (4.47). The matrix field $G_{\mu\nu}(x)$ can be written in terms of $G_\mu^m(x)$, using (4.44) and (4.32):

$$G_{\mu\nu} = G_{\mu\nu}^k T_k, \tag{4.51}$$

$$G_{\mu\nu}^k = \partial_\mu G_\nu^k - \partial_\nu G_\mu^k + f_{kmn}G_\mu^m G_\nu^n. \tag{4.52}$$

According to (4.45) and (4.47), $G_{\mu\nu}^k$ transforms in the adjoint representation of the group. The combination

$$G_{\mu\nu}^k G^{k\mu\nu} = \frac{1}{\rho}\,\mathrm{Tr}\left(G_{\mu\nu}G^{\mu\nu}\right) \tag{4.53}$$

is gauge invariant (ρ has been defined in (4.33)). Notice that the right-hand side does not depend on the representation chosen for Ω.

The action (4.21) can now be written in more detail,

$$S = -\int d^4x \left[\frac{1}{4g^2}G_{\mu\nu}^k G^{k\mu\nu} + \bar{\psi}\gamma^\mu\left(\partial_\mu - iT_m G_\mu^m\right)\psi + \bar{\psi}m\psi\right]. \tag{4.54}$$

Since $G_{\mu\nu}^{k}G^{k\mu\nu}$ contains terms of higher order than quadratic in G_μ^m, the non-Abelian gauge field is self-coupled. The coupling to the ψ field is completely determined by the generators T_m, i.e. by the representation Ω under which the ψ's transform. For the quark fields, Ω is the defining representation $T_m \rightarrow t_m$.

4.3 Lattice gauge field

We shall mimic the steps leading to the QCD action (4.54) with lattice derivatives,[1] except for choosing to work in Euclidean space–time. The QCD action has a straightforward generalization to $SU(n)$ gauge groups with $n \neq 3$. We shall assume the gauge group $\mathcal{G} = U(1)$ or $SU(n)$. The case of the non-compact group $\mathcal{G} = R$ will be discussed later.

Let the fermion field ψ_x be associated with the sites $x_\mu = m_\mu a$ of the lattice, analogously to the scalar field. Under local gauge transformations it transforms as

$$\psi'_x = \Omega_x \psi_x, \tag{4.55}$$

where as before Ω_x is an irreducible representation of the gauge group $\mathcal{G}$. Since the lattice derivative

$$\partial_\mu \psi_x = (\psi_{x+a\hat{\mu}} - \psi_x)/a \tag{4.56}$$

contains ψ both at x and at $x + a\hat{\mu}$, we try a covariant derivative of the form

$$D_\mu \psi_x = \frac{1}{a}(\psi_{x+a\hat{\mu}} - \psi_x) - i(\tilde{C}_{\mu x}\psi_x + C_{\mu x}\psi_{x+a\hat{\mu}}). \tag{4.57}$$

Here $C_{\mu x}$ and $\tilde{C}_{\mu x}$ are supposed to compensate for the lack of gauge covariance of the lattice derivative, analogously to the matrix gauge potential $G_\mu(x)$. The covariant derivative has to satisfy

$$D'_\mu \psi'_x = \Omega_x D_\mu \psi_x, \tag{4.58}$$

or

$$\frac{1}{a}(\psi'_{x+a\hat{\mu}} - \psi'_x) - i(\tilde{C}'_{\mu x}\psi'_x + C'_{\mu x}\psi'_{x+a\hat{\mu}})$$

$$= \Omega_x \left[\frac{1}{a}(\psi_{x+a\hat{\mu}} - \psi_x) - i(\tilde{C}_{\mu x}\psi_x + C_{\mu x}\psi_{x+a\hat{\mu}}) \right] \tag{4.59}$$

$$= \Omega_x \left[\frac{1}{a}(\Omega^\dagger_{x+a\hat{\mu}}\psi'_{x+a\hat{\mu}} - \Omega^\dagger_x \psi'_x 0 - i(\tilde{C}_{\mu x}\Omega^\dagger_x \psi'_x + C_{\mu x}\Omega^\dagger_{x+a\hat{\mu}}\psi'_{x+a\hat{\mu}}) \right].$$

Comparing coefficients of ψ'_x and $\psi'_{x+a\hat{\mu}}$ gives

$$\tilde{C}'_{\mu x} = \Omega_x \tilde{C}_{\mu x} \Omega^\dagger_x, \tag{4.60}$$

$$C'_{\mu x} = \Omega_x C_{\mu x} \Omega^\dagger_{x+a\hat{\mu}} + \frac{i}{a}(\Omega_x \Omega^\dagger_{x+a\hat{\mu}} - 1) \tag{4.61}$$

$$= \Omega_x C_{\mu x} \Omega^\dagger_{x+a\hat{\mu}} + \Omega_x i \partial_\mu \Omega^\dagger_x. \tag{4.62}$$

It is consistent to set

$$\tilde{C}_{\mu x} \equiv 0. \tag{4.63}$$

For $C_{\mu x}$ we then find the transformation rule (4.62), which resembles the transformation behavior of the continuum gauge potentials quite closely. By analogy with the continuum theory we try for the field strength the form

$$
\begin{aligned}
C_{\mu\nu x} &= D_\mu C_{\nu x} - D_\nu C_{\mu x} \\
&= \frac{1}{a}(C_{\nu,x+a\hat{\mu}} - C_{\nu x}) - i C_{\mu x} C_{\nu,x+a\hat{\mu}} - (\mu \leftrightarrow \nu).
\end{aligned} \tag{4.64}
$$

We find that $C_{\mu\nu x}$ indeed transforms homogeneously,

$$C'_{\mu\nu x} = \Omega_x C_{\mu\nu x} \Omega^\dagger_{x+a\hat{\mu}+a\hat{\nu}}, \tag{4.65}$$

and consequently $\mathrm{Tr}\,(C_{\mu\nu x} C^\dagger_{\mu\nu x})$ is gauge invariant ((4.62) implies that $C_{\mu\nu x}$ cannot be Hermitian in general).

The question is now that of how to parameterize the matrix $C_{\mu x}$ in a way consistent with the transformation rule (4.62). This can be answered by looking at the case $C_{\mu x} = 0$, as in the continuum in the previous section. Then

$$C'_{\mu x} = \Omega_x i \partial_\mu \Omega^\dagger_x = \frac{i}{a}(\Omega_x \Omega^\dagger_{x+a\hat{\mu}} - 1), \tag{4.66}$$

which suggests that we write

$$C_{\mu x} = \frac{i}{a}(U_{\mu x} - 1), \tag{4.67}$$

where $U_{\mu x}$ is a unitary matrix of the same form as Ω_x, i.e. it is a group element in the same representation of the gauge group $\mathcal{G}$. This parameterization of $C_{\mu x}$ is indeed consistent with the transformation rule (4.62), since it gives

$$U'_{\mu x} = \Omega_x U_{\mu x} \Omega^\dagger_{x+a\hat{\mu}}. \tag{4.68}$$

To connect with the gauge potentials $G^k_\mu(x)$ in the continuum we write

$$U_{\mu x} = e^{-iaG_{\mu x}}, \quad G_{\mu x} = G^k_{\mu x} T_k, \tag{4.69}$$

and identify

$$G^k_{\mu x} = G^k_\mu(x). \tag{4.70}$$

More precisely, let $G^k_\mu(x)$ be smooth gauge potentials in the continuum which we evaluate at the lattice points $x_\mu = m_\mu a$. Then $aG_\mu(x) \to 0$ as $a \to 0$, and by construction

$$C_{\mu x} \to G_\mu(x), \quad C_{\mu\nu x} \to G_{\mu\nu}(x), \tag{4.71}$$

where $G_{\mu\nu}(x)$ is the continuum form (4.48).

A possible lattice-regulated gauge-theory action is now given by

$$S = -\sum_x \left[\frac{1}{2}(\bar\psi_x \gamma_\mu D_\mu \psi_x - \bar\psi_x \gamma_\mu D^\dagger_{\mu x} \psi_x) + \bar\psi_x m \psi_x \right.$$
$$\left. + \frac{1}{4\rho g^2} \mathrm{Tr}\,(C_{\mu\nu x} C^\dagger_{\mu\nu x}) \right], \tag{4.72}$$

with ρ the representation-dependent constant defined in (4.33). Evidently, upon inserting $\psi_x = \psi(x)$, $\bar\psi_x = \bar\psi(x)$ and $G^k_{x\mu} = G^k_\mu(x)$ with smooth functions $\psi(x)$, $\bar\psi(x)$ and $G^k_\mu(x)$, this action reduces to the continuum form (4.54) in the limit $a \to 0$. The action (4.72) is not yet satisfactory in its fermion part: it describes too many fermions in the scaling region – this is the notorious phenomenon of 'fermion doubling'. We shall come back to this in a later chapter.

The transformation property (4.68) can be written in a more suggestive form by using the notation

$$U_{x,x+a\hat\mu} \equiv U_{\mu x}, \quad U_{x+a\hat\mu,x} \equiv U^\dagger_{\mu x}, \tag{4.73}$$

because then

$$U'_{x,y} = \Omega_x U_{x,y} \Omega^\dagger_y, \quad y = x + a\hat\mu. \tag{4.74}$$

This notation suggests that it is natural to think of $U_{x,x+a\hat\mu}$ as belonging to the *link* $(x, x+a\hat\mu)$ of the lattice, rather than having four $U_{1x}, \ldots, U_{4x}$ belonging to the site x, as illustrated in figure 4.1.

In fact, there is a better way of associating $U_{x,y}$ with the continuum gauge field $G_\mu(x)$: by identifying $U_{x,y}$ with the parallel transporter from y to x along the link (x, y). The parallel transporter $U(C_{xy})$ along a path C_{xy} from y to x is defined by the path-ordered product (in the continuum)

$$U(C_{xy}) = P \exp\left[-i \int_{C_{xy}} dz_\mu\, G_\mu(z) \right], \tag{4.75}$$

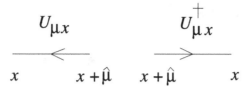

Fig. 4.1. Illustration of $U_{\mu x}$ and $U_{\mu x}^{\dagger}$.

where P denotes the path ordering. The path-ordered product can be defined by dividing the path into N segments $(z_n, z_n + dz_n)$, $n = 0, \ldots, N - 1$, and taking the ordered product,

$$U(C_{xy}) = \lim_{N \to \infty} \exp\left[-i \int_{z_{N-1}}^{x} dz_{\mu}\, G_{\mu}(z)\right] \cdots \exp\left[-i \int_{z_n}^{z_{n+1}} dz_{\mu}\, G_{\mu}(z)\right]$$
$$\cdots \exp\left[-i \int_{y}^{z_1} dz_{\mu}\, G_{\mu}(z)\right]$$
$$= \lim_{N \to \infty} [1 - i\, dz_{0\mu}\, G_{\mu}(z_0)] \cdots [1 - i\, dz_{N-1\,\mu}\, G_{\mu}(z_{N-1})]. \quad (4.76)$$

Under a gauge transformation $G'_{\mu}(z) = \Omega(z)G_{\mu}(z)\Omega^{\dagger}(z) + \Omega(z)i\partial_{\mu}\Omega^{\dagger}(z)$ we have

$$1 - i\, dz_{n\mu}G'_{\mu}(z_n) = \Omega(z_n)[\Omega^{\dagger}(z_n) - i\, dz_{n\mu}\, G_{\mu}(z_n)\Omega^{\dagger}(z_n)$$
$$+ dz_{n\mu}\, \partial_{\mu}\Omega^{\dagger}(z_n)] \quad (4.77)$$
$$= \Omega(z_n)[1 - i\, dz_{n\mu}\, G_{\mu}(z_n)]\Omega^{\dagger}(z_{n+1}) + O(dz^2),$$

such that all the Ω's cancel out in $U(C_{xy})$ except at the end points,

$$U'(C_{xy}) = \Omega(x)U(C_{xy})\Omega^{\dagger}(y). \quad (4.78)$$

Hence, $U(C_{xy})$ parallel transports vectors under the gauge group at y to vectors at x along the path C_{xy}. It is known that this way of associating $U_{x,y}$ with the continuum gauge field via $U(C_{xy})$ leads to smaller discretization errors in the action than does use of (4.69) and (4.70). For our lattice theory, however, the basic variables are the $U_{x,x+a\hat{\mu}} \equiv U_{\mu x}$, one for each link $(x, x + a\hat{\mu})$.

Expressing everything in terms of $U_{x,x+a\hat{\mu}}$ simplifies things and makes the transformation properties more transparent:

$$D_{\mu}\psi_x = \frac{1}{a}(U_{x,x+a\hat{\mu}}\psi_{x+a\hat{\mu}} - \psi_x), \quad (4.79)$$

$$C_{\mu\nu x} = \frac{i}{a^2}(U_{x,x+a\hat{\mu}}U_{x+a\hat{\mu},x+a\hat{\mu}+a\hat{\nu}} - U_{x,x+a\hat{\nu}}U_{x+a\hat{\nu},x+a\hat{\mu}+a\hat{\nu}}),$$

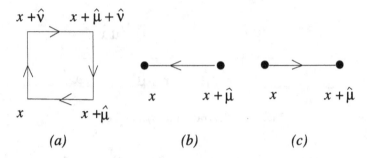

Fig. 4.2. Illustration of the terms in the action $\mathrm{Tr}\,U_{\mu\nu x}$ (a), $\bar{\psi}_x U_{\mu x}\psi_{x+a\hat{\mu}}$ (b) and $\bar{\psi}_{x+a\hat{\mu}}U^{\dagger}_{\mu x}\psi_x$ (c).

$$\mathrm{Tr}\,(C_{\mu\nu x}C^{\dagger}_{\mu\nu x}) = \frac{1}{a^4}\,\mathrm{Tr}\,(2 - U_{\mu\nu x} - U^{\dagger}_{\mu\nu x}),$$

$$U_{\mu\nu x} = U^{\dagger}_{\nu\mu x} = U_{x,x+a\hat{\mu}}U_{x+a\hat{\mu},x+a\hat{\mu}+a\hat{\nu}}U_{x+a\hat{\mu}+a\hat{\nu},x+a\hat{\nu}}U_{x+a\hat{\nu},x}.$$

We see in (4.79) how the covariant derivative involves parallel transport. The action can be written as

$$S = -\sum_{x\mu\nu} \frac{1}{2g^2\rho a^4}\,\mathrm{Tr}\,(1 - U_{\mu\nu x})$$

$$-\sum_{x\mu} \frac{1}{2a}(\bar{\psi}_x\gamma_\mu U_{\mu x}\psi_{x+a\hat{\mu}} - \bar{\psi}_{x+a\hat{\mu}}\gamma_\mu U^{\dagger}_{\mu x}\psi_x), \qquad (4.80)$$

which is illustrated in figure 4.2. The arrows representing $U_{\mu x}$ and $U^{\dagger}_{\mu x}$ are chosen such that they flow from ψ to $\bar{\psi}$, which conforms to a convention for the Feynman rules in the weak-coupling expansion.

We continue with the theory without fermions. The elementary square of a hypercubic lattice is called a *plaquette*. It may be denoted by p ($p = (x, \mu, \nu; \mu < \nu)$ and the product of the U's around p is denoted by U_p. The gauge-field part of the action can then be written as

$$S(U) = \frac{1}{g^2\rho}\sum_p \mathrm{Re}\,\mathrm{Tr}\,U_p + \mathrm{constant}, \qquad (4.81)$$

in lattice units ($a = 1$). This action depends on the representation of the gauge group chosen for the U's, which in our derivation was dictated by the representation carried by ψ and $\bar{\psi}$. This is in contrast to the classical action which is independent of the group representation, as we saw in the previous section (below (4.46)).

To make the representation dependence explicit, we will from now on assume U to be in the defining representation of the gauge group.

Supposing that (4.81) refers to representation $D^r(U)$, we replace $\rho \to \rho_r$ and $\text{Tr}\, U_p \to \chi_r(U_p)$, with χ_r the character (trace) in the representation r. A more general lattice action may involve a sum over representations r,

$$S(U) = \sum_p \sum_r \beta_r \frac{\text{Re}\, \chi_r(U_p)}{\chi_r(1)}, \qquad (4.82)$$

$$\chi_r(U) \equiv \text{Tr}\, [D^r(U)], \qquad (4.83)$$

which reduces to the classical gauge-field action in the classical continuum limit, with

$$\frac{1}{g^2} = \sum_r \frac{\beta_r \rho_r}{d_r}, \qquad d_r = \chi_r(1), \qquad (4.84)$$

where d_r is the dimension of representation r. For example, in an action containing both the fundamental irrep f and the adjoint irrep a of the gauge group $SU(n)$, we have $d_f = n$, $\rho_f = 1/2$, $\rho_a = n$, $d_a = n^2 - 1$ (cf. appendices A.1 and A.2), and

$$1/g^2 = \beta_f/2n + \beta_a n/(n^2 - 1). \qquad (4.85)$$

The simplest lattice formulation of QCD has a plaquette action with only the fundamental representation. It is usually called the Wilson action [39].

4.4 Gauge-invariant lattice path integral

We continue with a pure gauge theory (i.e. containing only gauge fields). The dynamical variables $U_{\mu x}$ are in the fundamental representation of the gauge group $\mathcal{G}$ and the system is described by the gauge-invariant action $S(U)$. If the gauge group is compact we can define a lattice path integral by

$$Z = \int DU \exp[S(U)], \qquad DU = \prod_{x\mu} dU_{\mu x}. \qquad (4.86)$$

Here dU for a given link is a volume element in group space. For a compact group the total volume of group space is finite and therefore Z is well defined for a finite lattice. We want Z to be gauge invariant, so we want the integration measure DU to satisfy

$$DU = DU^\Omega, \qquad U_{\mu x}^\Omega = \Omega_x U_{\mu x} \Omega_{x+a\hat{\mu}}^\dagger. \qquad (4.87)$$

On a given link with link variable U, gauge transformations $U' = \Omega_1 U \Omega_2^\dagger$ are combinations of left and right translations in group space:

$$U' = \Omega U, \quad \text{left}, \tag{4.88}$$

$$U' = U\Omega, \quad \text{right}. \tag{4.89}$$

A measure that is invariant under such translations in group space is well known: the Hurewicz or Haar measure. It can be written in a form familiar from general relativity,

$$dU = \nu \sqrt{\det g} \prod_k d\alpha^k, \tag{4.90}$$

where the α^k are coordinates on group space, $U = U(\alpha)$, and g_{kl} is a metric on this space, of the form

$$g_{kl} = \frac{1}{\rho} \text{Tr} \left(\frac{\partial U}{\partial \alpha^k} \frac{\partial U^\dagger}{\partial \alpha^l} \right), \quad \rho = \frac{1}{2}. \tag{4.91}$$

The normalization constant ν will be chosen such that

$$\int dU = 1. \tag{4.92}$$

The metric (4.91) is covariant under coordinate transformations $\alpha^k = f^k(\alpha')$,

$$g_{kl} = g_{mn} \frac{\partial \alpha'^m}{\partial \alpha^k} \frac{\partial \alpha'^n}{\partial \alpha^l}. \tag{4.93}$$

The Jacobian factors of coordinate transformations cancel out in (4.90), such that $dU' = dU$. Since left and right translations are special cases of coordinate transformations, e.g. $U = \Omega^\dagger U'$ corresponds to $U(\alpha) = \Omega^\dagger U(\alpha')$, the measure is again invariant, $dU' = dU$, or

$$d(\Omega U) = d(U\Omega) = dU. \tag{4.94}$$

The above may be illustrated by the exponential parameterization. For the one-dimensional group $U(1)$ we have

$$U = \exp(i\alpha), \quad g_{kl} = 1, \quad \int dU = \int_{-\pi}^{\pi} \frac{d\alpha}{2\pi} = 1. \tag{4.95}$$

For the $(n^2 - 1)$-dimensional group $SU(n)$ we have

$$U = \exp(i\alpha^k t_k), \quad g_{kl} = S_{km} S_{lm}, \tag{4.96}$$

where we used (4.41), (4.43) and $\partial U^\dagger / \partial \alpha^k = -U^\dagger \, \partial U / \partial \alpha^k \, U^\dagger$, which follows from differentiating $UU^\dagger = 1$. An explicit form for S_{km} is given in (A.43) in appendix A.2.

This completes the definition of the partition function Z. We shall introduce gauge-invariant observables later. One such object we know already: the plaquette field $\operatorname{Tr} U_{\mu\nu x}$, or more simply $\operatorname{Tr} U_p$. It is a composite field in QCD, which will later be seen to describe 'bound states of glue' – glueballs. Expectation values are defined as usual, for example

$$\langle \operatorname{Tr} U_p \operatorname{Tr} U_{p'} \rangle = Z^{-1} \int DU \, \exp[S(U)] \operatorname{Tr} U_p \operatorname{Tr} U_{p'}. \tag{4.97}$$

We stress at this point that gauge fixing (which is familiar in the formal continuum approach) is not necessary with the non-perturbative lattice regulator, for a compact gauge group. The need for gauge fixing shows up again when we attempt to make a weak-coupling expansion.

4.5 Compact and non-compact Abelian gauge theory

Let us write the formulas obtained so far more explicitly for $U(1)$:

$$U_{\mu x} = \exp(-iaA_{\mu x}), \tag{4.98}$$

$$U_{\mu\nu x} = \exp[-ia(A_{\mu x} + A_{\nu x+a\hat{\mu}} - A_{\mu x+a\hat{\nu}} - A_{\nu x})] \tag{4.99}$$

$$= \exp(-ia^2 F_{\mu\nu x}), \tag{4.100}$$

$$F_{\mu\nu x} = \partial_\mu A_{\nu x} - \partial_\nu A_{\mu x}, \tag{4.101}$$

$$S = -\frac{1}{4g^2 a^4} \sum_{x\mu\nu} [2 - 2\cos(a^2 F_{\mu\nu x})], \tag{4.102}$$

$$\int DU = \prod_{x\mu} \int_{-\pi}^{\pi} \frac{d(aA_{\mu x})}{2\pi}. \tag{4.103}$$

Gauge transformations $\Omega = \exp(i\omega_x)$ are linear for the gauge potentials,

$$U'_{\mu x} = \Omega_x U_{\mu x} \Omega^\dagger_{x+a\hat{\mu}}, \tag{4.104}$$

$$aA'_{\mu x} = aA_{\mu x} + \omega_{x+a\hat{\mu}} - \omega_x, \mod(2\pi), \tag{4.105}$$

except for the $\mod(2\pi)$.

We used the fundamental representation in S. The more general form (4.82) is a sum over irreps r = integer, with $D^r(U) = \exp(-iraA) = \chi_r(U)$, $d_r = 1$, and $\rho_r = r^2$, which takes the form of a Fourier series:

$$S = \frac{1}{2a^4} \sum_{x\mu\nu} \sum_r \beta_r \cos(ra^2 F_{\mu\nu x}) + \text{constant}, \tag{4.106}$$

$$\frac{1}{g^2} = \sum_{r=-\infty}^{\infty} \beta_r r^2. \tag{4.107}$$

We could for example choose the β_r such that

$$S = -\frac{1}{4g^2} \sum_{x\mu\nu} [F_{\mu\nu x}^2 \mod(2\pi/a^2)]. \tag{4.108}$$

The above is called the compact $U(1)$ gauge theory. It is clear that there is also a non-compact version of the Abelian gauge theory, with gauge transformations $\omega_x \in R$ acting on $A_{\mu x}$ as in (4.105), but without the $\mod(2\pi)$, $aA_{\mu x}' = aA_{\mu x} + \omega_{x+a\hat{\mu}} - \omega_x$, with $F_{\mu\nu x}$ as in (4.101), and the simple action

$$S(A) = -\frac{1}{4g^2} \sum_{x\mu\nu} F_{\mu\nu x}^2. \tag{4.109}$$

In this case the gauge-invariant measure is given by

$$\int DA = \prod_{x\mu} \int_{-\infty}^{\infty} d(aA_{\mu x}). \tag{4.110}$$

However, the path integral

$$Z = \int DA \, \exp[S(A)] \tag{4.111}$$

is ill defined because it is divergent. The reason is that $\int DA$ contains also an integration over all gauge transformations, which are unrestrained by the gauge-invariant weight $\exp S(A)$. As a consequence Z is proportional to the volume of the gauge group $\int D\Omega$. For the non-compact group $\mathcal{G} = R$ this is $\prod_x \int_{-\infty}^{\infty} d\omega_x$, which is infinite. On the other hand, this divergence formally cancels out in expectation values of gauge-invariant observables and e.g. Monte Carlo computations based on (4.111) still make sense.

To define (4.111) for the non-compact formulation, gauge fixing is needed. A suitable partition function is now given by

$$Z = \int DA \, \exp\left[S(A) - \frac{1}{2g^2\xi} \sum_{x\mu} (\partial_\mu' A_{\mu x})^2 \right], \tag{4.112}$$

where $\partial_\mu' = -\partial_\mu^\dagger$ is the backward derivative, $\partial_\mu' A_{\mu x} = (A_{\mu x} - A_{\mu x - a\hat{\mu}})/a$. See problem 5(i) for more details.

4.6 Hilbert space and transfer operator

We shall show here that the path integral

$$Z = \int \left(\prod_{x\mu} dU_{\mu x} \right) e^{S(U)} \tag{4.113}$$

can be expressed as the trace of a positive Hermitian transfer operator $\hat{T}$ in Hilbert space,

$$Z = \operatorname{Tr} \hat{T}^N, \tag{4.114}$$

where N is the number of time slices, thus providing the quantum-mechanical interpretation.

This Hilbert space is set up in the coordinate representation. The coordinates are $U_{m\mathbf{x}}$, $m = 1, 2, 3$, corresponding to the spatial link variables. A state $|\psi\rangle$ has a wavefunction $\psi(U) = \langle U|\psi\rangle$ depending on the $U_{m\mathbf{x}}$. The basis states $|U\rangle$ are eigenstates of operators $(\hat{U}_{m\mathbf{x}})_{ab}$, where a and b denote the matrix elements ($a, b = 1, \ldots, n$ for $SU(n)$):

$$(\hat{U}_{m\mathbf{x}})_{ab}|U\rangle = (U_{m\mathbf{x}})_{ab}|U\rangle. \tag{4.115}$$

In a parameterization $U = U(\alpha)$ with real parameters α^k one may think of the usual coordinate representation for Hermitian operators $\hat{\alpha}^k$: $\hat{\alpha}^k|\alpha\rangle = \alpha^k|\alpha\rangle$, $\hat{U}_{ab} = U(\hat{\alpha})_{ab}$, and $|U\rangle \equiv |\alpha\rangle$. The Hermitian conjugate matrix $U^\dagger$ corresponds to the operator $\hat{U}^\dagger_{ba} = U^*_{ba}(\hat{\alpha})$. We continue with the notation $|U\rangle$ for the basis states. The basis is orthonormal and complete

$$\langle U'|U\rangle = \prod_{\mathbf{x},m} \delta(U'_{m\mathbf{x}}, U_{m\mathbf{x}}), \tag{4.116}$$

$$1 = \int \left(\prod_{\mathbf{x},m} dU_{m\mathbf{x}} \right) |U\rangle\langle U|, \tag{4.117}$$

such that

$$\langle \psi_1|\psi_2\rangle = \int \left(\prod_{\mathbf{x},m} dU_{m\mathbf{x}} \right) \psi_1^*(U)\psi_2(U). \tag{4.118}$$

The delta function $\delta(U', U)$ corresponds to the measure dU such that $\int dU\, \delta(U, U') = 1$, which can of course be made explicit in a parameterization $U(\alpha)$.

After this specification of Hilbert space the trace in $Z = \operatorname{Tr} \hat{T}^N$ can be written more explicitly as

$$\operatorname{Tr} \hat{T}^N = \prod_{n=0}^{N-1} \left[\int \left(\prod_{\mathbf{x},m} dU_{m\mathbf{x}n} \right) \langle U_{n+1} | \hat{T} | U_n \rangle \right], \qquad (4.119)$$

where n indicates the U variables in time slice n. Notice that the timelike link variables $U_{4\mathbf{x}n}$ are not indicated explicitly in (4.119); these are hidden in $\hat{T}$.

We now have to work the path integral (4.113) into the form (4.119). It is useful for later to allow for different lattice spacings in time and space, a_t and a. Using a notation in which the x_μ are in lattice units (i.e. $\mathbf{x}$ and $x_4 = n$ become integers), but keeping the a's, the pure-gauge part of the action (4.80), $\sum_{x\mu\nu} \operatorname{Tr} U_{\mu\nu x} / 2g^2 \rho a^4$, takes the form $a^3 a_t \sum_{\mathbf{x},n} [\sum_j 2 \operatorname{Re} \operatorname{Tr} U_{4j\mathbf{x}n} / g^2 a_t^2 + \sum_{i<j} 2 \operatorname{Re} \operatorname{Tr} U_{ij\mathbf{x}n} / g^2 a^2]$ ($\rho = 1/2$ in the fundamental representation), or

$$S = \frac{2}{g^2} \left(\frac{a}{a_t} \sum_{p_t} \operatorname{Re} \operatorname{Tr} U_{p_t} + \frac{a_t}{a} \sum_{p_s} \operatorname{Re} \operatorname{Tr} U_{p_s} \right), \qquad (4.120)$$

where p_s and p_t are spacelike and timelike plaquettes, $U_{p_s} = U_{ij\mathbf{x}n}$, $U_{p_t} = U_{4j\mathbf{x}n}$. All the lattice-distance dependence is in the ratio a_t/a. This dependence is really a coupling-constant dependence, one coupling $g^2 a/a_t$ for the timelike plaquettes and another $g^2 a_t/a$ for the spacelike plaquettes. Inspection of (4.113) with action (4.120) shows that $\hat{T}$ can be identified in the form

$$\hat{T} = e^{-\frac{1}{2} a_t \hat{W}} \, \hat{T}'_K \, e^{-\frac{1}{2} a_t \hat{W}}, \qquad (4.121)$$

$$\hat{W} = \frac{-2}{g^2 a} \sum_{p_s} \operatorname{Re} \operatorname{Tr} \hat{U}_{p_s}, \qquad (4.122)$$

with the operator $\hat{T}'_K$ given by the matrix elements

$$\langle U' | \hat{T}'_K | U \rangle = \prod_{\mathbf{x},m} \int dU_{4\mathbf{x}} \exp \left[\frac{2a}{g^2 a_t} \operatorname{Re} \operatorname{Tr} \left(U_{m\mathbf{x}} U_{4\mathbf{x}+\hat{m}} U'^\dagger_{m\mathbf{x}} U^\dagger_{4\mathbf{x}} \right) \right]. \qquad (4.123)$$

The way the $U_{4\mathbf{x}}$ enter in (4.123) can be viewed as a gauge transformation on $U_{m\mathbf{x}}$,

$$U^\Omega_{m\mathbf{x}} = \Omega_x U_{m\mathbf{x}} \Omega^\dagger_{x+\hat{m}}, \qquad (4.124)$$

with $\Omega_x = U^\dagger_{4\mathbf{x}}$. Equivalently we can view this as a gauge transformation on $U'_{m\mathbf{x}}$. There is an integral over all such gauge transformations.

We can write this in operator notation as follows. Define the gauge-transformation operator $\hat{D}(\Omega)$ by

$$\hat{D}(\Omega)|U\rangle = |U^{\Omega^\dagger}\rangle. \tag{4.125}$$

Then

$$\langle U|\hat{D}(\Omega)|\psi\rangle = \langle U^\Omega|\psi\rangle = \psi(U^\Omega). \tag{4.126}$$

This operator is a unitary representation of the gauge group of time-independent gauge transformations in Hilbert space. Define furthermore $\hat{P}_0$ by

$$\hat{P}_0 = \int \left(\prod_{\mathbf{x}} d\Omega_{\mathbf{x}}\right) \hat{D}(\Omega). \tag{4.127}$$

It follows that $\hat{T}'_K$ can be written as

$$\hat{T}'_K = \hat{T}_K \hat{P}_0 = \hat{P}_0 \hat{T}_K, \tag{4.128}$$

with $\hat{T}_K$ given by

$$\langle U'|\hat{T}_K|U\rangle = \prod_{\mathbf{x},m} \exp\left[\frac{2a}{g^2 a_t} \,\mathrm{Re}\,\mathrm{Tr}\,(U_{m\mathbf{x}} U'^\dagger_{m\mathbf{x}})\right]. \tag{4.129}$$

The operator $\hat{P}_0$ is the projector onto the gauge-invariant subspace of Hilbert space. This follows from the fact that $\hat{D}(\Omega)$ is a representation of the gauge group,

$$\hat{D}(\Omega_1)\hat{D}(\Omega) = \hat{D}(\Omega_1\Omega), \tag{4.130}$$

$$\hat{D}(\Omega_1)\hat{P}_0 = \int \left(\prod_{\mathbf{x}} d\Omega_{\mathbf{x}}\right) \hat{D}(\Omega_1\Omega) = \int \left(\prod_{\mathbf{x}} d(\Omega_{1\mathbf{x}}\Omega_{\mathbf{x}})\right) \hat{D}(\Omega_1\Omega)$$

$$= \hat{P}_0 \tag{4.131}$$

$$= \hat{P}_0 \hat{D}(\Omega_1), \tag{4.132}$$

$$\hat{P}_0^2 = \hat{P}_0. \tag{4.133}$$

We used the invariance of the integration measure in group space and the normalization $\int d\Omega_x = 1$. It follows that a state $|\psi\rangle$ of the form $|\psi\rangle = \hat{P}_0|\phi\rangle$ is gauge invariant, $\hat{D}(\Omega)|\psi\rangle = |\psi\rangle$. It also follows easily by taking matrix elements that $\hat{P}_0$ commutes with $\hat{W}$.

We shall show in the next section that $\hat{T}_K$ is a positive operator. It can therefore be written in the from

$$\hat{T}_K = e^{-a_t \hat{K}}, \tag{4.134}$$

with $\hat{K}$ a Hermitian operator.

Summarizing, the path integral leads naturally to a quantum-mechanical Hilbert space and a transfer operator

$$\hat{T} = \hat{P}_0 \, e^{-\frac{1}{2}a_t \hat{W}} \, e^{-a_t \hat{K}} \, e^{-\frac{1}{2}a_t \hat{W}} = e^{-\frac{1}{2}a_t \hat{W}} \, e^{-a_t \hat{K}} \, e^{-\frac{1}{2}a_t \hat{W}} \hat{P}_0 \, , \quad (4.135)$$

which is positive and defines therefore a Hermitian Hamiltonian $\hat{H}$,

$$\hat{T} = \hat{P}_0 \, e^{-a_t \hat{H}} = e^{-a_t \hat{H}} \, \hat{P}_0. \quad (4.136)$$

We recognize a kinetic part ($\hat{K}$) and potential part ($\hat{W}$), analogously to the example of the scalar field. The form (4.129) for the matrix elements of $\hat{T}_K$ shows a plaquette in the *temporal gauge* $U_{4x} = 1$. The path integral has automatically provided the supplementary condition that has to be imposed in this 'gauge': physical states must be gauge invariant (i.e. invariant under time-independent gauge transformations),

$$|\text{phys}\rangle = \hat{P}_0|\text{phys}\rangle, \quad \hat{D}(\Omega)|\text{phys}\rangle = |\text{phys}\rangle. \quad (4.137)$$

In the continuum this corresponds to the 'Gauss law' condition (cf. appendix B).

4.7 The kinetic-energy operator

As we can see from its definition (4.129), the kinetic-energy transfer operator $\hat{T}_K$ is a product of uncoupled link operators. So let us concentrate on a single link $(\mathbf{x}, \mathbf{x}+\hat{m})$ and simple states $|\psi\rangle$ for which $\psi(U)$ depends only on $U_{m\mathbf{x}}$. To simplify the notation we write $U = U_{m\mathbf{x}}$. Then the single-link kinetic transfer operator is given by

$$\langle U'|\hat{T}_{K1}|U\rangle = \exp\left[\kappa \, \text{Re} \, \text{Tr} \, (UU'^\dagger)\right], \quad (4.138)$$

$$\kappa = \frac{2a}{g^2 a_t}, \quad (4.139)$$

where the subscript 1 reminds us of the fact that we are dealing with a single link.

Realizing that $\text{Re} \, \text{Tr} \, (UU'^\dagger)$ may be taken as the distance between the points U and U' in group space, we note that (4.138) is analogous to the expression (2.15) in quantum mechanics, which also involved a translation in the coordinates. So we may expect to gain understanding here too by introducing translation operators. Left and right translations $\hat{L}(V)$ and $\hat{R}(V)$ can be defined by

$$\hat{L}(V)|U\rangle = |V^\dagger U\rangle, \quad (4.140)$$

$$\hat{R}(V)|U\rangle = |UV\rangle. \quad (4.141)$$

By comparing matrix elements we see that $\hat{T}_{K1}$ can be written as

$$\hat{T}_{K1} = \int dV \, \exp[\kappa \, \mathrm{Re} \, \mathrm{Tr} \, V] \, \hat{L}(V), \qquad (4.142)$$

$$= \int dV \, \exp[\kappa \mathrm{Re} \, \mathrm{Tr} \, V] \, \hat{R}(V). \qquad (4.143)$$

The eigenstates of the translation operators can be found among the eigenstates of the Laplacian on group space, as summarized in appendix A.3 (eq. (A.76)). The eigenfunctions are the finite-dimensional unitary irreducible representations (irreps) $D^r_{mn}(U)$ of the group,

$$U \rightarrow D^r_{mn}(U) = \langle U | rmn \rangle. \qquad (4.144)$$

Here r labels the irreps and m and n label the matrix elements. These unitary matrices form a complete orthogonal set of basis functions,

$$\int dU \, D^{r'}_{m'n'}(U)^* \, D^r_{mn}(U) = \delta_{r'r} \delta_{m'm} \delta_{n'n} \frac{1}{d_r}, \qquad (4.145)$$

$$\sum_{rmn} d_r D^r_{mn}(U) \, D^r_{mn}(U')^* = \delta(U, U'), \qquad (4.146)$$

where d_r is the dimension of the representation ($D^r(U)$ is a $d_r \times d_r$ matrix). A function $\psi(U)$ can be expanded as

$$\psi(U) = \langle U | \psi \rangle = \sum_{rmn} \psi_{rmn} d_r D^r_{mn}(U), \qquad (4.147)$$

with the inversion

$$\psi_{rmn} = \langle rmn | \psi \rangle = \int dU \, D^r_{mn}(U)^* \psi(U). \qquad (4.148)$$

The action of $\hat{T}_{K1}$ on $|\psi\rangle$ now follows from

$$\langle U | \hat{T}_{K1} | \psi \rangle = \int dV \, \exp(\kappa \, \mathrm{Re} \, \mathrm{Tr} \, V) \, \langle U | \hat{L}(V) | \psi \rangle \qquad (4.149)$$

$$= \sum_{rmn} \psi_{rmn} d_r \int dV \, \exp(\kappa \, \mathrm{Re} \, \mathrm{Tr} \, V) \, D^r(VU)_{mn}.$$

Using the group-representation property $D^r(VU) = D^r(V)D^r(U)$, the integral in the above expression, i.e. the complex conjugate of

$$c_{rmn} \equiv \int dV \, D^r_{mn}(V)^* \, \exp(\kappa \, \mathrm{Re} \, \mathrm{Tr} \, V), \qquad (4.150)$$

is the coefficient for the expansion of the exponential in irreps,

$$\exp(\kappa \, \mathrm{Re} \, \mathrm{Tr} \, V) = \sum_{rmn} d_r c_{rmn} D^r_{mn}(V), \qquad (4.151)$$

as follows from the orthogonality of the irreps. A change of variables
$V \to V^\dagger$ in (4.150) shows that $c_{rmn} = c^*_{rnm}$. Making a transformation
of variables $V \to WVW^\dagger$, with arbitrary group element W, gives the
relation

$$c_{rmn} = D^r_{mm'}(W)c_{rm'n'}D^r_{n'n}(W^\dagger). \tag{4.152}$$

Using Schur's lemma it follows that c_{rmn} can be written in the form

$$c_{rmn} = c_r\,\delta_{mn}, \tag{4.153}$$

with real c_r. Returning to (4.149) we get

$$\langle U|\hat{T}_{K1}|\psi\rangle = \sum_{rmn} \psi_{rmn}d_r c_r\, D^r(U)_{mn}. \tag{4.154}$$

Every irrep r is just multiplied by the number c_r. The irrep states $|rmn\rangle$
are eigenstates of $\hat{T}_{K1}$ with eigenvalue c_r,

$$\hat{T}_{K1}|rmn\rangle = c_r\,|rmn\rangle. \tag{4.155}$$

The relation (4.153) holds generally for expansion coefficients of functions on the group which are invariant under $V \to WVW^\dagger$, i.e. class
functions. These have a *character expansion*,

$$\exp(\kappa\,\mathrm{Re}\,\mathrm{Tr}\,V) = \sum_r d_r c_r \chi_r(V), \tag{4.156}$$

with

$$\chi_r(V) = \mathrm{Tr}\,D^r(V) = D^r_{mm}(V) \tag{4.157}$$

the character in the representation r. The characters are orthonormal,

$$\int dV\,\chi_r(V)^*\chi_s(V) = \delta_{rs}, \tag{4.158}$$

as follows from (4.145). Writing

$$d_r c_r = \int dV\,\exp\left[\frac{\kappa}{2}\chi_f(V) + \frac{\kappa}{2}\chi_f(V)^*\right]\chi_r(V)^*, \tag{4.159}$$

where f is the fundamental (defining) representation, we can show that
the c_r are positive. Expansion of the right-hand side of (4.159) in powers
of κ leads to

$$c_r = \sum_{n=0}^{\infty} \frac{(\kappa/2)^n}{n!} \sum_{k=0}^{n} \frac{n!}{k!(n-k)!} \int dV\,\chi_f(V)^{*k}\chi_f(V)^{n-k}\chi_r(V)^*. \tag{4.160}$$

Reducing the tensor product representation $D^{r_1} \cdots D^{r_k}$ to irreducible components, we see that

$$\int dV \, \chi_{r_1}(V) \cdots \chi_{r_k}(V) = n(r_1, \ldots, r_k) \qquad (4.161)$$

is the number of times the singlet irrep occurs. Since κ is positive the c_r are positive.

It follows that the eigenvalues of $\hat{T}_{K1}$ are positive, i.e. $\hat{T}_{K1}$ is a positive operator. The full kinetic transfer operator $\hat{T}_K$, being the product of single-link operators $\hat{T}_{K1}$, is also positive.

4.8 Hamiltonian for continuous time

In the Hamiltonian approach to lattice gauge-theory time is kept continuous while space is replaced by a lattice. Taking the formal limit $a_t \to 0$ we get the appropriate Hamiltonian from

$$\hat{T} = \hat{P}_0 \exp\left[-a_t \hat{H} + O(a_t^2)\right]. \qquad (4.162)$$

Some work is required for $\hat{T}_K$ as $a_t \to 0$ since it depends explicitly on a/a_t through $\kappa = 2a/g^2 a_t$. Consider again the form (4.142) for one link,

$$\hat{T}_{K1} = \int dV \, \exp[\kappa \operatorname{Re} \operatorname{Tr} V] \, \hat{L}(V). \qquad (4.163)$$

Since $\kappa \to \infty$ as $a_t \to 0$ we can evaluate this expression with the saddle-point method. The highest saddle point is at $V = 1$. It is convenient to use the exponential parameterization,

$$V = \exp(i\alpha^k t_k), \qquad (4.164)$$

$$\operatorname{Re} \operatorname{Tr} V = d_f - \tfrac{1}{4}\alpha^k \alpha^k + O(\alpha^4), \qquad (4.165)$$

$$dV = \prod_k d\alpha_k \, [1 + O(\alpha^2)], \qquad (4.166)$$

$$\hat{L}(V) = 1 + i\alpha^k \hat{X}_k(L) - \tfrac{1}{2}\alpha^k \alpha^l \hat{X}_k(L)\hat{X}_l(L) + O(\alpha^3), \quad (4.167)$$

where we have written the left translator $\hat{L}$ in terms of its generators $\hat{X}_k(L)$ (cf. appendix A.3). Gaussian integration over α^k gives

$$\hat{T}_{K1} = \text{constant} \times \left(1 - \frac{1}{\kappa}\hat{X}^2 + \cdots\right), \qquad (4.168)$$

$$\hat{X}^2 = \hat{X}_k(L)\hat{X}_k(L) = \hat{X}_k(R)\hat{X}_k(R), \qquad (4.169)$$

$$\text{constant} = \int \prod_k d\alpha_k \, \exp[\kappa(d_f - \alpha^2/4)]. \qquad (4.170)$$

Here the constant could have been avoided by changing the measure in the path integral by an overall constant.

The Hamiltonian can be written as

$$\hat{H} = \hat{K} + \hat{W} \tag{4.171}$$

$$= \frac{1}{a}\left[\frac{g^2}{2}\sum_{l_s}\hat{X}_{l_s}^2 + \frac{2}{g^2}\sum_{p_s}\operatorname{Re}\operatorname{Tr}\left(1 - \hat{U}_{p_s}\right)\right] + \text{constant},$$

where the l_s denote the spatial links and p_s the spatial plaquettes. In the coordinate representation $\hat{U} \to U$ and $\hat{X}^2$ becomes the covariant Laplacian on group space. The above Hamiltonian is known as the Kogut–Susskind Hamiltonian [40].

It is good to keep in mind that, with continuous time and a lattice in space, the symmetry between time and space is broken. It is necessary to renormalize the velocity of light, which amounts to introducing different couplings g_K^2 and g_W^2 for the kinetic and potential terms in the Hamiltonian (4.171).

The formal continuum limit $a \to 0$, $U_{\mu x} = \exp(-iaG_{\mu x}) \to 1 - iaG_\mu(x)+\cdots$ leads to the formal continuum Hamiltonian in the temporal gauge:

$$H = \int d^3x\left(\frac{g^2}{2}\Pi_k^p\Pi_k^p + \frac{1}{4g^2}G_{lm}^pG_{lm}^p\right) = \int d^3x\left(\frac{1}{2}E^2 + \frac{1}{2}B^2\right),$$

$$\tag{4.172}$$

where

$$\Pi_k^p = -i\frac{\delta}{\delta G_k^p}, \quad p = 1,\ldots,n^2-1, \quad k = 1,2,3, \tag{4.173}$$

$$G_{lm}^p = \partial_l G_m^p - \partial_m G_l^p + f_{pqr}G_l^qG_m^r, \tag{4.174}$$

and the conventional 'electric' and 'magnetic' fields are given by

$$E_k^p = -g\Pi_k^p, \quad B_k^p = \frac{1}{g}\epsilon_{klm}G_{lm}^p. \tag{4.175}$$

In the continuum the canonical quantization in the temporal gauge is often lacking in text books, because it is less suited for weak-coupling perturbation theory. A brief exposition is given in appendix B.

4.9 Wilson loop and Polyakov line

In the classical Maxwell theory an external current J^μ enters in the weight factor in the real-time path integral as

$$e^{iS} \to e^{iS+i \int d^4x \, J^\mu A_\mu}. \qquad (4.176)$$

For a line current along a path $z^\mu(\tau)$,

$$J^\mu(x) = \int d\tau \, \frac{dz^\mu(\tau)}{d\tau} \, \delta^4(x - z(\tau)), \qquad (4.177)$$

the phase $\exp(i \int J^\mu A_\mu)$ takes the form

$$\exp\left[i \int dz^\mu A_\mu(z)\right], \qquad (4.178)$$

where the integral is along the path specified by $z(\tau)$. The current is 'conserved' (i.e. $\partial_\mu J^\mu = 0$) for a closed path or a never-ending path. In classical electrodynamics one thinks of $z^\mu(\tau)$ as the trajectory of a point charge. Then $dz^\mu/d\tau$ is timelike. For a positive static point charge at the origin the phase is

$$\exp\left[i \int dz^0 \, A_0(\mathbf{0}, z^0)\right]. \qquad (4.179)$$

We may however choose the external current as we like and use also spacelike $dz/d\tau$. For a line current running along the coordinate 3-axis the phase is

$$\exp\left[i \int dz^3 \, A_3(0, 0, z^3, 0)\right]. \qquad (4.180)$$

The Euclidean form is obtained from the Minkowski form by the substitution $J^0 = -iJ_4$, $dz^0 = -i \, dz_4$, $A_0 = iA_4$. The phase remains a phase,

$$\exp\left[i \int \sum_{\mu=1}^{4} dz_\mu A_\mu(z)\right]. \qquad (4.181)$$

The source affects the places where the time components enter in the action and we have to take a second look at the derivation of the transfer operator. Consider therefore first in the compact $U(1)$ theory the path integral

$$Z(J) = \int DU \exp\left[S(U) + \sum_{x\mu} J_{\mu x} A_{\mu x}\right], \qquad (4.182)$$

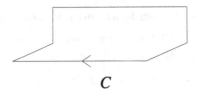

Fig. 4.3. A contour C specifying a line current or Wilson loop.

$$S(U) = \frac{1}{4g^2} \sum_{x\mu\nu} U_{\mu\nu x}, \qquad (4.183)$$

$$U_{\mu x} = \exp(-iA_{\mu x}), \qquad (4.184)$$

$$U_{\mu\nu x} = \exp[i(A_{\mu x+\hat{\nu}} - A_{\mu x} - A_{\nu x+\hat{\mu}} + A_{\nu x})]. \qquad (4.185)$$

We have written the source term in conventional Euclidean form as a real-looking addition to $S(U)$, but the current J_μ is purely imaginary. For a line current of unit strength over a closed contour C as illustrated in figure 4.3 we have

$$
\begin{aligned}
J_{\mu x} &= -i \quad \text{for links } (x, x + \hat{\mu}) \in C \\
&= +i \quad \text{for links } (x + \hat{\mu}, x) \in C \\
&= 0 \quad \text{otherwise.}
\end{aligned}
\qquad (4.186)
$$

This current is 'conserved',

$$\partial'_\mu J_{\mu x} = \sum_\mu (J_{\mu x} - J_{\mu x - \hat{\mu}}) = 0, \qquad (4.187)$$

and the integrand of the path integral is gauge invariant. The phase factor associated with the current can be written in another way,

$$\exp\left(\sum_{x\mu} J_{\mu x} A_{\mu x}\right) = \prod_{l \in C} U_l \equiv U(C), \qquad (4.188)$$

where l denotes a directed link, $U_l = U_{\mu x}$ for $l = (x, x + \hat{\mu})$. Such a product $U(C)$ of U's around a loop C is called a Wilson loop [39]. It is gauge invariant. The simplest Wilson loop is the plaquette $U_{\mu\nu x}$.

The Wilson-loop form of the interaction with an external line source generalizes easily to non-Abelian gauge theories. For a source in irrep r we have

$$\text{Tr } D^r(U(C)) = \chi_r(U(C)), \qquad (4.189)$$

with $D^r(U(C))$ the ordered product of the link matrices $D^r(U_l)$ along the loop C. Denoting the links l by the pair of neighbors (x, y), we have for example in the fundamental representation

$$U(C) = \text{Tr}\,(U_{x_1 x_2} U_{x_2 x_3} \cdots U_{x_n x_1}). \tag{4.190}$$

The gauge invariance is obvious: the gauge transformations cancel out pairwise in the product along the closed loop.

Consider now the derivation of the transfer operator. For the parts of C where it runs in spacelike directions it represents an operator in Hilbert space through $U_l \to \hat{U}_l$ as before. What about the timelike links? Suppose that between two time slices there are only two such links, say the links $(y, y + \hat{4})$ and $(z + \hat{4}, z)$. Then, for these time slices, (4.123) is modified to

$$\langle U'|\hat{T}'_K|U\rangle = \prod_{\mathbf{x},m} \int dU_{4\mathbf{x}}\,\exp[\cdots]\,D^r_{mn}(U_{4\mathbf{y}}) D^r_{pq}(U^\dagger_{4\mathbf{z}}), \tag{4.191}$$

where $\exp[\cdots]$ is the same as in (4.123) and the indices m, n, p, q hook up to the other D^r's of the Wilson loop. We see that the operator $\hat{P}_0$ defined in (4.127) is replaced by

$$\hat{P}_0 \to \int \prod_{\mathbf{x}} dV_{\mathbf{x}}\,\hat{D}(V) D^r_{mn}(V_{\mathbf{y}}) D^r_{pq}(V^\dagger_{\mathbf{z}}),' \tag{4.192}$$

where we used the notation $\Omega^\dagger_{\mathbf{x}} = V_{\mathbf{x}} = U_{4\mathbf{x}}$, $d\Omega = dV$. The gauge-transformation operator $\hat{D}(V)$ is the product of operators $\hat{D}(V_{\mathbf{x}})$ at sites $\mathbf{x}$. With the notation

$$\hat{P}^{r\mathbf{x}}_{mn} = \int dV_x\, D^r_{mn}(V_{\mathbf{x}})\hat{D}(V_{\mathbf{x}}), \tag{4.193}$$

the right-hand side of (4.192) can be written as

$$\hat{P}'_0\,\hat{P}^{r\mathbf{y}}_{mn}\,\hat{P}^{\bar{r}\mathbf{z}}_{pq}, \tag{4.194}$$

with

$$\hat{P}'_0 = \prod_{\mathbf{x} \neq \mathbf{y},\mathbf{z}} \hat{P}^{0\mathbf{x}} \tag{4.195}$$

the projector onto the gauge-invariant subspace except at $\mathbf{y}$ and $\mathbf{z}$. The irrep $\bar{r}$ is the Hermitian conjugate of the irrep r. The operator $\hat{P}^{r\mathbf{x}}_{mn}$ projects onto the subspace transforming at $\mathbf{x}$ in the irrep r in the following way. Let $|skl\rangle$ be an irrep state for some link $(\mathbf{u}, \mathbf{v})$,

$$\langle U|skl\rangle = D^s_{kl}(U), \quad U = U_{\mathbf{u},\mathbf{v}}; \tag{4.196}$$

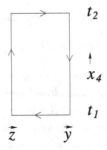

Fig. 4.4. A rectangular timelike Wilson loop.

then, for $\mathbf{x} = \mathbf{u}$,

$$\sum_n \langle U | \hat{P}^{r\mathbf{u}}_{mn} | snl \rangle = \int dV \, D^r_{mn}(V) D^s_{nl}(VU) = \delta_{r\bar{s}} D^s_{ml}(U),$$

$$\sum_n \hat{P}^{\bar{s}\mathbf{u}}_{mn} | snl \rangle = | sml \rangle; \tag{4.197}$$

similarly, for $\mathbf{x} = \mathbf{v}$,

$$\sum_m \langle U | \hat{P}^{r\mathbf{v}}_{mn} | skm \rangle = \int dV \, D^r_{mn}(V) D^s_{km}(UV^\dagger) = \delta_{rs} D^s_{kn}(U),$$

$$\sum_m \hat{P}^{s\mathbf{v}}_{mn} | skm \rangle = | skn \rangle. \tag{4.198}$$

The $\hat{P}^r$ are Hermitian projectors in the following sense:

$$(\hat{P}^{r\mathbf{x}}_{mn})^\dagger = \hat{P}^{r\mathbf{x}}_{nm}, \tag{4.199}$$

$$\sum_n \hat{P}^{r\mathbf{x}}_{mn} \hat{P}^{r\mathbf{x}}_{nq} = \hat{P}^{r\mathbf{x}}_{mq}. \tag{4.200}$$

Consider next a Wilson loop of the form shown in figure 4.4. In the $U(1)$ case this corresponds to two charges that are static at times between t_1 and t_2, a charge $+1$ at $\mathbf{z}$ and a charge -1 at $\mathbf{y}$:

$$J_4(\mathbf{x}, x_4) = -i[\delta_{\mathbf{x},\mathbf{y}} - \delta_{\mathbf{x},\mathbf{z}}], \quad t_1 < x_4 < t_2. \tag{4.201}$$

In the $SU(n)$ case the interpretation is evidently that we have a source in irrep $\bar{r}$ at $\mathbf{y}$ and a source in irrep r at $\mathbf{z}$. If r is the defining representation of the gauge group $SU(3)$ we say that we have a static quark at $\mathbf{z}$ and an antiquark at $\mathbf{y}$. The path-integral average of this Wilson loop

$$W(C) = \frac{1}{Z} \int DU \, \exp[S(U)] \, \chi_r(U(C)), \tag{4.202}$$

$$Z = \int DU \, \exp[S(U)] = \operatorname{Tr} \hat{T}^N, \tag{4.203}$$

can be expressed as

$$W(C) = \frac{1}{Z} \operatorname{Tr} \left[\hat{T}^{N-t} \, D_{kl}^r(\hat{U}^\dagger) \, (\hat{T}')^t \, \hat{P}_{lm}^{\bar{r}\mathbf{z}} \, D_{mn}^r(\hat{U}) \, \hat{P}_{nk}^{r\mathbf{y}} \right]. \tag{4.204}$$

Here C is the rectangular loop shown in figure 4.4, $t = t_2 - t_1$, $\hat{U}$ is the operator corresponding to the product of U's at time t_1 and similarly for $\hat{U}^\dagger$ at t_2, and $\hat{T}'$ is the transfer operator with $\hat{P}_0'$ (cf. (4.195)). In the zero-temperature limit $N \to \infty$, the trace in (4.204) is replaced by the expectation value in the ground state $|0\rangle$, $\hat{T}|0\rangle = \exp(-E_0)\,|0\rangle$. Inserting intermediate states $|n\rangle$, which are eigenstates of $\hat{T}' \, \hat{P}_{lm}^{\bar{r}\mathbf{z}} \, \hat{P}_{nk}^{r\mathbf{y}}$ with eigenvalues $\exp(-E_n')$, gives the representation

$$W(C) = \sum_n R_n \, e^{-(E_n' - E_0)t}, \quad N = \infty, \tag{4.205}$$

where R_n and E_n' depend on $\mathbf{y}$ and $\mathbf{z}$. For large times t the lowest energy level E_0' will dominate. This is the energy of the ground state $|0'\, rmn\rangle$ in that sector of Hilbert space which corresponds to the static sources at $\mathbf{y}$ and $\mathbf{z}$. By definition, the difference $E_0' - E_0$ is the *potential* V:

$$W(C) \overset{t \to \infty}{\to} R_0 \, e^{-Vt}, \quad V = V^r(\mathbf{y}, \mathbf{z}), \quad R_0 = R_0(\mathbf{y}, \mathbf{z}), \tag{4.206}$$
$$R_0 = \sum_{mn} \langle 0| D_{mn}^r(\hat{U}^\dagger)|0'\, rmn\rangle \langle 0'\, rmn| D_{mn}^r(\hat{U})|0\rangle.$$

Hence, we have found a formula for the static potential (e.g. for a quark–antiquark pair) in terms of the expectation value of a Wilson loop.

Another interesting quantity is the Polyakov line [41], which is a string of U's closed by periodic boundary conditions in the Euclidean time direction. (In case of closure by periodic boundary conditions in the spatial direction, this is often called a Wilson line.) For example, the situation illustrated in figure 4.5 corresponds to a single static quark, a source which is always switched on. The expectation value $W(L)$ of the Polyakov line operator at $\mathbf{x}$, e.g. in the defining representation

$$\operatorname{Tr} U(L) = \operatorname{Tr} \left(U_{4\mathbf{x},0} U_{4\mathbf{x},1} \cdots U_{4\mathbf{x},N-1} \right), \tag{4.207}$$

can be written as

$$W(L) = \langle \operatorname{Tr} U(L) \rangle = \frac{1}{Z} \operatorname{Tr} \left[(\hat{T}')^N \sum_m \hat{P}_{mm}^{r\mathbf{x}} \right]. \tag{4.208}$$

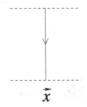

Fig. 4.5. A Polyakov line.

It is the free energy of a static quark at inverse temperature N. For temperature going to zero it behaves as

$$W(L) \propto e^{-\epsilon N}, \quad N \to \infty, \tag{4.209}$$

with ϵ the self-energy of a static quark.

4.10 Problems

(i) *The case $SU(2)$*

    (a) Work out the metric $g_{kl} = 2 \operatorname{Tr}\left[(\partial U/\partial \alpha_k)(\partial U^\dagger/\partial \alpha_l)\right]$ using the exponential parameterization $U = \exp(i\alpha^k \tau_k/2)$.

    (b) Determine the normalization constant ν in $dU = \nu\sqrt{\det g}\, d\alpha_1\, d\alpha_2\, d\alpha_3$ such that $\int dU = 1$.

    (c) Find the characters $\chi_j(U) = \operatorname{Tr} D^j(U)$ as a function of α^k ($j = \frac{1}{2}, 1, \frac{3}{2}, \ldots$).

    (d) Check the orthogonality relation $\int dU\, \chi_j(U)\chi_{j'}^*(U) = \delta_{jj'}$.

    (e) Verify for a one-link state that $_1\langle U|\hat{X}_k(L)|\psi\rangle_1 = X_k(L)\psi_1(U)$.

    (f) Verify that $X^2(L) = X^2(R)$.

(ii) *Two-dimensional $SU(n)$ gauge-field theory*

    Consider two-dimensional $SU(n)$ gauge theory with action

$$S = \sum_p L(U_p), \tag{4.210}$$

$$L(U) = \frac{1}{g^2} \sum_r \kappa_r \operatorname{Re} \chi_r(U), \tag{4.211}$$

$$\sum_r \kappa_r \rho_r = 1, \tag{4.212}$$

and periodic boundary conditions in space. In ordinary units g has the dimension of mass (in two dimensions), such that in lattice units $g \to 0$ in the continuum limit. The transfer operator is given by

$$\hat{T} = \hat{T}_K \hat{P}_0, \quad \hat{T}_K = \prod_l \hat{T}_{Kl}, \qquad (4.213)$$

where l, $l = 0, \ldots, N - 1$ labels the spatial links $(x, x + 1)$, $x = 0, \ldots, N - 1$. Since there is only one space direction, the link variables in the spatial direction may be denoted by U_x. Consider the wavefunction

$$\psi_{\{r,m,n\}}(U) = \prod_x D^{r_x}_{m_x n_x}(U_x), \qquad (4.214)$$

which is just a product of irreps r_x at each x.
(a) Show that

$$P_0\psi_{\{r,m,n\}}(U) \equiv \langle U|\hat{P}_0|\psi\rangle \qquad (4.215)$$
$$= d_{r_0}^{-N} \text{Tr}\left[D^{r_0}(U_0)D^{r_1}(U_1)\cdots D^{r_{N-1}}(U_{N-1})\right]$$
$$\times \delta_{r_0 r_1}\cdots\delta_{r_{N-1}r_0}\,\delta_{n_0 m_1}\delta_{n_1 m_2}\cdots\delta_{n_{N-1}m_0}.$$

Hence, the gauge-invariant component is non-zero only if all irreps are equal, say r, and it is a Wilson line in the spatial direction.
(b) Show that the energy spectrum of the system is given by

$$E_r = -N\left[\ln a_0 + \ln\left(\frac{\langle\chi_r\rangle_1}{d_r}\right)\right], \qquad (4.216)$$

where

$$\langle\chi_r\rangle_1 = \frac{\int dU\, e^{L(U)}\chi_r(U)}{\int dU\, e^{L(U)}}. \qquad (4.217)$$

(c) Show for $g \to 0$, using a saddle-point expansion about $U = 1$, that

$$\frac{\langle\chi_r\rangle_1}{d_r} \to 1 - \tfrac{1}{2}C_2^r g^2 + O(g^4), \qquad (4.218)$$

where C_2^r is the value of the quadratic Casimir operator in the representation r. This result holds independently of the detailed choice of κ_r's, as long as they satisfy the constraint (4.212),
(d) Restoring the lattice spacing a, $L = Na$, deduce from the result above that, in the continuum limit, the energy spectrum takes the universal form

$$E_r - E_0 = \tfrac{1}{2}g^2 C_2^r L. \qquad (4.219)$$

(iii) *Glueball masses and string tension*

Simple glueball operators may be defined in terms of the plaquette field $\operatorname{Tr} U_p$, where $p = (\mathbf{x}, m, n)$ denotes a spacelike plaquette. When this operator acts on the ground state (vacuum state) it creates a state with the quantum numbers of the plaquette. Similarly, a string state may be created by the operator $U_{\mathbf{x},\mathbf{y}} = \prod_{l \in C} U_l$, where the links l belong to an open contour from $\mathbf{x}$ to $\mathbf{y}$. The string state defined this way is not gauge invariant at $\mathbf{x}$ and $\mathbf{y}$; it has to be interpreted as a state with external sources at these points.

Using the transfer-operator formalism, derive to leading order a strong-coupling formula for the glueball mass corresponding to the plaquette, and for the string mass corresponding to $U_{\mathbf{x},\mathbf{y}}$. Use lattice units $a = a_t = 1$. Note that the potential-energy factors $\exp(-a_t \hat{W}/2)$ in the transfer operator may be neglected to leading order in $1/g^2$.

5

$U(1)$ and $SU(n)$ gauge theory

In this chapter we make a first exploration of $U(1)$ and $SU(n)$ 'pure gauge theories' (i.e. without electrons or quarks etc.), the static potential and the glueball masses.

5.1 Potential at weak coupling

According to (4.206) the static potential $V(r)$ in a gauge theory is given by the formula

$$V(r) = - \lim_{t \to \infty} \frac{1}{t} \ln W(r,t), \qquad (5.1)$$

where $W(r,t)$ is a rectangular $r \times t$ Wilson loop in a lattice of infinite extent in the time direction (figure 5.1). We shall first evaluate this formula for free gauge fields and then give the results of the first non-trivial order in the weak-coupling expansion. This will illustrate that (5.1) indeed gives the familiar Coulomb potential plus corrections.

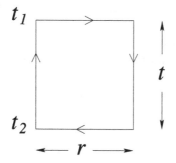

Fig. 5.1. A rectangular Wilson loop for the evaluation of the potential.

First we consider the compact $U(1)$ gauge theory (4.182), in which the external source $J_\mu(x)$ specified in (4.186) serves to introduce the Wilson loop. In this case (5.1) can be rewritten as

$$V(r) = -\lim_{t \to \infty} \frac{1}{t} \ln \left[\frac{Z(J)}{Z(0)} \right]. \tag{5.2}$$

The weak-coupling expansion can be obtained by substituting $U_\mu(x) = \exp[-igaA_\mu(x)]$ into the action,

$$S = -\sum_{x\mu\nu} \left(\frac{1}{4}[F_{\mu\nu x}]^2 - \frac{1}{48}g^2 a^2 [F_{\mu\nu x}]^4 + \cdots \right), \tag{5.3}$$

$$F_{\mu\nu x} = \partial_\mu A_{\nu x} - \partial_\nu A_{\mu x}, \tag{5.4}$$

and expanding the path integral in the gauge coupling g. The first term in (5.3) is the usual free Maxwell action (non-compact $U(1)$ theory). The other terms are interaction terms special to the compact $U(1)$ theory.

As usual, gauge fixing is necessary in the weak-coupling expansion. This can be done on the lattice in the same way as in the continuum formulation. We shall not go into details here (cf. problem (i)), and just state that the free part of S (the part quadratic in A_μ) leads in the Feynman gauge to the propagator

$$D_{\mu\nu}(p) = \delta_{\mu\nu} \frac{a^2}{\sum_\mu (2 - 2\cos ap_\mu)},$$

$$= \delta_{\mu\nu} \frac{1}{p^2}, \quad ap \to 0. \tag{5.5}$$

This is similar to the boson propagator (2.111). In position space

$$D_{\mu\nu}(x - y) \equiv D_{xy}^{\mu\nu} = \delta_{\mu\nu} \int_{-\pi/a}^{\pi/a} \frac{d^4p}{(2\pi)^4} e^{ip(x-y)} \frac{a^2}{\sum_\mu (2 - 2\cos ap_\mu)}$$

$$\to \delta_{\mu\nu} \frac{1}{4\pi^2(x-y)^2}, \quad (x-y)^2/a^2 \to \infty. \tag{5.6}$$

The large-x behavior of $D_{\mu\nu}(x)$ corresponds to the small-p behavior of $D_{\mu\nu}(p)$. This can be shown with the help of the saddle-point method for evaluating the large-x behavior.

To leading order in g^2, $Z(J)$ is given by

$$Z(J) = e^{\frac{1}{2}g^2 \sum_{xy} J_\mu(x) D_{\mu\nu}(x-y) J_\nu(y)} Z(0), \tag{5.7}$$

and

$$V(r) = -\frac{1}{t}\frac{1}{2}g^2 \sum_{xy} J_\mu(x) D_{\mu\nu}(x - y) J_\nu(y), \quad t \to \infty. \tag{5.8}$$

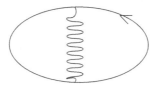

Fig. 5.2. Diagram illustrating $\frac{1}{2}g^2 \sum J\,D\,J$.

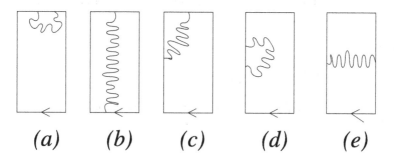

Fig. 5.3. Typical contributing diagrams.

This expression leads to the diagram in figure 5.2. With the currents J flowing according to figure 5.1, the following types of contributions can be distinguished (figure 5.3). Diagram (d) is a self-energy contribution,

$$
\frac{1}{2}g^2 \sum_{(d)} J\,D\,J = \frac{1}{2}g^2(i)^2 \sum_{x_4,y_4=-t/2}^{t/2} D_{44}(\mathbf{0}, x_4 - y_4), \qquad (5.9)
$$

where the times t_1 and t_2 in figure 5.1 have been taken as $\pm t/2$. We may first sum over y_4. For $t \to \infty$ this summation converges at large y_4 and becomes independent of x_4. The summation over y_4 sets p_4 in the Fourier representation for D to zero (cf. (2.90)),

$$
\frac{1}{2}g^2 \sum_{(d)} J\,D\,J \sim -\frac{1}{2}g^2 t \sum_{y_4=-\infty}^{\infty} D_{44}(\mathbf{0}, x_4 - y_4)
$$

$$
= -\frac{1}{2}g^2 t \int_{-\pi/a}^{\pi/a} \frac{d^4p}{(2\pi)^4} e^{ip_4(x_4-y_4)} \frac{a^2}{\sum_\mu (2 - 2\cos ap_\mu)}
$$

$$
= -\frac{1}{2}g^2 t \int_{-\pi/a}^{\pi/a} \frac{d^3p}{(2\pi)^3} \frac{a^2}{\sum_{j=1}^3 (2 - 2\cos ap_j)}
$$

$$
= -\frac{1}{2}g^2 t v(\mathbf{0}), \qquad (5.10)
$$

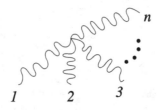

Fig. 5.4. Vertices in the compact $U(1)$ theory.

where

$$v(\mathbf{x}) = \int_{-\pi/a}^{\pi/a} \frac{d^3p}{(2\pi)^3} e^{i\mathbf{px}} \frac{a^2}{\sum_{j=1}^{3}(2 - 2\cos a p_j)} \qquad (5.11)$$

is the lattice-regularized Coulomb potential. Its numerical value at the origin is given by

$$av(\mathbf{0}) = 0.253 \cdots. \qquad (5.12)$$

The contribution of type (e) is given by

$$\frac{1}{2}g^2 \sum_{(e)} JDJ \sim \frac{1}{2}g^2 i(-i) \int_{-t/2}^{t/2} dx_4\, dy_4 \frac{1}{4\pi^2[(x_4 - y_4)^2 + r^2]}, \qquad (5.13)$$

where we assumed $r/a \gg 1$ such that the asymptotic form (5.6) is valid and the summations over x_4 and y_4 may be replaced by integrations. Proceeding as for diagram (d) we get

$$\frac{1}{2}g^2 \sum_{(e)} JDJ \sim \frac{1}{2}g^2 t \int_{-\infty}^{\infty} dy_4 \frac{1}{4\pi^2[(x_4 - y_4)^2 + r^2]}$$

$$= \frac{1}{2}g^2 t \frac{1}{4\pi r}. \qquad (5.14)$$

From these example calculations it is clear that the diagrams of types (a), (b) and (c) do not grow linearly with t. Remembering that there are two contributions of types (d) and (e) (related by interchanging x and y) we find for the potential to order g^2

$$V(\mathbf{x}) = g^2[v(\mathbf{0}) - v(\mathbf{x})], \qquad (5.15)$$

as expected.

Let us now briefly consider higher-order corrections in the compact $U(1)$ theory. The series (5.3) for S leads to interaction vertices of the type shown in figure 5.4, which are proportional to $(ag)^{n-2}$. Their effect vanishes in the continuum limit, unless the powers of a are compensated

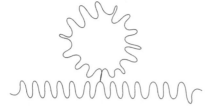

Fig. 5.5. A self-energy diagram in the compact $U(1)$ theory.

by powers of a^{-1} coming from divergent loop diagrams. An example of this is the self-energy diagram figure 5.5, which leads to a 'vacuum-polarization tensor' (cf. problem (ii))

$$\Pi_{\mu\nu}(p) = -\tfrac{1}{4}g^2(\delta_{\mu\nu}p^2 - p_\mu p_\nu) + O(a^2), \qquad (5.16)$$

and a modified propagator

$$D'^{-1}_{\mu\nu} = p^2\delta_{\mu\nu} + O(a^2) + \Pi_{\mu\nu}(p), \qquad (5.17)$$

$$D'_{\mu\nu}(p) = Z(g^2)\delta_{\mu\nu}\frac{1}{p^2} + \text{terms} \propto p_\mu p_\nu, \qquad (5.18)$$

$$Z(g^2) = [1 - \tfrac{1}{4}g^2 + O(g^4)]^{-1}. \qquad (5.19)$$

The terms $\propto p_\mu p_\nu$ do not contribute to the Wilson loop because of gauge invariance, as expressed by 'current conservation' $\partial'_\mu J_{\mu x} = 0$. Further analysis leads to the conclusion that there are no other effects of the self-interaction in the weak-coupling-expansion continuum limit. Note that $Z(g^2)$ is finite, i.e. it does not diverge as $a \to 0$.

We conclude that in the compact $U(1)$ theory the potential is given by

$$V(r) = -g^2 Z(g^2)\frac{1}{4\pi r} + \text{constant} + O(a^2), \to \infty, \qquad (5.20)$$

which is just a Coulomb potential. To make contact with the free Maxwell theory we identify the fine-structure constant α,

$$\alpha = \frac{e^2}{4\pi} = \frac{g^2 Z(g^2)}{4\pi}. \qquad (5.21)$$

The compact $U(1)$ theory is equivalent to the free Maxwell field at weak coupling.

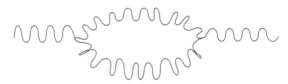

Fig. 5.6. Gluon self-energy contribution to the Wilson loop.

We now turn to the $SU(n)$ gauge theory. A calculation to order g^4 gives in this case the result for the magnitude of the force, $F(r)$, neglecting $O(a^2)$:

$$F(r) = \frac{\partial V(r)}{\partial r} = \frac{1}{4\pi r^2} C_2 \left\{ g^2 + \frac{11n}{48\pi^2} g^4 \left[\ln\left(\frac{r^2}{a^2}\right) + c \right] + O(g^6) \right\}.$$
$$(5.22)$$

Here C_2 is the value of the quadratic Casimir operator in the representation of the Wilson loop and c is a numerical constant which depends on lattice details. Some aspects of the calculation are described in [43]. The logarithm in (5.22) comes from the Feynman gauge self-energy contribution shown in figure 5.6, which is not present in the $U(1)$ theory. The formula (5.22) exhibits the typical divergencies occuring in perturbation theory. It diverges logarithmically as $a \to 0$. This problem is resolved by expressing physically measurable quantities in terms of each other. Here we shall choose an intuitive definition of a renormalized coupling constant g_R at some reference length scale d, by writing

$$F(d) = \frac{C_2 g_R^2}{4\pi d^2}.$$
$$(5.23)$$

This g_R is defined independently of perturbation theory. Its expansion in g^2 follows from (5.22),

$$g_R^2 = g^2 + \frac{11n}{48\pi^2} \left[\ln\left(\frac{d^2}{a^2}\right) + c \right] g^4 + \cdots,$$
$$(5.24)$$

which may be inverted,

$$g^2 = g_R^2 - \frac{11n}{48\pi^2} \left[\ln\left(\frac{d^2}{a^2}\right) + c \right] g_R^4 + \cdots.$$
$$(5.25)$$

The original parameter g in the action has to depend on a if we want to get a g_R independent of a. This dependence is here known incompletely: we cannot take the limit $a \to 0$ in (5.25) because then the coefficient of g_R^4 blows up (and similarly for the higher-order coefficients). The limit

$a \to 0$ will be discussed in the following sections. Insertion into (5.22) leads to the form

$$F(r) = \frac{1}{4\pi r^2} C_2 \left[g_R^2 + \frac{11n}{48\pi^2} g_R^4 \ln\left(\frac{r^2}{d^2}\right) + O(g_R^6) \right], \tag{5.26}$$

from which all dependence on a has disappeared to this order in g_R. The renormalizability of QCD implies that all divergences can be removed in this way to all orders in perturbation theory.

5.2 Asymptotic freedom

The perturbative form (5.26) is useless for $r \to 0$ or $r \to \infty$, since then the logarithm blows up. It is useful only for r of order d, the distance scale used in the definition of the renormalized coupling constant g_R. So let us take $d = r$ from now on. Then $g_R = g_R(r)$. We can extract more information from the weak-coupling expansion by considering renormalization-group beta functions, defined by

$$\beta_R(g_R) = -r\frac{\partial}{\partial r}g_R, \tag{5.27}$$

$$\beta(g) = -a\frac{\partial}{\partial a}g. \tag{5.28}$$

It is assumed here that g_R can be considered to depend only on r and not on a – its dependence on a is compensated by the dependence on a of g. Then the r- and a-dependence on the right-hand side of (5.27) and (5.28) can be converted into a g_R- and g-dependence, respectively, using (5.25) and (5.24), giving

$$\beta_R(g_R) = -\frac{11n}{48\pi^2}g_R^3 + \cdots, \tag{5.29}$$

$$\beta(g) = -\frac{11n}{48\pi^2}g^3 + \cdots. \tag{5.30}$$

Actually the first *two* terms in the expansions

$$\beta(g) = -\beta_1 g^3 - \beta_2 g^5 - \beta_2 g^7 - \cdots, \tag{5.31}$$

$$\beta_R(g_R) = -\beta_{R1} g_R^3 - \beta_{R2} g_R^5 - \beta_{R3} g_R^7 - \cdots \tag{5.32}$$

of the two beta functions are equal. The argument for this is as follows. Let

$$g_R = F(t, g), \tag{5.33}$$

$$t = \ln\left(\frac{r^2}{a^2}\right), \quad g = g(a), \quad g_R = g_R(r), \tag{5.34}$$

Make a scale transformation $a \to \lambda a$, $r \to \lambda r$, which does not affect t, and differentiate with respect to λ, setting $\lambda = 1$ afterwards. Then $\partial/\partial\lambda = a\,\partial/\partial a = r\,\partial/\partial r$, and

$$-\beta_{\mathrm{R}}(g_{\mathrm{R}}) = \frac{\partial g_{\mathrm{R}}}{\partial\lambda} = \frac{\partial F}{\partial g}\left(\frac{\partial g}{\partial\lambda}\right) = -\frac{\partial F}{\partial g}\beta(g). \qquad (5.35)$$

Inserting the expansions for $\beta(g)$ and

$$g_{\mathrm{R}} = g + F_1(t)g^3 + F_2(t)g^5 + \cdots, \qquad (5.36)$$

$$g = g_{\mathrm{R}} - F_1(t)g_{\mathrm{R}}^3 + \cdots, \qquad (5.37)$$

gives

$$\begin{aligned}
-\beta_{\mathrm{R}}(g_{\mathrm{R}}) &= [1 + 3F_1 g^2 + O(g^4)][\beta_1 g^3 + \beta_2 g^5 + O(g^7)] \\
&= [1 + 3F_1 g_{\mathrm{R}}^2 + O(g_{\mathrm{R}}^4)][\beta_1 g_{\mathrm{R}}^3 - 3\beta_1 F_1 g_{\mathrm{R}}^5 + \beta_2 g_{\mathrm{R}}^5 + O(g_{\mathrm{R}}^7)] \\
&= \beta_1 g_{\mathrm{R}}^3 + \beta_2 g_{\mathrm{R}}^5 + O(g_{\mathrm{R}}^7). \qquad (5.38)
\end{aligned}$$

Any coupling constant related to g by a series of the type (5.36) has the same beta function, so we may take the coefficient β_2 from calculations in the continuum using dimensional regularization,[†]

$$\beta_2 = \frac{102}{121}\beta_1^2, \quad \beta_1 = \frac{11n}{48\pi^2}. \qquad (5.39)$$

The remarkable fact in these formulas is that the beta functions are *negative* in a neighborhood of the origin, implying that the couplings become smaller as the length scale decreases. This property is called asymptotic freedom. As we shall see, it implies that $g \to 0$ in the continuum limit. We come back to this in a later section. It suggests furthermore that perturbation theory in the renormalized coupling g_{R} becomes reliable at short distances, provided that a 'running g_{R}' can be used at the appropriate length or momentum scale. In the case of the potential $V(r)$ there is only one relevant length scale, r, and we can use the r-dependence of $g_{\mathrm{R}}(r)$ to our advantage, as will now be shown.

The precise dependence of $g_{\mathrm{R}}(r)$ for small r follows by integrating the differential equation (5.27),

$$\frac{\partial g_{\mathrm{R}}}{\partial\ln r} = -\beta_{\mathrm{R}}(g_{\mathrm{R}}),$$

$$-\ln r = \int^{g_{\mathrm{R}}} \frac{dx}{\beta_{\mathrm{R}}(x)}$$

† Other authors write $\beta_{0,1}$ for our $\beta_{1,2}$.

$$= \int^{g_R} dx \left[\frac{-1}{\beta_1 x^3} + \frac{\beta_2}{\beta_1^2 x} + O(x) \right]$$

$$= \frac{1}{2\beta_1 g_R^2} + \frac{\beta_2}{\beta_1^2} \ln g_R + \text{constant} + O(g_R^2). \tag{5.40}$$

The integration constant can be partially combined with $\ln r$ to form a dimensionless quantity $\ln(r\Lambda_V)$ in a way that has become standard:

$$-\ln(r^2 \Lambda_V^2) = \frac{1}{\beta_1 g_R^2} + \frac{\beta_2}{\beta_1^2} \ln(\beta_1 g_R^2) + O(g_R^2). \tag{5.41}$$

Note the 'ln β_1 convention'. Note also that Λ_V can be defined precisely only if the β_2 term is taken into account – the $O(g_R^2)$ term no longer involves a constant term. This formula can be inverted so as to give g_R as a function of r,

$$\beta_1 g_R^2 = \frac{1}{s} - \frac{\beta_2}{\beta_1^2} \frac{1}{s^2} \ln s + O(s^{-3} \ln s), \tag{5.42}$$

$$s = -\ln(r^2 \Lambda_V^2). \tag{5.43}$$

Inserting this into the force formula (5.23) for $d = r$ gives

$$F(r) = \frac{C_2}{4\pi r^2} \frac{\beta_1^{-1}}{s + (\beta_2/\beta_1^2)s^{-1} \ln s + O(s^{-2} \ln s)}. \tag{5.44}$$

So the short-distance behavior of the potential can be reliably computed ('renormalization-group improved') in QCD by means of the weak-coupling expansion. However, this expansion tells us nothing about the long-distance behavior, because $g_R(r)$ *increases* as r increases, making the first few terms of the weak-coupling expansion irrelevant in this regime.

A second important implication of asymptotic freedom is the application of the renormalization-group equation to the bare coupling g. Integration of (5.28) leads to the analog of (5.41) for the bare coupling,

$$-\ln(a^2 \Lambda_L^2) = \frac{1}{\beta_1 g^2} + \frac{\beta_2}{\beta_1^2} \ln(\beta_1 g^2) + O(g^2), \tag{5.45}$$

where we introduced the 'lattice lambda scale' Λ_L. The analog of (5.42),

$$\beta_1 g^2 \approx 1/|\ln(a^2 \Lambda_L^2)|, \tag{5.46}$$

shows that the bare coupling vanishes in the continuum limit $a \to 0$. This means that the critical point of the theory (the one that is physically relevant, in case there is more than one) is known: it is $g = 0$.

The inverse of (5.45) can be written as

$$\Lambda_L^2 = \frac{1}{a^2}(\beta_1 g^2)^{-\beta_2/\beta_1^2} e^{-1/\beta_1 g^2}[1 + O(g^2)]. \qquad (5.47)$$

This equation is sometimes accompanied by the phrase 'dimensional transmutation': the pure gauge theory has no dimensional parameters (such as mass terms) in its classical action and we may think of transforming the bare coupling g into the dimensional lambda scale via the arbitrary regularization scale $1/a$. As we shall see later, all physical quantities with a dimension are proportional to the appropriate power of Λ_L (as in (1.4)).

The Λ_V and Λ_L are examples of the QCD lambda scales which set the physical scale of the theory. They are all proportional and their ratios can be calculated in one-loop perturbation theory. Let us see how this is done for the ratio Λ_V/Λ_L. The one-loop relation (5.25) can be rewritten as

$$\frac{1}{\beta_1 g^2} = \frac{1}{\beta_1 g_R^2} + \left[\ln\left(\frac{d^2}{a^2}\right) + c\right] + O(g_R^2). \qquad (5.48)$$

Inserting this relation into (5.47) and letting a and d go to zero with d/a fixed, such that g and g_R go to zero, gives

$$\Lambda_L^2 = \frac{e^{-c}}{d^2}(\beta_1 g_R^2)^{-\beta_2/\beta_1^2} e^{-1/\beta_1 g_R^2}[1 + O(g_R^2)] \qquad (5.49)$$

$$= \Lambda_V^2 e^{-c}. \qquad (5.50)$$

Hence the ratio is determined by the constant c, which depends on the details of the regularization.

A comparison of lambda scales on the lattice and in the continuum was done some time ago [45, 46, 47]. The relation with the popular MS-bar scheme (modified minimal subtraction scheme) in dimensional renormalization is

$$\frac{\Lambda_{\overline{MS}}}{\Lambda_L} = \exp[(1/16n - 0.0849780\,n)/\beta_1] \qquad (5.51)$$

$$= 19.82, \quad SU(2) \qquad (5.52)$$

$$= 28.81, \quad SU(3). \qquad (5.53)$$

A calculation [48] of the constant c in the MS-bar scheme then gave the relation to the potential scheme Λ_V, ($\gamma = 0.57 \cdots$ is Eulers's constant)

$$\frac{\Lambda_V}{\Lambda_L} = \exp[\gamma - 1 - (1/16n - 0.095884\,n)/\beta_1] \tag{5.54}$$

$$= 20.78, \quad \text{for } SU(2) \tag{5.55}$$

$$= 30.19 \text{ for } SU(3). \tag{5.56}$$

5.3 Strong-coupling expansion

The strong-coupling expansion is an expansion in powers of $1/g^2$. It has the advantage over the weak-coupling expansion that it has a non-zero radius of convergence. A lot of effort has been put into using it as a method of computation, similarly to the high-temperature or hopping expansion for scalar field theories, see e.g. [6, 44]. One has to be able to match on to coupling values where the theory exhibits continuum behavior. This turns out to be difficult for gauge theories. However, a very important aspect of the strong-coupling expansion is that it gives insight into the qualitative behavior of the theory, such as confinement and the particle spectrum. There are sophisticated methods for organizing the strong-coupling expansion, but here we give only a minimal outline of the basic ideas.

We start again with the compact $U(1)$ theory. Let p be the plaquette (x, μ, ν), $\mu < \nu$. We write the compact $U(1)$ action in the form

$$S = \sum_p L(U_p) + \text{constant,} \tag{5.57}$$

$$L(U_p) = \frac{1}{2g^2}(U_p + U_p^*), \tag{5.58}$$

$$U_p = U_{\mu\nu}(x) = U_{\nu\mu}(x)^*. \tag{5.59}$$

In the path integral we expand $\exp S$ in powers of $1/g^2$. First consider

$$\exp\left[\frac{1}{2g^2}(U_p + U_p^*)\right] = \sum_{m,n=0}^{\infty} \frac{1}{m!n!}\left(\frac{1}{2g^2}\right)^{m+n} U_p^m U_p^{*n}. \tag{5.60}$$

Since $U_p^* = U_p^{-1}$ we put $m = n + k$ and sum over n and k, $k = 0, \pm 1, \pm 2, \ldots$, which gives

$$\sum_{n=0}^{\infty} \left(\frac{1}{n!}\right)^2 \left(\frac{1}{2g^2}\right)^{2n} + (U_p + U_p^{-1}) \sum_{n=0}^{\infty} \frac{1}{(n+1)!n!} \left(\frac{1}{2g^2}\right)^{2n+1}$$

$$+ \cdots + (U_p^k + U_p^{-k}) \sum_{n=0}^{\infty} \frac{1}{(n+k)!n!} \left(\frac{1}{2g^2}\right)^{2n+k} + \cdots. \qquad (5.61)$$

Recognizing the modified Bessel function I_k,

$$I_k(x) = I_{-k}(x) = \sum_{n=0}^{\infty} \frac{1}{(n+k)!n!} \left(\frac{x}{2}\right)^{2n+k}, \qquad (5.62)$$

we find

$$e^{L(U_p)} = \sum_{k=-\infty}^{\infty} I_k\left(\frac{1}{g^2}\right) U_p^k. \qquad (5.63)$$

This is actually an expansion of $\exp L(U_p)$ in irreducible representations of the group $U(1)$, labeled by the integer k. It is useful to extract an overall factor,

$$e^{L(U_p)} = f \sum_k a_k U_p^k, \qquad (5.64)$$

$$a_k(1/g^2) = \frac{I_k(1/g^2)}{I_0(1/g^2)}, \qquad (5.65)$$

$$f(1/g^2) = I_0(1/g^2). \qquad (5.66)$$

The coefficients a_k are of order $(1/g^2)^k$.

Consider now the expansion of the partition function $Z = \int DU \exp S$. Using (5.57) and (5.64) we get a sum of products of U_p^k's,

$$Z = \int DU \sum (\text{coefficient}) \prod_p U_p^k. \qquad (5.67)$$

Each U_p^k is a product $U_1^k U_2^k U_3^{-k} U_4^{-k}$ of the four link variables $U_1, \ldots, U_4$ of the plaquette p, raised to the power k. A given link variable belongs to $2d$ plaquettes (in d dimensions). For each link there is an integration $\int dU$ over the group manifold, which for the group $U(1)$ is simply given by

$$\int d\bar{U}\, U^r = \int_0^{2\pi} \frac{d\theta}{2\pi} e^{ir\theta} = \delta_{r,0}, \qquad (5.68)$$

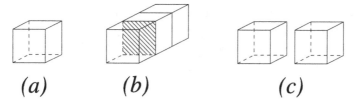

(a)   **(b)**   **(c)**

Fig. 5.7. Simple diagrams contributing to the partition function. The hatched area in (b) belongs to the closed surface. Diagram (c) is disconnected.

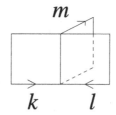

Fig. 5.8. Conservation of flux in three dimensions: $k + l + m = 0$.

where r is an integer. Hence the group integration projects out the trivial ($r = 0$) representation. Now r is the sum of the k's belonging to the plaquettes impinging on the link under consideration. It follows that, after integration, the non-vanishing terms in (5.67) can be represented by diagrams consisting of plaquettes forming closed surfaces, as in figure 5.7. We can interpret this as follows. Each plaquette carries an amount of electric or magnetic flux (depending on its being timelike or spacelike; recall that it corresponds to a miniature line current), labeled by k. The integration over the link variables enforces conservation of flux, as illustrated in figure 5.8. If the surface is not closed, then $\int dU^r = 0$ along each link of its boundary.

Diagram (a) in figure 5.7 represents the leading contribution to Z,

$$Z = f^{Vd(d-1)/2} \left[1 + \frac{1}{3!} Vd(d-1)(d-2) \sum_{k \neq 0} (a_k)^6 + \cdots \right], \quad (5.69)$$

where V is the number of lattice sites, $Vd(d-1)/2$ is the number of plaquettes, $Vd(d-1)(d-2)/3!$ is the number of ways the cube can be embedded in the d-dimensional hypercubic lattice ($d \geq 3$) and 6 is the number of faces of the cube.

The expansion can be arranged as an expansion for $\ln Z$ containing only connected diagrams, called polymers.

For a general gauge theory the derivation of the strong-coupling expansion is similar. One writes

$$L(U_p) = \frac{\beta}{\chi_f(1)} \, \text{Re} \, \chi_f(U_p), \tag{5.70}$$

where $\chi_f(U_p)$ is the character of U_p in the fundamental representation. Recall (we encountered this before in section 4.7) that these characters are orthonormal,

$$\int dU \, \chi_r(U)\chi_s(U)^* = \delta_{rs}, \tag{5.71}$$

and complete for class functions $F(U)$ (which satisfy $F(U) = F(VUV^{-1})$). Next $\exp L$ is written as a character expansion,

$$e^{L(U_p)} = f + f \sum_{r \neq 0} d_r a_r \chi_r(U_p), \tag{5.72}$$

where $r = 0$ denotes the trivial representation $U_p \to 1$ and $d_r = \chi_r(1)$ is the dimension of the representation r. The expansion coefficients are given by

$$f = \int dU \, e^{L(U)}, \tag{5.73}$$

$$d_r a_r = \frac{\int dU \, e^{L(U)}\chi_r^*}{\int dU e^{L(U)}}. \tag{5.74}$$

For the group $U(1)$, $r = 0, \pm 1, \pm 2, \ldots$, $\beta = 1/g^2$, $\chi_r(U) = \exp(ir\theta)$ and we recover (5.65) from the integral representation of the Bessel functions

$$I_r(x) = \frac{1}{\pi} \int_0^\pi d\theta \, \cos(k\theta) \, e^{x \cos\theta}. \tag{5.75}$$

For the group $SU(n)$, $\chi_f(1) = n$ and

$$\beta = 2n/g^2. \tag{5.76}$$

The leading β-dependence of $a_f(\beta)$ is easily found,

$$f(\beta) = \int dU \, e^{(\beta/2n)(\chi_f + \chi_f^*)}$$
$$= 1 + O(\beta^2), \tag{5.77}$$
$$n a_f(\beta) = f(\beta)^{-1} \int dU \, e^{(\beta/2n)(\chi_f + \chi_f^*)}\chi_f^*$$

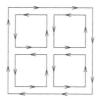

Fig. 5.9. A small Wilson loop with compensating plaquettes.

$$= \frac{\beta}{2n} + O(\beta^2), \;\; n > 2 \tag{5.78}$$

$$= \frac{\beta}{n} + O(\beta^2), \;\; n = 2. \tag{5.79}$$

For $SU(2)$ the characters are real. In terms of g^2,

$$a_{\mathrm{f}} = \frac{1}{g^2} + \cdots, \;\; n = 2, \tag{5.80}$$

$$= \frac{1}{ng^2} + \cdots, \;\; n = 3, 4, \ldots. \tag{5.81}$$

Up to group-theoretical complications (which can be formidable) the strong-coupling expansion for general gauge groups follows that of the $U(1)$ case. The graphs are the same, but the coefficients differ.

5.4 Potential at strong coupling

We now turn to the expectation value of the rectangular Wilson loop $\langle U(C) \rangle$, from which the potential can be calculated. The links on the curve C contain explicit factors of U that have to be compensated by plaquettes from the expansion of $\exp S$, otherwise the integration over U gives zero. Figure 5.9 shows a simple example. The contribution of this diagram is (the Wilson loop is taken in the fundamental representation of $U(1)$)

$$[a_1(1/g^2)]^4, \tag{5.82}$$

which is the leading contribution for this curve C. Recall that a_1 is given by

$$a_1(1/g^2) = \frac{I_1(1/g^2)}{I_0(1/g^2)} = \frac{1}{2g^2} - \frac{1}{2}\left(\frac{1}{2g^2}\right)^3 + \cdots. \tag{5.83}$$

In higher orders disconnected diagrams appear. It can be shown, however, that disconnected diagrams may be discarded: they cancel out

Fig. 5.10. Leading diagrams for a large Wilson loop.

between the numerator and denominator of $\langle U(C)\rangle$. The expansion can be rewritten as a sum of connected diagrams. Figure 5.10 illustrates the leading terms for a large Wilson loop,

$$W(r,t) = a_1^A + 2(d-2)Aa_1^{A+4} + \cdots, \tag{5.84}$$

where A is the area of the loop, in lattice units $A = rt$. Boundary corrections are also in the $\cdots$. The higher orders correspond to 'decorations' of the minimal surface.

The potential $V(r)$ follows now from (5.1) and $A = rt$,

$$
\begin{aligned}
V(r) &= \frac{1}{t}\ln W(r,t) \\
&= -[\ln a_1 + 2(d-2)a_1^4 + \cdots]\, r.
\end{aligned} \tag{5.85}
$$

For $r \to \infty$, $A \to \infty$ and the boundary corrections become negligible. Hence, the potential is *linearly confining* at large distances,

$$V(r) \approx \sigma r, \quad r \to \infty, \tag{5.86}$$

$$\sigma = -\ln a_1 - 2(d-2)a_1^4 + \cdots. \tag{5.87}$$

At strong coupling the compact $U(1)$ theory is confining.

For other gauge theories the calculation of the leading contribution to a Wilson loop in the fundamental representation goes similarly. A useful formula here is

$$\int dU\, \chi_r(UV)\chi_s(W^\dagger U^\dagger) = \frac{\delta_{rs}}{d_r}\chi_r(VW^\dagger), \tag{5.88}$$

which follows from

$$\int dU\, \mathcal{D}^r_{mn}(U)\mathcal{D}^{r'}_{m'n'}(U)^* = \delta_{rr'}\delta_{mm'}\delta_{nn'}\frac{1}{d_r}, \tag{5.89}$$

seen earlier in (4.145). The use of this formula is illustrated in figure 5.11. Successive integration in the simple Wilson-loop example in figure 5.9 is illustrated in figure 5.12. Each arrow in figure 5.12 denotes the result of 'integrating out a link'. The equality signs symbolize $UU^\dagger = 1$. Note that the factors d_r in (5.88) cancel out with those in (5.72). Hence

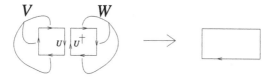

Fig. 5.11. Integration of a link variable.

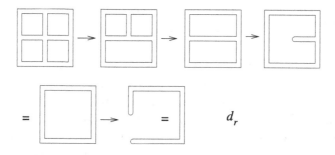

Fig. 5.12. Integrating the leading contribution to a 2×2 Wilson loop.

the numerical value of the diagram is $d_r a_r(\beta)^4$, for a Wilson loop in representation r.

Another way to see this is as follows: in figure 5.9 there are $n_l = 12$ links, $n_p = 4$ plaquettes and $n_s = 9$ sites. Integrating over each link gives a factor $d_r^{-n_l}$ by (5.89) and contracting the Kronecker deltas at each site gives a factor $d_r^{n_s}$. Each plaquette has a factor $d_r^{n_p}$ by (5.72). For a simple surface without handles, the Euler number is

$$-n_l + n_s + n_p = 1$$

$$\Rightarrow \text{leading contribution} = (d_r)^{-n_l+n_s}[d_r a_r(\beta)]^{n_p} = d_r a_r(\beta)^{n_p}. \quad (5.90)$$

For a Wilson loop in the fundamental representation of the $SU(n)$ theory the first few terms in the expansion for the string tension

$$\sigma = -\ln a_f(\beta) - 2(d-2)a_f(\beta)^4 + \cdots. \quad (5.91)$$

are similar to the $U(1)$ result (5.87). In higher orders the $a_r(\beta)$ corresponding to other irreps enter. The final result may then be re-expressed by expansion in powers of $1/ng^2$.

(a)   *(b)*

Fig. 5.13. Flux lines for sources in the fundamental (a) and the adjoint (b) representation.

5.5 Confinement versus screening

In the previous section we saw that the $U(1)$ and $SU(n)$ potentials are confining in the strong-coupling region. From the derivation we can see that this is true for external charges (Wilson loops) in the fundamental representation of any compact gauge group. However, external charges in the *adjoint* representation of $SU(n)$ are *not* confined. This is because the charges in the adjoint representation can be *screened* by the gauge field. A adjoint source is like a quark–antiquark pair, as illustrated intuitively in figure 5.13. We now show how this happens at strong coupling.

Let U denote the fundamental representation (as before) and R the adjoint representation. The latter can be constructed from U and $U^\dagger$,

$$R_{kl} = 2\,\mathrm{Tr}\,(U^\dagger t_k U t_l), \qquad (5.92)$$

where the t_k are the generators in the fundamental representation. Since R is an irrep,

$$\int dU\, R_{kl}(U) = 0. \qquad (5.93)$$

To compensate the R's on the links of the adjoint Wilson loop

$$\mathrm{Tr}\,R(C) = \mathrm{Tr}\,\prod_{l\in C} R_l \qquad (5.94)$$

by the plaquettes from the expansion of exp S, we may draw a Wilson surface and find in the same way as in the previous section the seemingly leading contribution

$$d_a a_a(\beta)^A, \quad d_a = n^2 - 1, \qquad (5.95)$$

with A the minimal surface spanned by C. However, there is a more economical possibility for large A, illustrated in figure 5.14. The tube of plaquettes is able to screen the adjoint loop. To evaluate this contribution we unfold the tube as in figure 5.15. The links in the interior

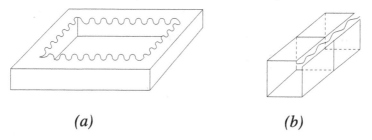

(a)   *(b)*

Fig. 5.14. Diagram contributing to a Wilson loop in the adjoint representation; (b) is a close up of a piece of the circumference in (a). The wavy line indicates the adjoint representation.

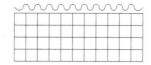

Fig. 5.15. Unfolding the tube of plaquettes. The horizontal and vertical boundaries are to be identified.

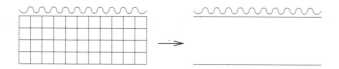

Fig. 5.16. Integrating out the interior.

can be integrated out as in figure 5.12, as illustrated in figure 5.16. The first step gives a factor $d_f a_f(\beta)^{N_p}$ with N_p the number of plaquettes ($d_f = n$). The second step gives an additional factor $1/d_f$. There remains the integration over the links of the Wilson loop, which leads to integrals of the type (for $n \geq 3$) (cf. (A.93) in appendix A.4)

$$\int dU \, U^a_b U^{q\dagger}_p R_{kl} = \frac{1}{d_a} 2 (t_k)^a_p (t_l)^q_b, \quad n > 2, \tag{5.96}$$

as illustrated in figure 5.17. So we get a trace of the form

$$2 d_a^{-1} (t_k)^a_p (t_l)^q_b \, 2 d_a^{-1} (t_l)^b_q (t_m)^r_c \cdots (t_k)^p_a = 1, \tag{5.97}$$

since $2 \, \text{Tr} \, (t_k t_k) = n^2 - 1 = d_a$. This leads to a factor

$$a_f(\beta)^{4P}, \tag{5.98}$$

where P is the perimeter of the (large) adjoint loop in lattice units:

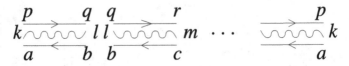

Fig. 5.17. Link variables on the adjoint loop.

$P = 2(r + t)$, and the factor 4 in the exponent reflects the fact that there are four plaquettes per unit length.

The leading contributions of the perimeter and area type in the $SU(n)$ theory are given by

$$W_a(r,t) \sim (n^2 - 1)(a_a)^{rt} + \cdots + 2(d-1)(d-2)(a_f)^{8(r+t)} + \cdots, \quad (5.99)$$

which by (5.1) leads to a potential

$$V(r) = \sigma_{\text{eff}}\, r, \quad r \le \frac{V(\infty)}{\sigma_{\text{eff}}}$$
$$= V(\infty), \quad r \ge \frac{V(\infty)}{\sigma_{\text{eff}}}, \quad (5.100)$$

with

$$\sigma_{\text{eff}} = -\ln a_a + \cdots, \quad (5.101)$$
$$V(\infty) = -8\ln a_f + \cdots, \quad (5.102)$$
$$a_f = (ng^2)^{-1} + \cdots,$$
$$a_a = \frac{n^2}{n^2 - 1}(ng^2)^{-2} + \cdots. \quad (5.103)$$

(This behavior of a_a follows easily from (5.74) and (5.96).)

At large distances the potential approaches a constant. The sharp crossover from linear to constant behavior (at $r \approx 4$) is an artifact of our simplistic strong-coupling calculation. Still, the calculation suggests that there is an intermediate region where the potential is approximately linear with some effective string tension σ_{eff}, although strictly speaking the string tension, defined by $\sigma = V(r)/r$, $r \to \infty$, vanishes for adjoint sources.

To decide whether static charges in an irreducible representation r can be screened by the gauge field, we consider the generalization of (5.96),

$$I = \int dU\, \mathcal{D}^s_{mn}(U)\, \mathcal{D}^s_{m'n'}(U)^*\, \mathcal{D}^r_{kl}(U), \quad (5.104)$$

where s denotes the irreps of the two screening plaquettes. If the integral I is zero, the source cannot be screened, and vice-versa. Let Z_k denote an element of the center of $SU(n)$, i.e. $Z_k \in SU(n)$ commutes with all group elements and it is represented in the fundamental representation as a multiple of the identity matrix,

$$(Z_k)_b^a = e^{ik2\pi/n} \delta_b^a, \quad k = 0, 1, \ldots, n-1. \tag{5.105}$$

Irreps r can be constructed from a tensor product $U \otimes U \cdots U \otimes U^\dagger \cdots \otimes U^\dagger$, say p times U and q times $U^\dagger$, so r can be assigned an integer $\nu(r) = p - q \bmod n$, from the way it transforms under $U \to Z_1 U$:

$$\mathcal{D}_{kl}^r(Z_1 U) \to e^{i\nu(r)2\pi/n} \mathcal{D}_{kl}^r(U). \tag{5.106}$$

The integer $\nu(r)$ is called the *n-ality* of the representation (triality for $n = 3$). Making the change of variables $U \to Z_1 U$ in (5.104) gives

$$I = e^{i\nu(r)2\pi/n} I, \tag{5.107}$$

and we conclude that $I = 0$ if the n-ality $\nu(r) \neq 0$. Sources with non-zero n-ality are confined; sources with zero n-ality are not confined. In QCD, static quarks have non-zero triality and are confined.

5.6 Glueballs

The particles of the pure gauge theory are called glueballs. They may be interpreted as bound states of gluons. Gluons appear as a sort of photons in the weak-coupling expansion and, because of asymptotic freedom, they manifest themselves as effective particle-like excitations at high energies. However, gluons do not exist as free particles because of confinement, as we shall see.

Masses of particles can be calculated from the long-distance behavior of suitable fields. These are gauge-invariant fields constructed out of the link variables $U_{\mu x}$, such as Wilson loops, with the quantum numbers of the particles being studied. The transfer-matrix formalism shows that an arbitrary state can be created out of the vacuum by application of a suitable combination of spacelike Wilson loops. The simplest of these is the plaquette field $\operatorname{Tr} U_{mnx}$, $m, n = 1, 2, 3$. The plaquette–plaquette expectation value (4.97) can be calculated easily at strong coupling. The relevant diagrams consist of tubes of plaquettes, as in figure 5.18. Since there are four plaquettes per unit of time, the glueball mass is given by $m = -4\ln a_f(\beta) + \cdots$. The higher-order corrections correspond to

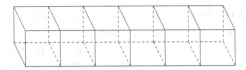

Fig. 5.18. The leading strong-coupling diagram for the plaquette–plaquette correlator. Time runs horizontally.

diagrams decorating the basic tube of figure 5.18, which will also cause the tube to perform random walks.

The plaquette can be decomposed into operators with definite quantum numbers under the symmetry group of the lattice, and such operators can in turn be embedded into representations of the continuum rotation group of spin zero, one and two. To be more precise, the quantum numbers J^{PC} (J = spin, P = parity, C = charge-conjugation parity) excited by the plaquette are 0^{++}, 1^{+-} and 2^{++}, which may be called scalar (S), axial vector (A) and tensor (T). The description of glueballs with other quantum numbers requires more complicated Wilson loops. The terms in the strong-coupling series

$$m_j = -4\ln u + \sum_k m_j^k u^k, \quad j = S, A, T, \quad u \equiv a_{\mathrm{f}}(\beta), \qquad (5.108)$$

have been calculated to order u^8 for gauge groups $SU(2)$ and $SU(3)$ [91, 92]. See [10] for details.

Since the strong-coupling diagrams are independent of the (compact) gauge group (but their numerical values are not), also the $U(1)$ and e.g. $Z(n)$ gauge theories† have a particle content at strong coupling similar to that of glueballs.

5.7 Coulomb phase, confinement phase

We have seen that all gauge theories with a compact gauge group such as $U(1)$, $SU(n)$ and $Z(n)$ have the property of confinement at strong coupling, and the emerging particles are 'glueballs'. On the other hand, we have also given arguments, for $U(1)$ and $SU(n)$, that the weak-coupling expansion on the lattice gives the usual universal results for renormalized quantities found with perturbation theory in the continuum.

In particular the compact $U(1)$ theory at weak coupling is not confining and it contains no glueballs but simply the photons of the free

† $Z(n)$ is the discrete group consisting of the center elements (5.105) of $SU(n)$.

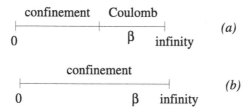

Fig. 5.19. Phase diagram of the compact $U(1)$ gauge theory (a) and the $SU(n)$ gauge theory, $n = 2, 3$ (b).

Maxwell theory. The physics of the compact $U(1)$ theory is clearly different in the weak- and strong-coupling regions. This can be understood from the fact that there is a phase transition as a function of the bare coupling constant (figure 5.19). One speaks of a Coulomb phase at weak coupling and a confining phase at strong coupling. In the Coulomb phase the static potential has the standard Coulomb form $V = -g_R^2/4\pi r+$constant, whereas in the confinement phase the potential is linearly confining at large distances, $V \approx \sigma r$. There is a phase transition at a critical coupling $\beta_c \equiv 1/g_c^2 \approx 1.01$, at which the string tension $\sigma(\beta)$ vanishes; see for example [95].

The Wilson loop serves as an order field in pure gauge theories. Consider a rectangular $r \times t$ Wilson loop C, with perimeter $P = 2(r+t)$ and area $A = rt$. When the loop size is scaled up to infinity, the dominant behavior is a decay according to a perimeter law or an area law:

$$W(C) \sim e^{-\epsilon P}, \quad \text{Coulomb phase,} \tag{5.109}$$
$$W(C) \sim e^{-\sigma A}, \quad \text{confinement phase.} \tag{5.110}$$

Here ϵ may be interpreted as the self-energy of a particle tracing out the path C in (Euclidean) space–time, and σ is the string tension experienced by a particle.

There is no phase transition in the $SU(2)$ and $SU(3)$ models with the standard plaquette action in the fundamental representation in the whole region $0 < \beta < \infty$ ($\beta = 2n/g^2$). This conclusion is based primarily on numerical evidence (see e.g. the collection of articles in [5]) and it is also supported by analytic mean-field calculations (see e.g. [6] for a review). The absence of a phase transition, combined with confinement at strong coupling, may be interpreted as evidence for confinement also in the weak-coupling region.

Fig. 5.20. Qualitative phase diagram of mixed-action $SU(n)$ gauge theory for $n = 2, 3$.

It should be kept in mind that the phase structure of a theory is not universal and depends on the action chosen. Only the scaling region near a critical point is supposed to have universal properties. For example, in $SU(n)$ gauge theory with an action consisting of a term in the fundamental representation and a term in the adjoint representation,

$$S = \sum_p [\beta_f d_f^{-1} \mathrm{Re\,Tr}\, U_p + \beta_a d_a^{-1} \mathrm{Re\,Tr}\, D^a(U_p)], \qquad (5.111)$$

the phase diagram in the β_f–β_a coupling plane looks schematically like figure 5.20. This figure shows two connected phase regions; the one relevant for QCD is the region connected to the weak-coupling region $\beta_f/2n + \beta_a n/(n^2 - 1) = 1/g^2 \to \infty$ (recall (4.85)). For $n > 3$ the phase boundary going downward in the south-east direction crosses the β_f axis. This implies that, for $n > 3$, the model with only the standard plaquette action in the fundamental representation shows a phase transition. It is, however, not a deconfining transition because we can go around it continuously through negative values of the adjoint coupling β_a.

The phase structure of lattice gauge theories is rich subject and for more information we refer the reader to [5] and [6], and [10].

5.8 Mechanisms of confinement

As we have seen in section 5.1, the calculation of the static potential from a Wilson loop to lowest order of perturbation in g^2 gives a Coulomb potential. In the compact $U(1)$ theory, higher orders did not change this result qualitatively, whereas in $SU(n)$ gauge theory, there are logarithmic corrections, that can be interpreted in terms of asymptotic freedom.

However, there is no sign of confinement in weak-coupling perturbation theory. This can be understood from the fact that we expect the string tension to depend on the bare coupling g^2 as

$$\sqrt{\sigma} = C_\sigma \Lambda_{\mathrm{L}} = C_\sigma \, \frac{1}{a} \, (\beta_1 g^2)^{-\beta_2/2\beta_1^2} e^{-1/\beta_1 g^2} [1 + O(g^2)], \qquad (5.112)$$

which has no weak-coupling expansion (all derivatives $\partial/\partial g^2$ vanish at $g = 0$). The physical region is at weak coupling, where the lattice spacing is small, so how can we understand confinement in this region?

Non-perturbative field configurations have long been suspected to do the job. Such configurations are fundamentally different from mere fluctuations on a zero or pure-gauge background. We mention here in particular *magnetic-monopole* configurations envisioned by Nambu [49], 't Hooft [50], and Polyakov [41], and $Z(n)$ *vortex* configurations put forward by 't Hooft [51] and Mack [52].

It can be shown that the confinement of the compact $U(1)$ theory is due to the fact that it is really a theory of photons interacting with magnetic monopoles (see e.g. the first reference in [53] for a review). These monopoles *condense* in the confinement phase in which the model behaves like a *dual superconductor*. In a standard type-II superconductor, electrically charged Cooper pairs are condensed in the ground state, which phenomenon causes magnetic-field lines to be concentrated into line-like structures, called Abrikosov flux tubes. Magnetic monopoles, if they were to exist, would be confined in such a superconductor, because the energy in the magnetic flux tube between a monopole and an antimonopole would increase linearly with the distance between them.

In a dual superconductor electric and magnetic properties are interchanged. The compact $U(1)$ model is a dual superconductor in the strong-coupling phase, in which the magnetically charged monopoles condense and the electric-field lines are concentrated in tubes, such that the energy between a pair of positively and negatively charged particles increases linearly with distance. In this way the model is an illustration of the dual-superconductor hypothesis as the explanation of confinement in QCD.

At weak coupling the monopoles decouple in the compact $U(1)$ model, because they are point particles that acquire a Coulomb self-mass of order of the inverse lattice spacing a^{-1}. However, in $SU(2)$ gauge theory, according to [53], there are 'fat' monopoles that have physical sizes and masses, and do not decouple at weak bare gauge coupling g^2. They remain condensed as $g^2 \to 0$ and continue to produce a non-zero string tension for all values of g^2. A similar mechanism is supposed to take place in $SU(n)$ gauge theory for $n > 2$.

The mechanism for confinement in $SU(n)$ gauge theory proposed by Mack is condensation of fat $Z(n)$ vortices. The latter cause an area-type decay of large Wilson loops in much the same way as in the $Z(n)$ gauge theory at strong coupling.

There seems to be more than one explanation of confinement, depending on the gauge one chooses to work in. This may seem disturbing, but, e.g. also in scattering processes, different reference frames (such as 'center of mass' or 'laboratory') lead to different physical pictures. Numerical simulations offer a great help in studying these fundamental questions. Lattice XX reviews are in [54], see also [55, 56, 57, 58, 59].

5.9 Scaling and asymptotic scaling, numerical results

We say that relations between physical quantities scale if they become independent of the correlation length ξ as it increases toward infinity. In practice this means once ξ is sufficiently large. In pure $SU(n)$ gauge theory the correlation length is given by the mass in lattice units of the lightest glueball, $\xi = 1/am$. For instance, glueball-mass ratios m_i/m_j are said to scale when they become approximately independent of ξ. Typically one expects corrections of order a^2,

$$m_i/m_j = r_{ij} + r'_{ij} a^2 m^2 + O(a^4). \tag{5.113}$$

For the usual plaquette action am is only a function of the bare gauge coupling g^2. We can write $m = C_m \Lambda_L$, with Λ_L the lambda scale introduced in (5.45) and C_m a numerical constant characterizing the glueball. The correlation length is then related to the gauge coupling by

$$\xi_m^{-2} = a^2 m^2 = C_m^2 \, a^2 \Lambda_L^2 = C_m^2 \, (\beta_1 g^2)^{-\beta_2/\beta_1^2} e^{-1/\beta_1 g^2} [1 + O(g^2)], \tag{5.114}$$

for sufficiently small g^2. Neglecting the $O(g^2)$, this behavior as a function of g^2 is called asymptotic scaling.

It turns out that asymptotic scaling is a much stronger property than scaling, in the sense that scaling may set in when the correlation length is only a few lattice spacings, whereas asymptotic scaling is not very well satisfied yet. In the usual range of couplings, which are of order $\beta = 2n/g^2 = 6$ for $SU(3)$ gauge theory with the plaquette action in the fundamental representation, once $\beta \geq 5.7$ or so, the correlation length appears to be sufficiently large and the $O(a^4)$ corrections small enough for scaling corrections to be under control. However, asymptotic scaling does not hold very well yet in this region. Apparently the $O(g^2)$ corrections in (5.114) cannot be neglected. This has led to a search for 'better' expansion parameters, i.e. 'improved' definitions of a bare coupling that may give better convergence, see e.g. [60]. Note that $\exp(-1/\beta_1 g^2)$ is a rapidly varying function of g^2 because $\beta_1 = 11n/48\pi^2 \approx 0.070$ ($n = 3$) is so small. Typically $\Delta\beta \approx 0.48$ corresponds to a reduction of a^2 by a factor of four near $\beta = 6$.

The potential $V(r)$ is a good quantity to test for scaling because it is relatively easy to compute and there are many values $V(r)$. As a measure of the correlation length we may take

$$\xi_\sigma(\beta) = 1/a\sqrt{\sigma}, \qquad (5.115)$$

where $a^2\sigma$ is the string tension in lattice units, which goes to zero as β approaches infinity. Assuming $\sqrt{\sigma} = 400$ MeV, for example (cf. section 1.1), the value of $a\sqrt{\sigma}$ give us the lattice distance a in units $(\text{MeV})^{-1}$. This can be used to express the potential in physical units as follows. The potential in lattice units can be written as

$$aV = v\left(\frac{r}{a}, \beta\right), \qquad (5.116)$$

where v is a function of the dimensionless variables r/a and β. Recall that V contains the unphysical self-energy of the sources, which is distance independent. Expressing the potential in physical units, as set by the string tension, gives

$$\frac{V}{\sqrt{\sigma}} = \xi_\sigma(\beta)\, v\left(\frac{r\sqrt{\sigma}}{\xi_\sigma(\beta)}, \beta\right) \equiv \tilde{V}(r\sqrt{\sigma}, \beta) + v_0(\beta). \qquad (5.117)$$

These relations 'scale' when $\tilde{V}$ becomes independent of β. Here $v_0(\beta)$ is the self-energy, which can be fixed by a suitable choice of the zero point of energy, e.g. $\tilde{V}(1) = 0$. In practice, after computing σ from the long-distance behavior $V \approx \sigma r + \text{constant} + O(r^{-1})$, the data points at various $\beta \geq 6$ can be made to form a single scaling curve by plotting $V/\sqrt{\sigma}$ versus $r\sqrt{\sigma}$ with a suitable vertical shift corresponding to $v_0(\beta)$.

However, the accuracy of such scaling tests is limited by the fact that σ is an asymptotic quantity defined in terms of the behavior of the potential at infinity. This problem may be circumvented by concentrating on the force $F = \partial V/\partial r$, in terms of which we can define a reference distance r_0 by

$$r_0^2 F(r_0) = 1.65. \qquad (5.118)$$

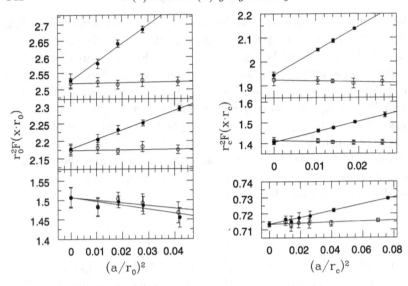

Fig. 5.21. Scaling of the $SU(3)$ force and the continuum limit at $x = r/r_0 = 0.4$, 0.5, and 0.9 (left), and $x = r/r_c = 0.5$, 0.6, and 1.5 (right) from top to bottom. The stronger/weaker dependence on a corresponds to r_1 defined in (5.119)/(5.120). From [62].

The choice 1.65 turns out to give $r_0 \approx 1/\sqrt{\sigma}$, which is in the intermediate-distance regime within which the potential and force can be computed accurately [61]. The force may be computed as

$$F(r_1) = [V(r + a) - V(r)]/a, \quad r_1 = r - a/2, \qquad (5.119)$$

and scaling tests can then be performed as above with $\sqrt{\sigma} \to 1/r_0$. There is another choice for r_1 that gives an improved definition of the force, leading to much smaller scaling violations in the small- and intermediate-distance region [61], namely

$$(4\pi r_1)^{-2} = [v(r_1, 0, 0) - v(r_1 - a, 0, 0)]/a, \qquad (5.120)$$

where $v(x, y, z)$ is the lattice Coulomb potential (5.11). The scaling test for the force avoids ambiguities from the Coulomb self-energy in the potential. Writing $r = xr_0$, or $r = xr_c$, where r_c is defined as in (5.118) with $1.65 \to 0.65$, a scaling analysis is carried out in [62] in the form $r_0^2 F(xr_0) = f_0(x) + f_0'(a/r_0)^2 + O(a^4)$, or with $r_0 \to r_c$, as shown in figure 5.21. The values of $(a/r_{0,c})^2$ correspond to β in the interval [5.7, 6.92].

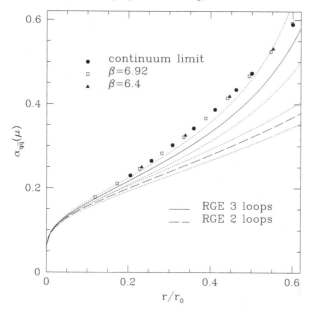

Fig. 5.22. The running coupling $\alpha_{q\bar{q}}(\mu) = g_R^2(\mu)/4\pi$, $\mu = 1/r$ plotted versus r/r_0 and compared with the dependence on r as predicted by the weak-coupling expansion for the renormalization-group beta function (the curves labeled RGE; dotted lines correspond to 1 σ uncertainties of $\Lambda_{\overline{MS}}(r_0)$. From [64].

In the small-distance regime the running of the coupling (5.23), i.e. $g_R^2(\mu) = 4\pi r^2 F(r)/C_2$, $\mu = 1/r$, can be compared with the prediction of the perturbative beta function, which is known to three-loop order. One could use the perturbative expansion (5.41) in which Λ_V, or equivalently $\Lambda_{\overline{MS}}$, appears as an integration constant. This scale in units of r_0, i.e. $r_0\Lambda_{\overline{MS}}$, has been determined independently in an elaborate non-perturbative renormalization-group computation [63]. Instead of using the perturbative expansion it is more accurate to integrate the two- or three-loop renormalization-group equation numerically. The result is shown in figure 5.22, where we see that perturbation theory works surprisingly well, when it is implemented in this way, up to quite large α's. In physical units $r_0 \approx 1/\sqrt{\sigma} \approx 0.5$ fm.

Note that knowledge of a non-perturbative Λ scale allows the *prediction* of α_s. Such a program has been pursued in full QCD in various ways [66] and the resulting α_s agrees well with the experimentally measured values, see also the review in [2].

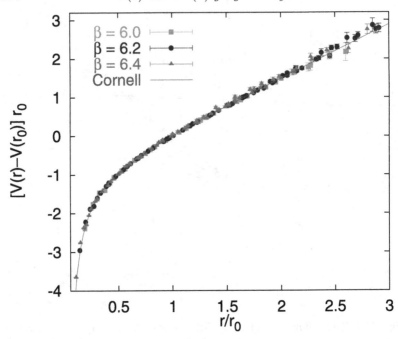

Fig. 5.23. The potential from two values of β. The curve labeled 'Cornell' is a fit of the form $-\frac{4}{3}\alpha/r + \text{constant} + \sigma r$ with constant α. From [65].

An overview of numerically computed potential is given in figure 5.23.

Glueball masses have by now also been computed with good accuracy in the $SU(n)$ models, using variational methods for determining the eigenvalues of the transfer matrix. It is particularly interesting to do this for varying n, since the theory simplifies in the large-n limit in the sense that only planar diagrams contribute [67]. The same is true in the strong-coupling expansion [68]. Figure 5.24 shows recent results for various n. We see that ratios with $\sqrt{\sigma}$ do indeed behave smoothly as a function of $1/n^2$ all the way down to $n = 2$.

Last, but not least, *analytic* computations in *finite volume* are theoretically very interesting and a comparison with numerical data is very rewarding. For a review, see [18].

5.10 Problems

(i) *Gauge fixing and the weak-coupling expansion*

Consider a partition function for a $U(1)$ or $SU(n)$ lattice gauge-

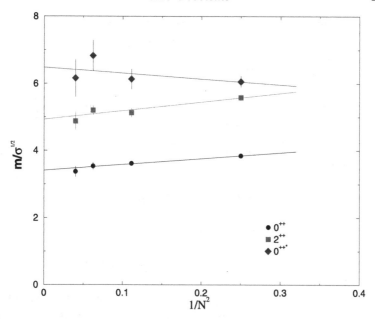

Fig. 5.24. Ratios of glueball masses with $\sqrt{\sigma}$, extrapolated to $a \to 0$ and infinite volume, as a function of $1/n^2$, for $n = 2, 3, 4, 5$. From [69].

field theory with gauge-invariant action $S(U)$,

$$Z = \int DU \exp[S(U)]. \tag{5.121}$$

The action may be the standard plaquette action

$$S(U) = -\frac{1}{2\rho g^2} \sum_{x\mu\nu} \operatorname{Tr}(1 - U_{\mu\nu x}), \tag{5.122}$$

it may also contain the effect of dynamical fermions in the form $\ln \det A(U)$, with A the 'fermion matrix', cf. section 7.1. Let $O(U)$ be a gauge-invariant observable, $O(U) = O(U^\Omega)$,

$$U^\Omega_{\mu x} = \Omega_x U_{\mu x} \Omega^\dagger_{x+\hat\mu}, \tag{5.123}$$

and

$$\langle O \rangle = \frac{\int DU \exp[S(U)] \, O(U)}{Z}. \tag{5.124}$$

be the average of O.

We want to evaluate the path integrals in the weak-coupling expansion and expect to have to use gauge fixing, as in the continuum. We can try to restrict the implicit integration over all gauge transformations in $\langle O \rangle$, loosely called gauge fixing, by adding an action $S_{\text{gf}}(U)$ to $S(U)$ that is not invariant under gauge transformations. For example,

$$S_{\text{gf}}(U) = -\frac{1}{\xi} \sum_x \frac{1}{2g^2 \rho} \operatorname{Tr}(\partial'_\mu \operatorname{Im} U_{\mu x})^2, \quad \operatorname{Im} U \equiv \frac{U - U^\dagger}{2i},$$

$$(5.125)$$

with $\partial'_\mu = -\partial^\dagger_\mu$ the backward derivative, $\partial'_\mu f_x = f_x - f_{x+\hat{\mu}}$. Let $\Delta(U)$ be defined by

$$\Delta(U)^{-1} = \int D\Omega \exp[S_{\text{gf}}(U^\Omega)], \qquad (5.126)$$

where $\int D\Omega$ is the integration over all gauge transformations. It is assumed that $\Delta(U)^{-1} \neq 0$.

(a) Show that the Faddeev–Popov measure factor $\Delta(U)$ is gauge invariant.

We insert $1 = \Delta(U) \int D\Omega \exp[S_{\text{gf}}(U^\Omega)]$ into the integrands in the above path-integral expression for $\langle O \rangle$ and make a transformation of variables $U \to U^{\Omega^\dagger}$. Using the gauge invariance of $S(U)$, $O(U)$, and $\Delta(U)$ we get

$$\begin{aligned}
\langle O \rangle &= \frac{\int D\Omega \int DU\, \Delta(U) \exp[S(U) + S_{\text{gf}}(U)]\, O(U)}{\int D\Omega \int DU\, \Delta(U) \exp[S(U) + S_{\text{gf}}(U)]} \\
&= \frac{\int DU\, \Delta(U) \exp[S(U) + S_{\text{gf}}(U)]\, O(U)}{\int DU\, \Delta(U) \exp[S(U) + S_{\text{gf}}(U)]}.
\end{aligned} \qquad (5.127)$$

In the weak-coupling expansion we expand about the saddle points with maximum action. We assume this maximum to be given by $U_{\mu x} = 1$. There will in general be more maxima. For example, without dynamical fermions, $U_{\mu x} = U$ (i.e. independent of x and μ) and $U_{\mu x} = Z_\mu$, with Z_μ an element of the center of the gauge group, give the same value of the plaquette action as does $U_{\mu x} = 1$. Intuitively we expect constant modes to be important for finite-size effects, but not important in the limit that the space–time volume goes to infinity. Restricting ourselves here to the latter case, we shall not integrate over constant modes and expand about $U_{\mu x} = 1$, writing

$$U_{\mu x} = \exp(-ig A^k_{\mu x} t_k). \qquad (5.128)$$

The evaluation of the integral (5.126) defining $\Delta(U)$ is also done perturbatively. Because of the factor $1/g^2$ in the gauge-fixing action, we only need to know $\Delta(U)$ for small $\partial'_\mu \operatorname{Im} \operatorname{Tr} U_{\mu x}$. The integral (5.126) has a saddle point at $\Omega_x = 1$, but there are in general many more saddle points Ω_x, called Gribov copies, with $S_{\mathrm{gf}}(U^\Omega) = S_{\mathrm{gf}}(U)$. The study of Gribov copies is complicated. One can give arguments that the correct weak-coupling expansion is obtained by restriction to the standard choice $\Omega_x = 1$, and this is what we shall do in the following. This means that, for the perturbative evaluation of $\Delta(U)$, we can write

$$\Omega_x = \exp(ig\omega_x^k t_k) \tag{5.129}$$

and expand in $g\omega_x$. In perturbation theory we may just as well simplify the gauge-fixing action and use

$$S_{\mathrm{gf}} = -\frac{1}{2\xi^2} \sum_x \partial'_\mu A^k_{\mu x} \partial'_\nu A^k_{\nu x}. \tag{5.130}$$

(In the neighborhood of the identity, $A^k_{\mu x}$ and ω^k_x are well defined in terms of $U_{\mu x}$ and Ω_x.)

We extend the initially compact integration region over $A^k_{\mu x}$ and ω^k_x to the entire real line $(-\infty, \infty)$. The error made in doing so is expected to be of order $\exp(-\mathrm{constant}/g^2)$, and therefore negligible compared with powers of g, as $g \to 0$. A typical example is given by

$$\int_{-\pi}^{\pi} dx\, e^{-x^2/g^2} = \int_{-\infty}^{\infty} dx\, e^{-x^2/g^2} + O(e^{-\pi^2/g^2}). \tag{5.131}$$

(b) For a $U(1)$ gauge theory show that (5.130) leads to a Faddeev–Popov factor that is independent of U,

$$\Delta(U) = \mathrm{constant} \times \det(\Box), \quad \Box_{xy} = \partial'_\mu \partial_\mu \bar{\delta}_{xy}, \tag{5.132}$$

with the constant independent of A_μ.

(ii) *Weak-coupling expansion in compact QED*

We consider first the bosonic theory given by the action

$$S(U) = -\frac{1}{2g^2} \sum_{x\mu\nu} (1 - U_{\mu\nu x}) \tag{5.133}$$

and use (5.130) for gauge fixing. The bare vertex functions $-V$ are given by

$$S_A + S_{\mathrm{gf}} = -\sum_n \frac{1}{n!} \sum_{x_1 \cdots x_n} V^{x_1 \cdots x_n}_{\mu_1 \cdots \mu_n} A_{\mu_1 x_1} \cdots A_{\mu_n x_n}. \tag{5.134}$$

(a) Show that, in momentum space, for even $n \geq 2$ (by convention the momentum-conserving periodic delta function is omitted in the definition of the Fourier transform of $V_{\mu_1 \cdots \mu_n}^{x_1 \cdots x_n}$),

$$V_{\mu_1 \cdots \mu_n}(k_1 \cdots k_n) = -\tfrac{1}{2}(g^2)^{n/2-1} \sum_{\alpha\beta} T_{\mu_1}^{\alpha\beta}(k_1) \cdots T_{\mu_n}^{\alpha\beta}(k_n)$$

$$- \delta_{n,2} \frac{1}{\xi} K_{\mu_1}(k_1) K_{\mu_2}(k_2), \qquad (5.135)$$

where

$$K_\mu(k) = \frac{1}{i}(e^{ik_\mu} - 1), \quad K_\mu^*(k) = -K_\mu(-k), \qquad (5.136)$$

$$T_\mu^{\alpha\beta}(k) = K_\alpha^*(k)\delta_{\beta\mu} - K_\beta^*(k)\delta_{\alpha\mu}. \qquad (5.137)$$

(b) Show that the photon propagator $D_{\mu\nu}(k)$ is given by

$$D_{\mu\nu}(k) = \left(\delta_{\mu\nu} - \frac{K_\mu(k)K_\nu^*(k)}{|K(k)|^2}\right) \frac{1}{|K(k)|^2} + \xi \frac{K_\mu(k)K_\nu^*(k)}{|K(k)|^4}.$$

$$(5.138)$$

The Feynman gauge corresponds to $\xi = 1$.

(c) Derive (5.19), for arbitrary ξ.

6

Fermions on the lattice

In this chapter we introduce the path integral for Fermi fields. We shall discuss the species-doubling phenomenon – the fact that a naively discretized Dirac fermion field leads to more particle excitations than expected and desired, two remedies for this, which go under the names 'Wilson fermions' and 'staggered fermions', the interpretation of the path integral in Hilbert space, and the construction of the transfer operator. Integration over 'anticommuting numbers', the 'Grassmann variables' and the relation with creation and annihilation operators in fermionic Hilbert space is reviewed in appendix C.

6.1 Naive discretization of the Dirac action

In continuous Minkowski space–time the action for a free fermion field can be written as (see appendix D for an introduction)

$$S = -\int d^4x \left[\tfrac{1}{2}(\bar{\psi}(x)\gamma^\mu \partial_\mu \psi(x) - \partial_\mu \bar{\psi}(x)\gamma^\mu \psi(x)) + m\bar{\psi}(x)\psi(x) \right], \quad (6.1)$$

or, exhibiting the Dirac indices α, β, ... (but suppressing the label x for brevity),

$$S = -\int d^4x \left[(\gamma^\mu)_{\alpha\beta} \tfrac{1}{2}(\bar{\psi}_\alpha \partial_\mu \psi_\beta - \partial_\mu \bar{\psi}_\alpha \psi_\beta) + m\bar{\psi}_\alpha \psi_\alpha \right]. \quad (6.2)$$

The ψ and $\bar{\psi}$ are anticommuting objects, so-called Grassmann variables, e.g. $\psi_\alpha(x)\bar{\psi}_\beta(y) = -\bar{\psi}_\beta(y)\psi_\alpha(x)$. The integrand in (6.1) is Hermitian, treating ψ and ψ^+,

$$\psi^+ = \bar{\psi}\beta, \quad \beta = i\gamma^0, \quad (6.3)$$

as Hermitian conjugates, e.g. $(\psi_\alpha(x)\psi_\beta^+(y))^\dagger = \psi_\beta^+(y)\psi_\alpha(x)$. Note, however, that ψ and ψ^+ are independent 'variables' (which is why we use the superscript $+$ instead of $\dagger$). The Dirac matrices have the following properties:

$$\gamma^0 = -\gamma_0, \quad \gamma_0 = -\gamma_0^\dagger, \quad \gamma_0^2 = -1, \tag{6.4}$$

$$\gamma^k = \gamma_k = \gamma_k^\dagger, \quad \gamma_k^2 = 1, \quad k = 1, 2, 3. \tag{6.5}$$

implying that $\beta = \beta^\dagger$ and $\beta^2 = 1.\dagger$ Replacing the derivative operators by discrete differences,

$$\partial_\mu \psi(x) \to \frac{1}{a_\mu} [\psi(x + a_\mu \hat\mu) - \psi(x)], \tag{6.6}$$

we obtain from (6.1) a lattice version

$$S = -\sum_{x,\mu} \frac{1}{2a_\mu} [\bar\psi(x)\gamma^\mu \psi(x + a_\mu \hat\mu) - \bar\psi(x + a_\mu \hat\mu)\gamma^\mu \psi(x)] - m \sum_x \bar\psi(x)\psi(x). \tag{6.7}$$

Recall that a_μ is the lattice spacing in the μ direction. We shall occasionally only need the spacing in the time direction, a_0, to be different from the spatial lattice spacing $a_k = a$, $k = 1, 2, 3$.

The path integral for free fermions with anticommuting external sources η and $\bar\eta$ is now tentatively defined by

$$Z(\eta, \bar\eta) = \int D\bar\psi \, D\psi \, e^{i[S + \sum_x (\bar\eta\psi + \bar\psi\eta)]}, \tag{6.8}$$

where

$$D\bar\psi \, D\psi = \prod_{x,\alpha} d\bar\psi_{x\alpha} \, d\psi_{x\alpha} = \prod_{x\alpha} d\psi_{x\alpha}^+ \, d\psi_{x\alpha}. \tag{6.9}$$

We assumed the action to be rewritten in terms of dimensionless ψ_x and $\bar\psi_x$,

$$\psi_x = a^{3/2}\psi(x), \quad \bar\psi_x = a^{3/2}\bar\psi(x), \tag{6.10}$$

and similarly the symbols $d\psi_{x\alpha}^+$ and $d\psi_{x\alpha}$ are dimensionless. The last equality in (6.9) follows from the rule $d(T\psi) = (\det T)^{-1} d\psi$ (cf. appendix C) and $\det \beta = 1$. The $\psi_{x\alpha}$ and $\psi_{x\alpha}^+$ are independent generators of a Grassmann algebra. We recall also the definition of fermionic integration (cf. appendix C),

$$\int db = 0, \quad \int db\, b = 1, \tag{6.11}$$

where b is any of the $\psi_{x\alpha}$ or $\psi_{x\alpha}^+$. Before making the transition to imaginary time we need to make the dependence on a_0 explicit. So let n_μ

$\dagger$ We usually write just 1 for the unit matrix $\mathbb{1}$.

be the integers specifying the lattice site x, $x^0 = n_0 a_t$, $\mathbf{x} = \mathbf{n}a$, and let $\psi_n \equiv \psi_x$ and $\bar{\psi}_n \equiv \bar{\psi}_x$. Recalling that $\sum_x = a_0 a^3 \sum_n$ in our notational convention, the lattice action reads more explicitly

$$S = -\sum_n \left[\frac{1}{2}(\bar{\psi}_n \gamma^0 \psi_{n+\hat{0}} - \bar{\psi}_{n+\hat{0}} \gamma^0 \psi_n) \right.$$
$$\left. + \sum_{k=1}^{3} \frac{a_0}{2a}(\bar{\psi}_n \gamma^k \psi_{n+\hat{k}} - \bar{\psi}_{n+\hat{k}} \gamma^k \psi_n) + (a_0 m)\bar{\psi}_n \psi_n \right]. \quad (6.12)$$

Furthermore

$$\sum_x (\bar{\eta}\psi + \bar{\psi}\eta) \equiv \frac{a_0}{a} \sum_n (\bar{\eta}_{\alpha n} \psi_{\alpha n} + \bar{\psi}_{\alpha n} \eta_{\alpha n}), \quad (6.13)$$

with dimensionless $\eta_{\alpha n}$ and $\bar{\eta}_{\alpha n}$.

It follows from the rules of fermionic integration that the path integral for a finite space–time volume is a polynomial in $a_0 m$ and a_0/a. Hence, an analytic continuation to 'imaginary time' poses no problem:

$$a_0 = |a_0| \exp(-i\varphi), \quad \varphi: 0 \to \pi/2, \quad a_0 \to -i a_4, \quad (6.14)$$

with $a_4 = |a_0|$. This transforms the path integral into its Euclidean version ($iS \to S_\Im$, dropping the $\Im$),

$$Z = \int D\bar{\psi} D\psi \, e^{S + \sum_n (\bar{\eta}\psi + \bar{\psi}\eta)}, \quad (6.15)$$

$$S = -\sum_n \left[\sum_\mu \frac{a_4}{2a_\mu}(\bar{\psi}_n \gamma_\mu \psi_{n+\hat{\mu}} - \bar{\psi}_{n+\hat{\mu}} \gamma_\mu \psi_n) + a_4 m \, \bar{\psi}_n \psi_n \right],$$

where μ now runs from 1 to 4 (with $n_4 \equiv n_0$, $\hat{4} \equiv \hat{0}$), and

$$\gamma_4 = i\gamma^0 = \beta. \quad (6.16)$$

6.2 Species doubling

It turns out that the model described by the action in (6.15) yields $2^4 = 16$ Dirac particles (fermions with two charge and two spin states) instead of one. This is the species-doubling phenomenon. We shall infer it in this section from inspection of the fermion propagator and the excitation energy spectrum.

Using a matrix notation, writing

$$Z(\eta, \bar{\eta}) = \int D\bar{\psi} D\psi \, e^{-\bar{\psi}A\psi + \bar{\eta}\psi + \bar{\psi}\eta}, \quad (6.17)$$

where (in lattice units, $a = a_4 = 1$)

$$A_{xy} = \sum_{z\mu} \gamma_\mu \frac{1}{2} (\bar{\delta}_{x,z}\bar{\delta}_{y,z+\hat{\mu}} - \bar{\delta}_{x,z+\hat{\mu}}\bar{\delta}_{y,z}) + m \sum_z \bar{\delta}_{x,z}\bar{\delta}_{y,z}, \qquad (6.18)$$

the path integral is easily integrated (appendix C) to give

$$Z(\eta, \bar{\eta}) = \det A \, e^{\bar{\eta}A^{-1}\eta}. \qquad (6.19)$$

Here $A_{xy}^{-1} \equiv S_{xy}$ is the fermion propagator. It can be evaluated in momentum space, assuming infinite space–time,

$$A(k, -l) = \sum_{xy} e^{-ikx+ily} A_{xy} = S(k)^{-1}\bar{\delta}(k - l), \qquad (6.20)$$

$$S(k)^{-1} = \sum_\mu i\gamma_\mu \sin k_\mu + m, \qquad (6.21)$$

$$S(k) = \frac{m - i\gamma_\mu s_\mu}{m^2 + s^2}, \quad s_\mu = \sin k_\mu. \qquad (6.22)$$

Reverting to non-lattice units the propagator becomes

$$S(k) = \frac{m - i\sum_\mu \gamma_\mu \sin(ak_\mu)/a}{m^2 + \sum_\mu \sin^2(ak_\mu)/a^2}, \qquad (6.23)$$

for which the limit $a \to 0$ gives the continuum result

$$S(k) = \frac{m - i\gamma k}{m^2 + k^2} + O(a^2). \qquad (6.24)$$

The propagator has a pole at $k_4 = i\omega = i\sqrt{\mathbf{k}^2 + m^2}$ corresponding to a Dirac particle. The pole is near the zeros of the sine functions at the origin $ak_\mu = 0$. However, there are 15 more regions in the four dimensional torus $-\pi < ak_\mu \le \pi$ where the sine functions vanish, 16 in total:

$$S(k) = \frac{m - i\gamma_\mu^{(A)} p_\mu}{m^2 + p^2} + O(a), \quad k = k_A + p \qquad (6.25)$$

where the k_A is one of the 16 four-vectors

$$k_A = \frac{\pi_A}{a}, \quad \mathrm{mod}\ 2\pi \qquad (6.26)$$

with

$$\begin{aligned}
\pi_0 &= (0,0,0,0), \quad \pi_{1234} = (\pi,\pi,\pi,\pi), \\
\pi_1 &= (\pi,0,0,0), \quad \pi_2 = (0,\pi,0,0), \ldots, \quad \pi_4 = (0,0,0,\pi), \\
\pi_{12} &= (\pi,\pi,0,0), \ldots, \quad \pi_{34} = (0,0,\pi,\pi), \\
\pi_{123} &= (\pi,\pi,\pi,0), \ldots, \quad \pi_{234} = (0,\pi,\pi,\pi),
\end{aligned} \qquad (6.27)$$

and

$$\gamma_\mu^{(A)} = \gamma_\mu \cos \pi_{A\mu} = \pm \gamma_\mu. \tag{6.28}$$

Since the γ_μ^A differ only by a sign from the original γ_μ, they are equivalent to these by a unitary transformation. This transformation is easy to build up out of products of $\gamma_\rho \gamma_5$, where $\gamma_5 = i\gamma^0\gamma^1\gamma^2\gamma^3 = -\gamma_1\gamma_2\gamma_3\gamma_4$ is the Hermitian and unitary matrix which anticommutes with the γ_μ: $\gamma_\mu\gamma_5 = -\gamma_5\gamma_\mu$. So let

$$\{S_A\} = \{\mathbb{1}, S_\rho, S_\rho S_\sigma, S_\rho S_\sigma S_\tau, S_1 S_2 S_3 S_4\}, \quad S_\rho = i\gamma_\rho\gamma_5, \tag{6.29}$$

where $\rho \neq \sigma \neq \tau \neq \rho$ and $A \leftrightarrow \pi_A \leftrightarrow S_A$, e.g. $\pi_{23} \leftrightarrow S_{23} = S_2 S_3$. Then

$$\gamma_\mu^{(A)} = S_A^\dagger \gamma_\mu S_A, \tag{6.30}$$

and we have

$$S(k_A + p) = S_A^\dagger \frac{m - i\gamma_\mu p_\mu}{m^2 + p^2} S_A + O(a^2). \tag{6.31}$$

The transformations S_A are useful for the detailed interpretation of the zeros of the sine functions near $k_A \neq 0$ in terms of genuine particles [70]. Here we shall support the interpretation of the 15 additional particles – the species doublers – by deriving the spectrum of excitation energies above the energy of the ground state.

The excitation-energy spectrum is conveniently obtained from the time dependence of the propagator, analogously to the boson case:

$$S(\mathbf{x}, t) = \int_{-\pi}^\pi \frac{d^3 k}{(2\pi)^3} e^{i k x} \int_{-\pi}^\pi \frac{dk_4}{2\pi} e^{i k_4 t} \frac{m - i\gamma \mathbf{s} - i\gamma_4 \sin k_4}{m^2 + \mathbf{s}^2 + \sin^2 k_4}, \tag{6.32}$$

where we reverted to lattice units and used the notation $s_\mu = \sin k_\mu$. The k_4 integral can be performed by changing variables to

$$z = e^{i k_4}, \tag{6.33}$$

in terms of which $s_4^2 = 1 - (z^2 + z^{-2} + 2)/4$, and

$$S(\mathbf{x}, t) = -4 \int \frac{d^3 k}{(2\pi)^3} e^{i k x} \int \frac{dz}{2\pi i} z^t \frac{z(m - i\gamma \mathbf{s}) - \gamma_4(z^2 - 1)/2}{z^4 - 2fz^2 + 1},$$
$$f = 1 + 2(m^2 + \mathbf{s}^2). \tag{6.34}$$

The integral over z is over the unit circle in the complex plane, as shown in figure 6.1. The denominator of the integrand has four zeros, at $\pm z_+$ and $\pm z_-$, where $z_\pm$ are given by

$$(z_\pm)^2 = f \pm \sqrt{f^2 - 1}, \quad z_\pm = e^{\pm\omega}, \tag{6.35}$$

$$\cosh(2\omega) = f, \quad \sinh\omega = \sqrt{m^2 + \mathbf{s}^2}. \tag{6.36}$$

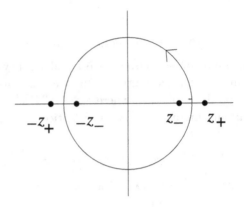

Fig. 6.1. Contour integration in the complex-z plane.

For $t > 0$ (t = integer) the two poles at $z = \pm z_-$ contribute, giving

$$S(\mathbf{x}, t) = \int \frac{d^3 k}{(2\pi)^3} \frac{e^{i\mathbf{kx} - \omega t}}{\sinh(2\omega)} (m - i\gamma \mathbf{s} + \gamma_4 \sinh \omega)$$

$$+ (-1)^t \int \frac{d^3 k}{(2\pi)^3} \frac{e^{i\mathbf{kx} - \omega t}}{\sinh(2\omega)} (m - i\gamma \mathbf{s} - \gamma_4 \sinh \omega). \quad (6.37)$$

Before interpreting this result we want to summarize it in terms of the variable k_4, for later use. In terms of k_4 the zeros of the denominator $m^2 + \mathbf{s}^2 + \sin^4 k_4$ at $z = z_\pm$ are at $k_4 = \mp i\omega$, and for $z = -z_\pm$ at $k_4 = \mp i\omega + \pi$ (mod 2π). The $k_4 = -i\omega$, $-i\omega + \pi$ poles are relevant for $t < 0$. The residues of the other poles are given by

$$e^{ik_4 t} (m - i\gamma \mathbf{s} - i\gamma_4 s_4)$$

$$= e^{-\omega t} (m - i\gamma \mathbf{s} + \gamma_4 \sinh \omega), \qquad k_4 = i\omega,$$

$$= (-1)^t e^{-\omega t} (m - i\gamma \mathbf{s} - \gamma_4 \sinh \omega), \quad k_4 = i\omega + \pi,$$

$$= e^{\omega t} (m - i\gamma \mathbf{s} - \gamma_4 \sinh \omega), \qquad k_4 = -i\omega,$$

$$= (-1)^t e^{\omega t} (m - i\gamma \mathbf{s} + \gamma_4 \sinh \omega), \quad k_4 = -i\omega + \pi. \quad (6.38)$$

We see that we cannot blindly perform the inverse Wick rotation on the lattice $k_4 \to ik^0$ and look for particle poles at $k^0 = \pm \omega$. We have to let $k_4 \to ik^0 + \varphi$, $\varphi \in [0, 2\pi)$: then $k^0 = \pm \omega$ corresponds to $e^{\mp \omega t} e^{i\varphi t}$, $t \gtrless 0$. In this case we have have poles at $\varphi = 0$ and $\varphi = \pi$. Recall that the Bose-field denominator $m^2 + 2 \sum_\mu (1 - \cos k_\mu)$ gives only a pole for $\varphi = 0$.

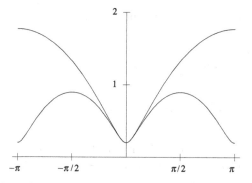

Fig. 6.2. Excitation-energy spectra for bosons (upper curve) and fermions (lower curve) on the lattice, in lattice units ($m = 0.2$).

We now interpret the result (6.37). From the time dependence of the propagator we identify the energy spectrum $\omega(\mathbf{k})$. Since there are two poles contributing for $t > 0$, there must be two fermion particles for every $\mathbf{k}$. One of them (the pole at $z = z_-$) has the usual $e^{-\omega t}$ factor. The other (at $-z_-$) has in addition the rapidly oscillating factor $(-1)^t$. Apparently, to obtain smooth behavior at large times (in lattice units) we have to take two lattice units as our basic time step. This is in accordance with the transfer operator interpretation of the path integral, in which in general *two* adjacent time slices are identified with the fermion Hilbert space [78, 79, 89], in which *two independent* operator Dirac fields $\hat{\psi}_{1,2}$ act, corresponding to the two particle poles. An exception is Wilson's fermion method [86, 87], which has no fermion doubling (for $r = 1$, see below).

So there is a doubling of fermion species due to the discretization of time. There is a further proliferation of particles due to the discretization of space. In figure 6.2 we compare the boson and fermion excitation-energy spectra

$$\cosh \omega = 1 + \tfrac{1}{2}\left[m^2 + 2\sum_{j=1}^{3}(1 - \cos k_j)\right], \quad \text{boson};\qquad (6.39)$$

$$\sinh \omega = \sqrt{m^2 + \sum_{j=1}^{3}\sin^2 k_j}, \quad \text{fermion}.\qquad (6.40)$$

We define a particle state to correspond to a local minimum of the energy surface $\omega(\mathbf{k})$. The minima are at $\mathbf{k} = \mathbf{k}_A$, $\mathbf{k}_A = 0$, $\boldsymbol{\pi}_1$, $\boldsymbol{\pi}_2$, $\boldsymbol{\pi}_3$, $\boldsymbol{\pi}_{12}$, $\boldsymbol{\pi}_{23}$, $\boldsymbol{\pi}_{31}$, $\boldsymbol{\pi}_{123}$, with rest energy given by $\omega_A \equiv \omega(\mathbf{k}_A)$, $\sinh \omega_A = m$. For $m \to 0$ (in lattice units) the spectrum is relativistic near $\mathbf{k} = \mathbf{k}_A$,

$$\omega \to \sqrt{m^2 + \mathbf{p}^2}, \quad m \to 0, \quad \mathbf{p} = \mathbf{k} - \mathbf{k}_A \to 0, \qquad (6.41)$$

and $\mathbf{p}$ can be interpreted as the momentum of the particle. From the time and space 'doubling' we count $2^4 = 16$ particles. Note that the wave vector $\mathbf{k}$ is just a label to identify the states and that the physical momentum interpretation has to be supplied separately.

One may wish to ignore the $k_A \neq 0$ particles. However, in an interacting theory this is not possible, because k_μ is conserved only modulo 2π. For example, two $k_A = 0$ particles may collide and produce two $k_A = \pi_1 = (\pi, 0, 0, 0)$ particles: $p_1 + p_2 = p_3 + \pi_1 + p_4 + \pi_1 = p_3 + p_4$ (mod 2π).

The phenomena related to fermions on a lattice touch on deep issues involving anomalies and topology. This is a vast and technically difficult subject and we shall give only a brief review in sections 8.4 and 8.6. In a first exploration we shall describe two important methods used for ameliorating the effects of species doubling in QCD-like theories: Wilson's method [71] and the method of Kogut–Susskind [72, 40] (in the Hamiltonian formulation). The latter is also known as the staggered-fermion method, in its generalization to Euclidean space–time (see for example [79, 74, 80]). For the hypercubic lattice the staggered-fermion method is equivalent to the 'geometrical' or Dirac–Kähler fermion method of Becher and Joos [81], provided that an appropriate choice is made of the couplings to the gauge fields.

We shall first describe Wilson's method and then briefly introduce the staggered-fermion method.

6.3 Wilson's fermion method

Wilson's method can be viewed as adding a momentum-dependent 'mass term' to the fermion action, which raises the masses of the unwanted doublers to values of the order of the cutoff, thereby decoupling them from continuum physics. For free fermions we replace the mass term in the action as follows,

$$m \sum_x \bar{\psi}_x \psi_x \rightarrow m \sum_x \bar{\psi}_x \psi_x + \frac{ar}{2} \sum_{x\mu} \partial_\mu \bar{\psi}_x \partial_\mu \psi_x \tag{6.42}$$

$$= m \sum_x \bar{\psi}_x \psi_x + \frac{ar}{2} \sum_{x\mu} \frac{1}{a^2} (\bar{\psi}_{x+a\hat{\mu}} - \bar{\psi}_x)(\psi_{x+a\hat{\mu}} - \psi_x)$$

$$= \left(m + \frac{4r}{a} \right) \sum_x \bar{\psi}_x \psi_x - \frac{r}{2a} \sum_{x\mu} (\bar{\psi}_{x+a\hat{\mu}} \psi_x + \bar{\psi}_x \psi_{x+a\hat{\mu}}).$$

The has the effect of replacing the mass m in the inverse propagator in momentum space by

$$m + r \sum_\mu (1 - \cos k_\mu) \equiv \mathcal{M}(k), \tag{6.43}$$

in lattice units. The propagator is then given by

$$S(k) = \frac{\mathcal{M}(k) - i\gamma_\mu \sin k_\mu}{\mathcal{M}^2(k) + \sum_\mu \sin^2 k_\mu}. \tag{6.44}$$

For $k = k_A + p$ and small p in lattice units this takes the form

$$S(p) = \frac{m_A - i\gamma_\mu^{(A)} p_\mu}{m_A^2 + p^2}, \tag{6.45}$$

$$m_A = m + 2n_A r, \quad n_A = 0, 1, \ldots, 4, \tag{6.46}$$

where n_A is the number of π's in k_A.

Hence, the mass parameters of the doubler ($n_A > 0$) fermions are of order one in lattice units as long as $r \neq 0$. These mass parameters m_A may be identified with the fermion masses if they are small in lattice units, i.e. for small m and r. For general r and momenta $\mathbf{p}$ the fermion energies differ from $\sqrt{m_A^2 + \mathbf{p}^2}$ and it is interesting to see what they actually are. We therefore look for the poles of the propagator as a function of k_4 and identify the energy ω from $k_4 = i\omega$ or $k_4 = i\omega + \pi$, as explained below (6.37). For simplicity we shall use the notation

$$s_\mu = \sin k_\mu, \quad c_\mu = \cos k_\mu, \quad s^2 = s_\mu s_\mu. \tag{6.47}$$

Separating the k_4 dependence, the denominator of the propagator can be written as

$$\mathcal{M}^2 + s^2 = 1 + \mathbf{s}^2 + \Sigma^2 - 2r\Sigma c_4 - (1 - r^2)c_4^2, \tag{6.48}$$

$$\Sigma = m + r + r \sum_{j=1}^{3} (1 - c_j), \tag{6.49}$$

which denominator vanishes for

$$\cosh \omega^\pm = \frac{\sqrt{\Sigma^2 + (1 - r^2)(1 + \mathbf{s}^2)} \pm r\Sigma}{1 - r^2}. \tag{6.50}$$

Here the plus sign corresponds to $k_4 = i\omega + \pi$ and the minus sign to $k_4 = i\omega$. The rest energies of the particles at $\mathbf{k} = \mathbf{k}_A$ follow from $\mathbf{s} = 0$ and

$$\Sigma = m + r + 2nr, \quad n = 0, 1, 2, 3 \text{ for } \mathbf{k}_A = 0, \, \pi_j, \, \pi_{jk}, \, \pi_{123}. \tag{6.51}$$

For $m = 0$ the particles have rest energy ω_n given by

$$\cosh \omega_n^\pm = \frac{\sqrt{r^2(1 + 2n)^2 + 1 - r^2} \pm r^2(1 + 2n)}{1 - r^2}. \tag{6.52}$$

Hence, only the wanted ($n = 0, -$ sign) fermion has rest energy zero and the doubler fermions have rest energies of order 1 in lattice units (energies of order of the cutoff). For $r \to 1$ the rest energies of the time-doublers (for which the $+$ sign applies) become infinite, $\omega^+ \to \infty$. The non-time-doubler rest energies become $\omega_n^- = \ln(1 + 2n)$ at $r = 1$. Actually, as r increases from 0 to 1 the doublers disappear before reaching $r = 1$ in the sense that the local minima of the energy surface at $k_A \neq 0$ disappear.

Wilson's choice is $r = 1$. It can be seen directly from (6.48) that in this case there is no species doubling because the inverse propagator is linear in $\cos k_4$. Re-installing the lattice spacing a, the particle energy can be found to contain errors of order a, to be compared with $O(a^2)$ for naive/staggered fermions or bosons,

$$\omega = \omega_0^- = \sqrt{m^2 + \mathbf{p}^2} + O(a). \tag{6.53}$$

The special significance of $r = 1$ can be seen in another way from the complete action, which has the form

$$S = \sum_{x\mu} \left(\bar{\psi}_x \frac{r - \gamma_\mu}{2} \psi_{x+\hat{\mu}} + \bar{\psi}_{x+\hat{\mu}} \frac{r + \gamma_\mu}{2} \psi_x \right) - M \sum_x \bar{\psi}_x \psi_x,$$
$$M = m + 4r. \tag{6.54}$$

The combinations

$$P_\mu^\pm = \frac{r \pm \gamma_\mu}{2}, \tag{6.55}$$

become orthogonal projectors for $r = 1$,

$$(P_\mu^\pm)^2 = P_\mu^\pm, \quad P_\mu^+ P_\mu^- = 0, \quad P_\mu^+ + P_\mu^- = 1. \tag{6.56}$$

Replacing derivatives by covariant derivatives we obtain the expression for the fermion action coupled to a lattice gauge field $U_{\mu x}$,

$$S_{\mathrm{F}} = \sum_{x\mu} (\bar{\psi}_x P_\mu^- U_{\mu x} \psi_{x+\hat{\mu}} + \bar{\psi}_{x+\hat{\mu}} P_\mu^+ U_{\mu x}^\dagger \psi_x) - \sum_x \bar{\psi}_x M \psi_x, \quad (6.57)$$

or, temporarily reintroducing the lattice spacing a,

$$S_{\mathrm{F}} = -\sum_x \left[\frac{1}{2} (\bar{\psi}\gamma_\mu D_\mu \psi - D_\mu \bar{\psi}\gamma_\mu \psi) + \bar{\psi}m\psi + a\frac{r}{2} D_\mu \bar{\psi} D_\mu \psi \right], \quad (6.58)$$

where $D_\mu \bar{\psi}(x) = [\bar{\psi}(x + \hat{\mu}a)U_{\mu x}^\dagger - \bar{\psi}(x)]/a$, etc., we rearranged the summation over x, and $m = M - 4r/a$ is sometimes called the bare fermion mass.

In the QCD case M is a diagonal matrix in flavor space and r is usually chosen flavor-independent, mostly $r = 1$. A parameterization introduced by Wilson follows from rescaling $\psi \to M^{-1/2}\psi$, $\bar{\psi} \to \bar{\psi}M^{-1/2}$. For one flavor this gives the form

$$S = -\sum_x \bar{\psi}_x \psi_x + \kappa \sum_{x\mu} [\bar{\psi}_x (r - \gamma_\mu)U_{\mu x}\psi_{x+\hat{\mu}} + \bar{\psi}_{x+\hat{\mu}}(r + \gamma_\mu)U_{\mu x}^\dagger \psi_x],$$

$$(6.59)$$

where

$$\kappa = \frac{1}{2M}, \quad (6.60)$$

is Wilson's hopping parameter (it is flavor dependent). This κ is analogous to the hopping parameter in the scalar field models. We may interpret $-\sum_x \bar{\psi}_x \psi_x$ as belonging to the integration measure in the path integral.

For free fermions the continuum limit means $m \to 0$ in lattice units, which implies a critical value for the hopping parameter

$$\kappa \to \kappa_{\mathrm{c}} = 1/8r, \quad M \to M_{\mathrm{c}} = 4r. \quad (6.61)$$

At this critical value there is somehow a cancellation of the $\bar{\psi}\psi$-like terms, such that the fermions acquire zero mass. With the gauge field present the effective strength of the hopping term is reduced by the 'fluctuating' unitary $U_{\mu x}$. We then expect $M_{\mathrm{c}} < 4$ and $\kappa_{\mathrm{c}} > 1/8r$, for given gauge coupling g. However, in the QCD case we know already that g itself should go to zero in the continuum limit, because of asymptotic freedom, implying $U_{\mu x} \to 1$ in a suitable gauge and (6.61) should still be valid (κ_{c} is of course gauge independent). However, at gauge coupling of order one we can be deep in the scaling region of QCD and we may expect an effective κ_{c} substantially larger than $1/8r$. Since there are no free quarks in QCD we cannot define κ_c as the value at which the quark mass vanishes. We shall see later that it may be defined as the value at which the pion mass vanishes.

6.4 Staggered fermions

Starting with the naive fermion action we make the unitary transformation of variables (in lattice units)

$$\psi_x = \gamma^x \chi_x, \quad \bar{\psi}_x = \bar{\chi}_x (\gamma^x)^\dagger, \qquad (6.62)$$

$$\gamma^x \equiv (\gamma_1)^{x_1} (\gamma_2)^{x_2} (\gamma_3)^{x_3} (\gamma_4)^{x_4}. \qquad (6.63)$$

Because

$$\left[(\gamma^x)^\dagger \gamma_\mu \gamma^{x+\hat{\mu}} \right]_{\alpha\beta} = \left[\gamma^x \gamma_\mu (\gamma^{x+\hat{\mu}})^\dagger \right]_{\alpha\beta} = \eta_{\mu x} \delta_{\alpha\beta}, \qquad (6.64)$$

where

$$\eta_{1x} = 1, \quad \eta_{2x} = (-1)^{x_1}, \quad \eta_{3x} = (-1)^{x_1+x_2}, \quad \eta_{4x} = (-1)^{x_1+x_2+x_3}, \qquad (6.65)$$

this transformation has the effect of removing the gamma matrices from the naive fermion action, which acquires the form

$$S = -\sum_{\alpha=1}^{4} \left[\sum_{x\mu} \eta_{\mu x} \frac{1}{2} (\bar{\chi}_x^\alpha \chi_{x+\hat{\mu}}^\alpha - \bar{\chi}_{x+\hat{\mu}}^\alpha \chi_x^\alpha) + m \sum_x \bar{\chi}_x^\alpha \chi_x^\alpha \right]. \qquad (6.66)$$

In this representation the Dirac spinor labels α on $\bar{\chi}$ and χ are like internal symmetry labels and the action is just a sum of four identical terms, one for each value of the Dirac index. Hence, one of these should suffice in describing fermion particles. It can indeed be shown that taking χ and $\bar{\chi}$ as one-component fields leads to $16/4 = 4$ Dirac particles in the continuum limit. In QCD all these fermions are interpreted as quark flavors. Inserting the 'parallel transporters' $U_{\mu x}$ then leads to a gauge-invariant staggered-fermion action

$$S_F = -\sum_{x\mu} \eta_{\mu x} \frac{1}{2} \left[\bar{\chi}_{ax} (U_{\mu x})_{ab} \chi_{bx+\hat{\mu}} - \bar{\chi}_{ax+\mu} (U_{\mu x}^\dagger)_{ab} \chi_{bx} \right] - \sum_x m \bar{\chi}_{ax} \chi_{ax}, \qquad (6.67)$$

where we have made all indices on χ and $\bar{\chi}$ explicit (a and b are color indices) – there are e.g. no spin or flavor indices for $\bar{\chi}$ and $\bar{\chi}$. Analysis in weak-coupling perturbation theory leads to the conclusion that this action describes QCD with four mass-degenerate flavors in the continuum limit [73, 74] (the mass degeneracy of the quarks can be lifted by adding other terms to the action). The action has an interesting symmetry group [76], which is important for the construction of composite fields with the quantum numbers of hadrons [75, 77]. In the scaling region this symmetry group enlarges to the group in the continuum (including 'anomalies').

A further reduction by a factor of two is possible by assigning $\bar{\chi}_x$ only to the even sites and χ_x only to the odd sites [78, 79, 80]. Even and odd sites are defined by $\epsilon_x = 1$ and -1, respectively, with

$$\epsilon_x = (-1)^{x_1+x_2+x_3+x_4}. \tag{6.68}$$

In this formulation we may as well omit the bar on $\bar{\chi}_x$ since no confusion between even and odd sites is possible. Then a minimal action with only one Grassmann variable per site is given by

$$S = -\sum_{x\mu} \eta_{\mu\,x} \tfrac{1}{2} \chi_x \chi_{x+\hat{\mu}}, \tag{6.69}$$

in case of zero fermion mass. This method leads essentially to four Majorana fermions, which are equivalent to two Dirac fermions or eight Weyl fermions. Non-zero mass requires one-link or multilink couplings, since $\chi_x^2 = 0$.

Staggered fermions are technically rather specialized and we shall not emphasize them in this book. For an application of the method (6.69) to numerical simulations of the Higgs–Yukawa sector of the Standard Model see [36].

6.5 Transfer operator for Wilson fermions

It will now be shown that the fermion partition function with Wilson fermions can for $r = 1$ be written in the form

$$Z = \operatorname{Tr} \hat{T}^N, \tag{6.70}$$

where $\hat{T}$ is a positive transfer operator in Hilbert space and N is the number of time slices. A transfer operator was first given by Wilson [86] and a study of its properties was presented in [87]. The construction below is slightly different. (A general construction for $r \neq 1$ is sketched in [89], which is easily adapted to naive or staggered fermions. See also [79, 88, 90] and references therein.) To identify $\hat{T}$ we first assume that the gauge field is external and and write $\operatorname{Tr} \hat{T}^N$ in the Grassmann representation,

$$
\operatorname{Tr} \hat{T}^N = \int da_1^+ \, da_1 \cdots da_N^+ \, da_N \, e^{-a_N^+ a_N} \, T(a_N^+, a_{N-1}) \, e^{-a_{N-1}^+ a_{N-1}}
$$
$$
\times \, T(a_{N-1}^+, a_{N-2}) \cdots T(a_{k+1}^+, a_k) \, e^{-a_k^+ a_k} \, T(a_k^+, a_{k-1}) \cdots
$$
$$
\times \, e^{-a_2^+ a_2} \, T(a_2^+, a_1) \, e^{-a_1^+ a_1} \, T(a_1^+, -a_N). \tag{6.71}
$$

The minus sign in the last factor corresponds to the same sign in (C.68) in appendix C. It implies that there are antiperiodic boundary conditions in the path integral, i.e. there is a change of sign in the couplings in the action between time slices 0 and $N-1$. The expression above is to be compared with

$$
Z_F = \int D\bar\psi D\psi \, \exp S_F, \tag{6.72}
$$

where S_F is the fermion part of the action. We have seen in the pure-gauge case that the integration over the timelike links $U_{4\mathbf{x}}$ leads to the projector on the gauge-invariant subspace of Hilbert space, together with a transfer operator in the temporal 'gauge' $U_{4\mathbf{x}} = 1$. We therefore set $U_{4\mathbf{x}} = 1$ and write the fermion action in the form (using lattice units, $t \equiv x_4$ and $\mathbf{x}$ are integers)

$$
S_F = \sum_t \left(-\psi_t^+ \frac{1-\beta}{2} \psi_{t+1} + \psi_{t+1}^+ \frac{1+\beta}{2} \psi_t \right)
$$
$$
- \sum_t \psi_t^+ \beta A_t \psi_t - \epsilon \sum_t \psi_t^+ D_t \psi_t. \tag{6.73}
$$

Here a matrix notation is used with

$$
A_{\mathbf{xy},t} = M \delta_{\mathbf{x},\mathbf{y}} - \epsilon \sum_{j=1}^{3} \frac{1}{2} (U_{\mathbf{xy},t} \, \delta_{\mathbf{x}+\hat{j},\mathbf{y}} + \mathbf{x} \leftrightarrow \mathbf{y}), \tag{6.74}
$$

$$
D_{\mathbf{xy},t} = \sum_{j=1}^{3} \alpha_j \frac{1}{2i} (U_{\mathbf{xy},t} \, \delta_{\mathbf{x}+\hat{j},\mathbf{y}} - \mathbf{x} \leftrightarrow \mathbf{y}), \tag{6.75}
$$

and $\alpha_j = i\gamma_4\gamma_j$ and $\beta = \gamma_4$ are Dirac's matrices, and furthermore

$$
\epsilon = a_4/a. \tag{6.76}
$$

We recognize the projectors

$$
P^\pm \equiv P_4^\pm = (1 \pm \beta)/2 \tag{6.77}
$$

for Wilson parameter $r = 1$. They reduce the number of $\psi_{t\pm1}^+ \psi_t$ couplings by a factor of two compared with the naive fermion action.

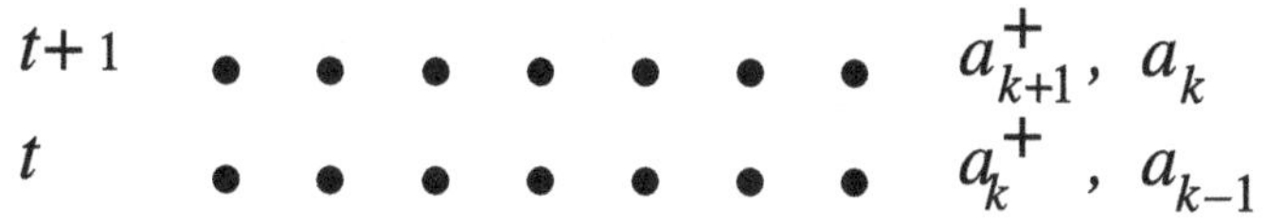

$$t+1 \quad \bullet \; \bullet \; \bullet \; \bullet \; \bullet \; \bullet \; \bullet \quad a_{k+1}^{+}, \; a_k$$

$$t \quad \bullet \; \bullet \; \bullet \; \bullet \; \bullet \; \bullet \; \bullet \quad a_k^{+}, \; a_{k-1}$$

Fig. 6.3. The association of time slices with Hilbert space for Wilson fermions ($r = 1$).

The association of time slices t with Hilbert-space slices k is as follows (for each $\mathbf{x}$, T denotes transposition):

$$P^{+}\psi_t = (a_k^{+}P^{+})^{\mathrm{T}}, \quad \psi_t^{+}P^{-} = a_k^{+}P^{-}, \tag{6.78}$$

$$P^{-}\psi_{t+1} = P^{-}a_k, \quad \psi_{t+1}^{+}P^{+} = (P^{+}a_k)^{\mathrm{T}}, \tag{6.79}$$

$$A_t = A_k, \quad D_t = D_k \tag{6.80}$$

as illustrated in figure 6.3. With this notation the action can be written as

$$S_{\mathrm{F}} = -\sum_k a_k^{+}a_k + \sum_k a_k^{+}A_k a_{k-1}$$

$$- \epsilon \sum_k (a_{k-1}P^{+}D_k P^{-}a_{k-1} + a_k^{+}P^{-}D_k P^{+}a_k^{+}). \tag{6.81}$$

Here we have used $\beta D = -D\beta$, such that

$$D = (P^{+} + P^{-})D(P^{+} + P^{-}) = P^{+}DP^{-} + P^{-}DP^{+}, \tag{6.82}$$

and abused the notation by leaving out the transposition symbol T. Comparison with $\mathrm{Tr}\,\hat{T}^N$ in the form (6.71) gives the (Grassmannian) transfer-matrix elements

$$T_{\mathrm{F}}(a_k^{+}, a_{k-1}) = \exp(-\epsilon a_k^{+}P^{-}D_k P^{+}a_k^{+}) \, \exp(a_k^{+}A_k a_{k-1})$$

$$\times \exp(-\epsilon a_{k-1}P^{+}D_k P^{-}a_{k-1}). \tag{6.83}$$

Using the rules listed above (C.68) in appendix C this translates into operator form as

$$\hat{T}_{\mathrm{F}} = e^{-\epsilon \hat{a}^{\dagger}P^{-}DP^{+}\hat{a}^{\dagger}} \, e^{\hat{a}^{\dagger}\ln(A)\hat{a}} \, e^{-\epsilon \hat{a}P^{+}DP^{-}\hat{a}}. \tag{6.84}$$

Here D and A depend in general on the gauge-field configuration in a time slice.

Consider now first the case of free fermions, $U_{\mathbf{xy}} = 1$. Then $\hat{T}$ is clearly a positive operator provided that A is positive, i.e. a Hermitian matrix with only positive eigenvalues. In momentum space we get the eigenvalues

$$A(\mathbf{p}) = M - \epsilon \sum_{j=1}^{3} \cos p_j, \tag{6.85}$$

which shows that $A > 0$ for

$$M > 3\epsilon. \tag{6.86}$$

With dynamical gauge fields we have to take into account in (6.84) the transfer operator for the gauge field $\hat{T}_U$. The complete transfer operator can be taken as

$$\hat{T} = \hat{T}_F^{1/2} \, \hat{T}_U \, \hat{T}_F^{1/2} \, \hat{P}_0, \tag{6.87}$$

$$\hat{T}_F = e^{-\epsilon \hat{a}^\dagger P^- \hat{D} P^+ \hat{a}^\dagger} \, e^{\hat{a}^\dagger \ln \hat{A} \, \hat{a}} \, e^{-\epsilon \hat{a} P^+ \hat{D} P^- \hat{a}} \tag{6.88}$$

where we have also put in the projector P_0 on the gauge-invariant subspace. Since A has lowest eigenvalues when the link variables are unity, the condition (6.86) remains sufficient in general for positivity of $\hat{T}_F$.

We can now use (6.80) in reverse and define operator fields $\hat{\psi}$ and $\hat{\psi}^\dagger$, for each spatial site $\mathbf{x}$, by

$$P^+ \hat{\psi} = (\hat{a}^\dagger P^+)^{\mathrm{T}}, \quad \hat{\psi}^\dagger P^- = \hat{a}^\dagger P^-, \tag{6.89}$$

$$P^- \hat{\psi} = P^- \hat{a}, \quad \hat{\psi}^\dagger = (P^+ \hat{a})^{\mathrm{T}}. \tag{6.90}$$

In terms of these fields the fermion transfer operator takes the explicitly charge-conserving form

$$\hat{T}_F = e^{-\epsilon \hat{\psi}^\dagger P^- \hat{D} P^+ \hat{\psi}} \, e^{-\hat{\psi}^\dagger \beta \ln \hat{A} \, \hat{\psi}} \, e^{\mathrm{Tr} \, P^+ \ln \hat{A}} \, e^{-\epsilon \hat{\psi}^\dagger P^+ D P^- \hat{\psi}}. \tag{6.91}$$

Notice the Dirac-sea factor $\exp(\mathrm{Tr} \, P^+ \ln A)$.

The continuous time limit $\hat{T} = 1 - \epsilon \hat{H} + O(\epsilon^2)$ can be taken if we let M depend on $\epsilon \to 0$ according to

$$M = 1 + \epsilon M_3, \tag{6.92}$$

such that A takes the form

$$A(U) = 1 + \epsilon \mathcal{M}_3(U) \tag{6.93}$$

$$\mathcal{M}_3 = M_3 - \sum_{j=1}^{3} \tfrac{1}{2}(U_{\mathbf{x},\mathbf{y}} \, \delta_{\mathbf{x}+\hat{\jmath},\mathbf{y}} + \mathbf{x} \leftrightarrow \mathbf{y}), \tag{6.94}$$

and we get the fermion Hamiltonian

$$\hat{H}_F = \hat{\psi}^\dagger [\mathcal{M}_3(\hat{U})\beta + D(\hat{U})]\hat{\psi}. \tag{6.95}$$

This may be called a Wilson–Dirac Hamiltonian on a spatial lattice.

In summary, we conclude that the Euclidean-lattice formulation of QCD using Wilson's fermion method has a good Hilbert-space interpretation, with a positive transfer operator.

6.6 Problems

The following exercises serve to clarify the continuum limit in QED and the phenomenon of species doubling by calculation of the photon self-energy at one loop in the weak-coupling expansion.

(i) *Vertex functions*

Consider the naive fermion action in QED

$$S_F = -\sum_{x\mu} \tfrac{1}{2}(\bar\psi_x \gamma_\mu e^{-igA_{\mu x}}\psi_{x+\hat\mu} - \bar\psi_{x+\hat\mu}\gamma_\mu e^{igA_{\mu x}}\psi_x). \tag{6.96}$$

The bare fermion–photon vertex functions are the derivatives of the action with respect to the fields. Taking out a minus sign and the momentum-conserving delta functions, let the vertex function $V_{\mu_1\cdots\mu_n}(p, q; k_1\cdots k_n)$ be defined by

$$S_F = -\sum_{uvx_1\cdots x_n} \frac{1}{n!} \bar\psi_u V_{\mu_1\cdots\mu_n}(u, v; x_1\cdots x_n)\psi_v A_{\mu_1 x_1}\cdots A_{\mu_n x_n}, \tag{6.97}$$

$$\sum_{uvx_1\cdots x_n} e^{-ipu+iqv-ik_1 x_1\cdots-ik_n x_n} V_{\mu_1\cdots\mu_n}(u, v; x_1\cdots x_n)$$
$$= V_{\mu_1\cdots\mu_n}(p, q; k_1\cdots k_n)\,\bar\delta(p - q + k_1 + \cdots k_n). \tag{6.98}$$

Show that $(p - q + k_1 + \cdots k_n = 0)$

$$V_{\mu_1\cdots\mu_n}(p, q; k_1\cdots k_n) = \sum_\mu \gamma_\mu \tfrac{1}{2}[(-ig)^n e^{iq_\mu} - (ig)^n e^{-ip_\mu}]$$
$$\times \delta_{\mu\mu_1}\cdots\delta_{\mu\mu_n}, \tag{6.99}$$

as illustrated in figure 6.4. The fermion propagator is given by

$$S(p)^{-1} = V(p, p), \quad S(p) = \frac{m - i\gamma_\mu \sin p_\mu}{m^2 + \sin^2 p}. \tag{6.100}$$

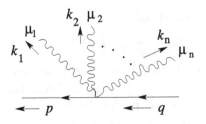

Fig. 6.4. Fermion vertex function $-V_{\mu_1 \cdots \mu_n}(p, q; k_1, \ldots, k_n)$.

(ii) *Ward–Takahashi identities*

The gauge invariance of S_F implies certain properties of the vertex functions, called Ward–Takahashi identities. Consider a small gauge transformation $\psi'_x = (1 + i\omega_x + O(\omega^2))\psi_x$, $\bar{\psi}'_x = (1 - i\omega_x + O(\omega^2))\bar{\psi}_x$, $A'_{\mu x} = A_{\mu x} + (1/g)\partial_\mu \omega_x$ (recall the definition of the forward and backward lattice derivatives, $\partial_\mu \omega_x = \omega_{x+\hat\mu} - \omega_x$ and $\partial'_\mu \omega_x = \omega_x - \omega_{x-\hat\mu}$). Collect the linear terms in ω_x in the invariance relation $0 = S_F(\psi', \bar{\psi}', A') - S_F(\psi, \bar{\psi}, A)$ and derive the Ward identities

$$
\begin{aligned}
0 = {} & \frac{1}{g} i \partial'_\mu V_{\nu\mu_1 \cdots \mu_n}(u, v; x, x_1, \ldots, x_n) \\
& + \delta_{ux} V_{\mu_1 \cdots \mu_n}(u, v; x_1, \ldots, x_n) \\
& - \delta_{vx} V_{\mu_1 \cdots \mu_n}(u, v; x_1, \ldots, x_n),
\end{aligned}
\tag{6.101}
$$

and the momentum-space version

$$
\begin{aligned}
0 = {} & \frac{1}{g} K^*_\mu(k) V_{\mu\mu_1 \cdots \mu_n}(p, q; k, k_1, \ldots, k_n) \\
& + V_{\mu_1 \cdots \mu_n}(p + k, q; k_1, \ldots, k_n) \\
& - V_{\mu_1 \cdots \mu_n}(p, q - k; k_1, \ldots, k_n),
\end{aligned}
\tag{6.102}
$$

where $K_\mu(k) = (e^{ik_\mu} - 1)/i$. In particular, for $n = 0$ and 1,

$$
K^*_\mu(k) V_\mu(p, q; k) = S(p)^{-1} - S(q)^{-1},
\tag{6.103}
$$

$$
K^*_\mu(k) V_{\mu\nu}(p, q; k, l) = V_\mu(p, p + l; l) - V_\nu(q - l, q; l)
$$

$(p - q + k = 0, \; p - q + k + l = 0)$.

(iii) *Photon self-energy*

We study the 'vacuum-polarization' diagrams in figure 6.5, which describe the photon self-energy vertex function $\Pi = \Pi^{(a)} + \Pi^{(b)}$

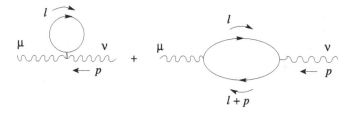

Fig. 6.5. Vacuum-polarization diagrams for $-\Pi_{\mu\nu}(p)$.

given by

$$\Pi_{\mu\nu}^{(a)}(p) = -g^2 \int_{-\pi}^{\pi} \frac{d^4l}{(2\pi)^4}\, \mathrm{Tr}\,[V_{\mu\nu}(l,-l,p,-p)S(l)], \qquad (6.104)$$

$$\Pi_{\mu\nu}^{(b)}(p) = g^2 \int_{-\pi}^{\pi} \frac{d^4l}{(2\pi)^4}\, \mathrm{Tr}\,[V_\mu(l,l+p)S(l+p)V_\nu(l+p,l)S(l)].$$

Use the identities (6.103) to show that the sum $\Pi_{\mu\nu} = \Pi_{\mu\nu}^{(a)} + \Pi_{\mu\nu}^{(b)}$ satisfies the Ward identity

$$K_\mu^*(p)\Pi_{\mu\nu}(p) = 0. \qquad (6.105)$$

(Note that the loop integrals in the lattice regularization are invariant under translation of the integration variable.)

(iv) *Continuum region and lattice-artifact region*
The calculation of the continuum limit of $\Pi_{\mu\nu}(p)$ can be done in the same way as for the scalar field in section 3.4 We split the integration region into a ball of radius δ around the origin $l = 0$ and the rest, where δ is so small that we may use the continuum form of the propagators and vertex functions.

Going over to physical units, $p \to ap$, $m \to am$, $\Pi \to \Pi a^{-2}$, show that in the scaling region limit $a \to 0$, $\delta \to 0$, $am/\delta \to 0$, $ap/\delta \to 0$ the contribution of this ball can be written as

$$-\frac{g^2}{2\pi^2}(\delta_{\mu\nu}p^2 - p_\mu p_\nu)\int_0^1 dx\, x(1-x)\ln[a^2(m^2 + x(1-x)p^2)] \qquad (6.106)$$

up to a second-degree polynomial in p.

Verify that the 15 fermion doublers in similar balls around non-zero $l = \pi_A$ give identical contributions, up to possible arbitrariness in the polynomials.

The region outside the 16 balls can contribute only a second-degree polynomial $T_{\mu\nu}(p)$ in a^{-1}, m and p in the continuum limit, because possible infrared divergences cannot develop in the outside regions.

Note that $\Pi_{\mu\nu}^{(a)}$ contributes only to the polynomial part of $\Pi_{\mu\nu}$ in the continuum limit (it is just a constant $\propto \delta_{\mu\nu}$). The reason is that there is no logarithmic contribution from the balls around $l = \pi_A$ because the vertex function $V_{\mu\nu}$ vanishes in the classical continuum limit.

(v) *Lattice symmetries*

The polynomial has to comply with the symmetries of the model, in particular cubic rotations $R^{(\rho\sigma)}$ in a plane (ρ, σ),

$$
\begin{aligned}
(R^{(\rho\sigma)}p)_\mu &\equiv R_{\mu\nu}^{(\rho\sigma)}p_\nu, \\
(R^{(\rho\sigma)}p)_\rho &= p_\sigma, \quad (R^{(\rho\sigma)}p)_\sigma = -p_\rho, \\
(R^{(\rho\sigma)}p)_\mu &= p_\mu, \quad \mu \neq \{\rho, \sigma\}
\end{aligned}
\tag{6.107}
$$

and inversions $I^{(\rho)}$,

$$
(I^{(\rho)}p)_\rho = -p_\rho, \quad (I^{(\rho)}p)_\mu = p_\mu, \quad \mu \neq \rho.
\tag{6.108}
$$

The polynomial $T_{\mu\nu}(p)$ has to be a tensor under these transformations,

$$
T_{\mu\nu}(R^{(\rho\sigma)}p) = R_{\mu\mu'}^{(\rho\sigma)} R_{\nu\nu'}^{(\rho\sigma)} T_{\mu'\nu'}(p),
\tag{6.109}
$$

$$
T_{\mu\nu}(I^{(\rho)}p) = I_{\mu\mu'}^{(\rho)} I_{\nu\nu'}^{(\rho)} T_{\mu'\nu'}(p).
\tag{6.110}
$$

Show using the lattice symmetries that the form of the polynomial is limited to

$$
c_1 a^{-2}\delta_{\mu\nu} + c_2 m^2 + c_3 p_\mu^2 \delta_{\mu\nu} + c_4 p_\mu p_\nu + c_5 p^2 \delta_{\mu\nu}.
\tag{6.111}
$$

(In the third term there is of course no summation over μ.)

(vi) *Constraints from the Ward identity*

Use the continuum limit of the Ward identity (6.105) to show finally that $\Pi_{\mu\nu}(p)$ has the continuum covariant form,

$$
\begin{aligned}
\Pi_{\mu\nu}(p) = {}&-16\frac{g^2}{2\pi^2}(\delta_{\mu\nu}p^2 - p_\mu p_\nu) \\
&\times \int_0^1 dx\, x(1-x) \ln[a^2(m^2 + x(1-x)p^2)] \\
&+ c(p^2\delta_{\mu\nu} - p_\mu p_\nu).
\end{aligned}
\tag{6.112}
$$

Note that the coefficient c_1 of the quadratic divergence is zero. This can of course also be verified by an explicit calculation, e.g.

for $p = 0$. In non-Abelian gauge theory such quadratic divergences are also absent, provided that the contribution from the integration (Haar) measure in the path integral is not forgotten. Note also that the coefficient c_3, of the term that is lattice covariant but not covariant under continuous rotations, is zero. Such cancellations will not happens in models in which vector fields are not gauge fields (no Ward identities). Then counterterms are needed in order to ensure covariance.

The numerical constant c can be obtained by a further careful analysis and numerical integration. It determines e.g. the ratio of lambda scales $\Lambda_{\overline{MS}}/\Lambda_L$ in the theory with (naive) dynamical fermions. On dividing by a factor of four we get the analogous result for four-flavor staggered fermions described by the $U(1)$ version of the action (6.67).

7

Low-mass hadrons in QCD

In this chapter we address the calculation of the properties of hadrons composed of the light quarks u, d and s.

7.1 Integrating over the fermion fields

The partition function for a fermion gauge theory

$$Z = \int DU\, D\bar{\psi} D\psi \, \exp(S_U + S_F) \qquad (7.1)$$

can be expressed in an alternative form involving only the gauge fields by first integrating out the fermion fields. We shall use Wilson fermions as an example. We can write S_F in matrix notation,

$$S_F = -\sum_{x\mu} \frac{1}{2}(\bar{\psi}_x \gamma_\mu U_{\mu x} \psi_{x+\hat{\mu}} - \bar{\psi}_{x+\hat{\mu}} \gamma_\mu U_{\mu x}^\dagger \psi_x)$$

$$- \sum_x \bar{\psi}_x M \psi_x + \sum_{x\mu} \frac{r}{2}(\bar{\psi}_x U_{\mu x} \psi_{x+\hat{\mu}} + \bar{\psi}_{x+\hat{\mu}} U_{\mu x}^\dagger \psi_x) \qquad (7.2)$$

$$\equiv -\bar{\psi} A \psi, \qquad (7.3)$$

$$A = \slashed{D} + M - W. \qquad (7.4)$$

The matrix $A = A(U)$ is called the fermion matrix. Its inverse is the fermion propagator $S(U)$ in a given gauge-field configuration,

$$S_{xy}(U) \equiv A_{xy}^{-1}(U) = \left[\frac{1}{M - W(U) + \slashed{D}(U)}\right]_{xy}. \qquad (7.5)$$

Since ψ and $\bar{\psi}$ occur only bilinearly in the action, we can perform the integration over these variables and evaluate fermion correlation

170

functions explicitly:

$$Z = \int DU \, \det[A(U)] \, \exp(S_U)$$

$$\langle \psi_x \bar{\psi}_y \rangle = Z^{-1} \int DU \, \det[A(U)] \, e^{S_U} \, S_{xy}(U)$$

$$\equiv \langle S_{xy} \rangle_U,$$

$$\langle \psi_u \bar{\psi}_v \psi_x \bar{\psi}_y \rangle = \langle S_{uv} S_{xy} - S_{uy} S_{xv} \rangle_U,$$

$$\langle \psi_u \psi_v \psi_w \bar{\psi}_x \bar{\psi}_y \bar{\psi}_z \rangle = \langle S_{ux} S_{vy} S_{wz} + 5 \text{ permutations} \rangle_U, \qquad (7.6)$$

etc. For clarity we indicated only the space–time indices $x, \ldots$ and suppressed the color, Dirac, and flavor indices a, α, and f of the fermion fields $\psi_x^{a\alpha f}$.

7.2 Hopping expansion for the fermion propagator

For Wilson fermions an expansion in the hopping parameter $\kappa = 1/2M$ has given useful results. Here we describe it for the propagator, for which it gives an intuitive representation in terms of a summation over random paths. We have seen this earlier for the scalar field in section 3.7. Let us define the hopping matrix H by

$$H_{x a \alpha, y b \beta} = \sum_\mu [(P_\mu^-)_{\alpha\beta} (U_{xy})_{ab} \, \delta_{x+\hat{\mu}, y} + (P_\mu^+)_{\alpha\beta} (U_{yx})_{ab} \, \delta_{y+\hat{\mu}, x}], \quad (7.7)$$

in terms of which

$$A_{x a \alpha f, y b \beta g} = M_f \, \delta_{fg} \, [1 - 2\kappa_f H_{x a \alpha, y b \beta}]. \qquad (7.8)$$

For a given flavor we have

$$S_{xy} = M^{-1} \left(\frac{1}{1 - 2\kappa H} \right)_{xy} = M^{-1} \sum_{L=0}^{\infty} (2\kappa)^L \, (H^L)_{xy}, \qquad (7.9)$$

where we suppressed again the non-space–time indices. The successive terms in this series can be represented as a sum over paths of length L, as illustrated in figure 7.1. The diagrams for $L + 1$ are obtained from those for L by application of H (by attaching the $L = 1$ diagrams). To each path C there corresponds a color factor $U_{xy}(C)$ and a spin factor

$$\Gamma(C) = \prod_{l \in C} P_l, \qquad (7.10)$$

$$P_l = P_\mu^-, \quad l = (x, x + \hat{\mu}) \qquad (7.11)$$

$$= P_\mu^+, \quad l = (x + \hat{\mu}, x). \qquad (7.12)$$

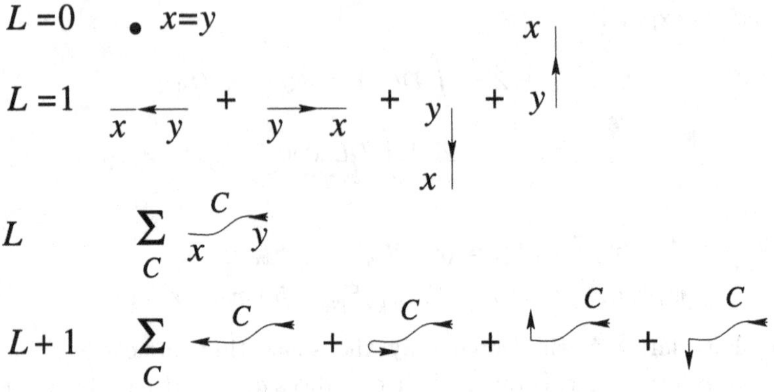

Fig. 7.1. Illustration of the terms in the hopping expansion of the propagator.

Note that, for $r = 1$, there are no paths with back-tracking because then $P_\mu^+ P_\mu^- = P_\mu^- P_\mu^+ = 0$.

For free fermions we can perform the summation in momentum space,

$$(H^L)_{xy} = \sum_p e^{ip(x-y)} H(p)^L, \tag{7.13}$$

$$H(p)\bar\delta_{pq} = \sum_{xy} e^{-ipx+iqy} H_{xy}, \tag{7.14}$$

$$H(p) = \sum_\mu (e^{ip_\mu} P_\mu^- + e^{-ip_\mu} P_\mu^+)$$

$$= \sum_\mu (r \cos p_\mu - i\gamma_\mu \sin p_\mu), \tag{7.15}$$

$$S(p) = \frac{1}{M - \sum_\mu (r \cos p_\mu - i\gamma_\mu \sin p_\mu)}. \tag{7.16}$$

So the summation over random paths with the particular weight factor (7.10) leads to the free Wilson fermion propagator. The maximum eigenvalue of $H(p)$ is $4r$ at $p = 0$, which means that the radius of convergence of the hopping expansion for free fermions is given by $|\kappa| < 1/8r$. For fixed κ and complex p the expansion diverges at the position of the particle pole in the propagator. In the interacting case the unitary $U_{\mu x}$ tend to reduce the maximum eigenvalue of H and the convergence radius is generically larger than $1/8r$, depending on the configuration of U's.

The hopping expansion for the propagator leads to an expansion of the fermion determinant in terms of closed paths,

$$\det A \rightarrow \det(1 - 2\kappa H) = \exp[\mathrm{Tr}\,\ln(1 - 2\kappa H)]$$

$$= \exp\left[-\sum_L \frac{(2\kappa)^L}{L}\,\mathrm{Tr}\,H^L\right]. \tag{7.17}$$

With each closed path C there is associated a Wilson loop $\mathrm{Tr}\,U(C)$ and a spin loop $\mathrm{Tr}\,\Gamma(C)$.

7.3 Meson and baryon propagators

The expressions (7.6) are well defined without gauge fixing. Since $\int DU$ contains an implicit integration over all gauge transformations, it projects on the gauge-invariant content of the integrand. Under a gauge transformation

$$U^{\Omega}_{\mu x} = \Omega_x U_{\mu x} \Omega^{\dagger}_{x+\hat{\mu}}, \tag{7.18}$$

$$A(U^{\Omega})_{xy} = \Omega_x A(U)_{xy} \Omega^{\dagger}_y, \tag{7.19}$$

$$A^{-1}(U^{\Omega})_{xy} = \Omega_x A^{-1}(U)_{xy} \Omega^{\dagger}_y, \tag{7.20}$$

or

$$S(U^{\Omega})_{xy} = \Omega_x S(U)_{xy} \Omega^{\dagger}_y. \tag{7.21}$$

Since $\int d\Omega_x\,\Omega_x = 0$, the gauge-invariant content of Ω_x is zero and it follows that the expectation value of the gauge-field-dependent fermion propagator is zero, unless $x = y$,

$$\langle S_{x a\alpha f, y b\beta g}\rangle_U \propto \delta_{xy}\,\delta_{ab}\,\delta_{fg}. \tag{7.22}$$

The fermions cannot propagate 'on their own' without gauge fixing. Of course, this does not mean that fermion propagation is a non-gauge-invariant phenomenon and it is also not an expression of confinement. It forces us to consider carefully what the gauge-invariant description of propagation means.

A simple gauge-invariant correlation function is of the type $\langle\bar{\psi}_x\psi_x\bar{\psi}_y\psi_y\rangle$. In QCD we call these mesonic, since $\bar{\psi}\gamma\psi$ combinations carry mesonic quantum numbers. More explicitly, we can define gauge-invariant meson and baryon fields

$$\mathcal{M}^{\alpha f}_{x\beta g} = \delta^b_a \, \psi^{a\alpha f}_x \bar{\psi}_{xb\beta g}, \tag{7.23}$$

$$\mathcal{B}^{\alpha_1 f_1 \alpha_2 f_2 \alpha_3 f_3}_x = \epsilon_{a_1 a_2 a_3} \psi^{a_1 \alpha_1 f_1}_x \psi^{a_2 \alpha_2 f_2}_x \psi^{a_3 \alpha_3 f_3}_x, \tag{7.24}$$

$$\bar{\mathcal{B}}_{x\alpha_1 f_1 \alpha_2 f_2 \alpha_3 f_3} = \epsilon^{a_1 a_2 a_3} \bar{\psi}_{xa_1 \alpha_1 f_1} \bar{\psi}_{xa_2 \alpha_2 f_2} \bar{\psi}_{xa_3 \alpha_3 f_3}. \tag{7.25}$$

The gauge invariance of the meson fields is obvious. For the baryon fields the effect of a gauge transformation is given by (suppressing non-color indices)

$$\epsilon_{abc} \psi^a \psi^b \psi^c \rightarrow \epsilon_{abc} \, \Omega^a_{a'} \Omega^b_{b'} \Omega^c_{c'} \, \psi^{a'} \psi^{b'} \psi^{c'}$$
$$= (\det \Omega) \, \epsilon_{a'b'c'} \, \psi^{a'} \psi^{b'} \psi^{c'}, \tag{7.26}$$

and invariance follows from $\det \Omega = 1$.

By taking special linear combinations, we can construct from $\mathcal{M}$, $\mathcal{B}$, and $\bar{\mathcal{B}}$ fields with the required quantum numbers. For example, for π^+ (which has spin zero) we can use the scalar field combination

$$\bar{d}_x i\gamma_5 u_x, \quad \bar{d}_x i\gamma_\mu \gamma_5 u_x, \tag{7.27}$$

where we made flavor explicit according to

$$u^{a\alpha}_x = \psi^{a\alpha u}_x, \quad d^{a\alpha}_x = \psi^{a\alpha d}_x, \tag{7.28}$$

etc. For the ρ^+ particle (which has spin one) we can use the vector and tensor fields (both containing spin 1)

$$\bar{d}_x i\gamma_\mu u_x, \quad \bar{d}_x i[\gamma_\mu, \gamma_\nu] u_x. \tag{7.29}$$

An example for the proton (spin $\frac{1}{2}$) is given by

$$\epsilon_{abc} \, (C^\dagger \gamma_5)_{\beta\gamma} \, u^{a\alpha}_x (u^{b\beta}_x d^{c\gamma}_x - d^{b\beta}_x u^{c\gamma}_x), \tag{7.30}$$

and for the Δ^{++} (spin $\frac{3}{2}$),

$$\epsilon_{abc} \, (C^\dagger \gamma_\mu)_{\beta\gamma} \, u^{a\alpha}_x u^{b\beta}_x u^{c\gamma}_x. \tag{7.31}$$

Here C is the charge-conjugation matrix; $\bar{\psi}^{(c)} = -(C^\dagger \psi)^{\mathrm{T}}$ is the charge conjugate of $\bar{\psi}$ (cf. (D.27) in appendix D). For example, for the Δ^{++} the last two u fields combine to give a vector (containing spin 1) of the form $\bar{\psi}^{(c)} \gamma_\mu \psi$, which, together with the first u field, contains spin $\frac{3}{2}$. More examples are in [10]; [82] gives group-theoretical details of the spin–flavor content of the baryon fields.

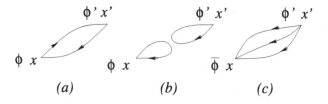

Fig. 7.2. Meson (a) and (b), and baryon (c) propagators.

Putting these hadron fields in the form $(A = \{a\alpha f\})$

$$\mathcal{M}_x(\Phi) = \Phi_B^A \, \mathcal{M}_{xA}^B, \tag{7.32}$$

$$\mathcal{B}_x(\Phi) = \bar{\Phi}_{ABC} \, \mathcal{B}_x^{ABC}, \tag{7.33}$$

$$\bar{\mathcal{B}}_x(\Phi) = \bar{\mathcal{B}}_{xABC} \, \Phi^{ABC}, \tag{7.34}$$

where Φ specifies the spin–flavor structure, we can write gauge-invariant meson correlation functions as

$$\langle \mathcal{M}_x(\Phi) \mathcal{M}_{x'}(\Phi') \rangle = -\langle \mathrm{Tr}\,(\Phi S_{xx'} \Phi' S_{x'x}) \rangle_U + \langle \mathrm{Tr}\,(\Phi S_{xx})\,\mathrm{Tr}\,(\Phi' S_{x'x'}) \rangle_U, \tag{7.35}$$

and the baryon correlation function

$$\langle \mathcal{B}_x(\Phi) \bar{\mathcal{B}}_{x'}(\Phi') \rangle = \bar{\Phi}_{ABC} \langle S_{A'xx'}^A S_{B'xx'}^B S_{C'xx'}^C \rangle_U \Phi'^{A'B'C'}, \tag{7.36}$$

as illustrated in figure 7.2. The contribution (b) to the meson correlation function is non-zero only for flavor-neutral fields, i.e. fields of the form $\bar{f}\gamma f$, $f = u, d, c, s, t, b$.

These composite field correlation functions describe bound states, the mesons and baryons; for this reason we also call them meson and baryon propagators. Consider first the meson propagator in figure 7.2(a). It is a sum over Wilson loops going through x and x'. This follows from the hopping expansion of the quark propagator which expresses $S_{xx'}$ as a sum over random paths C weighted by the Wilson line $U_{xx'}(C)$ and the spinor factor associated with the path. We can now intuitively understand the implication of confinement as expressed by the area law for Wilson loops. The exponential fall-off in the area law greatly suppresses the contribution of widely separated paths of the two propagators in figure 7.2(a). Contributions in which the two paths stay together dominate and, when they are together, they make a random walk, which implies the formation of a bound state. A similar story can be told for the baryon propagator in figure 7.2(c); the combined random

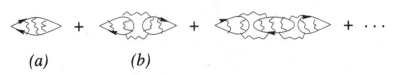

(a)   *(b)*

Fig. 7.3. Diagrams for flavor-neutral mesons with sea-quark loops and gluon lines also indicated.

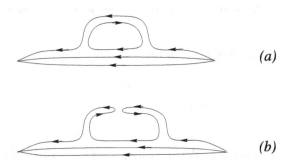

Fig. 7.4. Meson-loop corrections to the baryon propagator. The closeness of lines is to suggest binding by 'glue'.

walk of the three-quark propagators leads to a pole in momentum space corresponding to a bound baryon state.

So far we have concentrated on the fermion lines related to the hadron fields (the 'valence-quark' lines) and ignored the effect of the fermion determinant. Its hopping expansion leads to a sum of closed fermion lines called 'sea-quark' loops or 'vacuum loops', and we have to imagine such sea-quark loops everywhere in figure 7.2. This is particularly relevant for the case of flavor-neutral mesons for which diagram (b) contributes. Figure 7.3 shows diagrams (a) and (b) as the first two terms in an infinite series with the sea-quark loops included one by one. As a reminder of the presence of 'glue' implied by the average $\langle \cdots \rangle_U$ we have also shown some gluon lines in this figure. Figure 7.4 illustrates a meson-loop contribution to a baryon propagator: (a) uses a sea-quark loop but not (b), which is already included in diagram (c) of figure 7.2.

We were led by confinement to the intuitive picture of random walks for the composite hadron propagators. In a theory without confinement, such as QED, there is no area law and there will be relatively large contributions in the fermion–antifermion correlation function also for widely separated fermion paths. These will correspond to fermions propagating almost freely at large distances from each other. Of course, they will feel

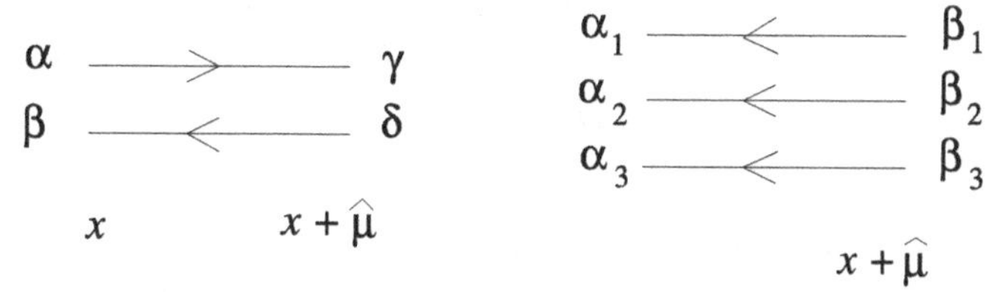

Fig. 7.5. Hopping matrix for the mesons and baryons at strong coupling.

the long range electromagnetic interactions, which may but need not lead to bound states. This is the gauge-invariant description of fermion propagation in QED.

7.4 Hadron masses at strong coupling

At bare gauge coupling $g = \infty$ the string tension diverges and there are only contributions to the Wilson paths with zero area. Neglecting vacuum fermion loops, it is an interesting approximation to take into account only simultaneous quark–antiquark hopping for mesons and three-quark hopping for baryons, as illustrated in figure 7.5. The inverse propagators can be written down explicitly and solved for the position of the pole in $p_4 = im$ at $\mathbf{p} = 0$, which determines the mass of the bound state, m. We can even derive effective actions describing the coupling constants [83, 84, 85, 109, 82] in terms of the meson and baryon fields (7.23)–(7.25). For example, the meson effective action has the form

$$S_{\text{eff}} = n_c \sum_x \text{Tr} \left[-\ln \mathcal{M}_x + M\mathcal{M}_x - \sum_\mu \mathcal{M}_x P_\mu^- \mathcal{M}_{x+\hat{\mu}} P_\mu^+ + O(\mathcal{M}^4) \right].$$

(7.37)

where n_c is the number of colors and now $\mathcal{M}$ is an *effective* field. For $r = 1$ it turns out that the low-lying states are the pions, rho mesons, nucleons, and deltas ($m_\pi \approx 140$, $m_\rho \approx 770$, $m_N \approx 940$ and $m_\Delta \approx 1232$ MeV). In the flavor-degenerate case $M_u = M_d = M$ the masses are given by

$$\cosh m_\pi = 1 + \frac{(M^2 - 4)(M^2 - 1)}{2M^2 - 3},$$

(7.38)

$$\cosh m_\rho = 1 + \frac{(M^2 - 3)(M^2 - 2)}{2M^2 - 3},$$

(7.39)

$$\exp m_\Delta = \frac{(M^3 - 3/2)(M^3 - 1/2)}{M^3 - 5/4}, \tag{7.40}$$

$$\exp m_N = \frac{M^3(M^3 - 2)}{M^3 - 5/4}. \tag{7.41}$$

On decreasing M from infinity toward zero or equivalently increasing the hopping parameter κ from zero upward we see that the pion mass vanishes at $M = M_c = 2$, whereas the other masses stay non-zero. At strong coupling the critical hopping parameter is $\kappa_c = 1/2M_c = \frac{1}{4}$. For weak coupling one can calculate $M_c(g) = 4 + O(g^2)$, and κ_c as defined by the vanishing of m_π will decrease toward $\frac{1}{8}$ as $g^2 \to 0$.

Although we know that the scaling region of QCD is at weak coupling, it is still interesting to compare these strong-coupling results with experiment. For small $M - M_c$,

$$m_\pi^2 = 4.8 m_q, \quad m_q \equiv M - M_c, \tag{7.42}$$

$$m_\rho = 0.894 + 1.97 m_q. \tag{7.43}$$

We can choose M such that m_π/m_ρ takes the experimental value 140 MeV/770 MeV, which gives $m_q = 0.0055$. This may be compared with $m_\rho = 0.894$ at $M = M_c$. Introducing the lattice distance, $am_\rho = 0.994$, means that the lattice cutoff is $1/a = 770/0.894 = 860$ MeV and the quark mass $m_q = 0.0055/a = 4.7$ MeV, which is remarkably close to the up–down quark mass found in numerical simulations ($m_{ud} \approx 4.5$ MeV (quenched), see section 7.5). Mass ratios not involving the pion can be approximated by taking $M = M_c$. Then we have the strong-coupling predictions

$$\frac{m_N}{m_\rho} = 1.7 \; (1.21), \quad \frac{m_\Delta}{m_N} = 1.01 \; (1.31), \quad M = M_c, \tag{7.44}$$

where the experimental values are given in parentheses. For $M \to \infty$ the baryon/meson mass ratio would be $\frac{3}{2}$. The results are not improved much by including $O(1/g^2)$ corrections, which are already hard to calculate [82]. The idea of using the strong-coupling expansion as a method for calculating the properties of hadrons has failed up to now because of its great complexity.

Other quantities such as the decay constants f_π and f_ρ, the π–π scattering amplitudes, and the splitting $m_\eta'^2 - (m_\eta^2 + m_{\pi^0}^2)/2$ in the neutral pseudoscalar meson sector, which is related to the notorious $U(1)$ problem, have also been calculated at strong coupling. Quantitatively these predictions are wrong of course, but they present an interesting caricature of hadron physics.

Table 7.1. *Low-mass hadrons: the baryon octet (N, Σ, Λ, Ξ), the baryon decuplet (Δ, Σ*, Ξ*, Ω) and the mesons*

State	Spin	Mass (MeV)	Valence-quark content
N	1/2	940	$u\bar{u}d$, udd
Σ	1/2	1193	uus, $(ud+du)s$, dds
Λ	1/2	1116	$(ud-du)s$
Ξ	1/2	1315	uss, dss
Δ	3/2	1232	uuu, uud, udd, ddd
Σ^*	3/2	1384	uus, uds, dds
Ξ^*	3/2	1532	uss, dss
Ω	3/2	1673	sss
π	0	135	$u\bar{d}$, $d\bar{u}$
K	0	498	$u\bar{s}$, $d\bar{s}$, $s\bar{u}$, $s\bar{d}$
ρ	1	768	$u\bar{d}$, $d\bar{u}$
K^*	1	896	$u\bar{s}$, $d\bar{s}$, $s\bar{u}$, $s\bar{d}$
π^0	0	135	$u\bar{u} - d\bar{u}$ (& $s\bar{s}$)
η	0	547	$u\bar{u} + d\bar{u} - 2s\bar{s}$
η'	0	958	$u\bar{u} + d\bar{u} + s\bar{s}$
ρ^0	1	768	$u\bar{u} - d\bar{u}$ (& $s\bar{s}$)
ω	1	783	$u\bar{u} + d\bar{d}$ (& $s\bar{s}$)
ϕ	1	1019	$s\bar{s}$ (& $u\bar{u}$ & $d\bar{d}$)

7.5 Numerical results

In table 7.1 the low-mass hadrons found experimentally are listed, with their valence-quark contents indicated. The electric charge of a state is just the sum of the charges of the quarks, which is $+\frac{2}{3}$ for u, $-\frac{1}{3}$ for d and s, and the opposite for the antiquarks indicated by a 'bar'. The flavor-neutral mesons (π^0, ..., ϕ) have mixed quark content, approximately as indicated (with small 'contaminations' in parentheses). The decuplet baryons are symmetric in their flavor content, whereas the octet has mixed symmetry. The neutral octet members Σ and Λ differ in the symmetry properties of their u, d flavor content. The primary aim of the numerical simulations is to recover this spectrum of hadron masses with essentially only three parameters: the Λ scale which corresponds to the gauge coupling and which sets the overall mass scale, the non-strange-quark mass in the approximation $m_u = m_d$ and the strange-quark mass m_s.

In numerical simulations the fermion determinant $\det A$ poses the greatest problem. The *quenched approximation* consists of the replacement $\det A \to 1$, while taking its effect on the effective gauge coupling into account by a change in the bare coupling. This means that only the valence-quark propagators are taken into account and the sea-quark loops are neglected. For this reason the approximation is also called the valence approximation.

The reliability of this approximation (which destroys the Hilbert-space interpretation of the fermion path integral) is hard to establish *a priori*. It helps to consider the generalization of the $SU(3)$ gauge group to $SU(n_c)$, with $n_c \to \infty$ [67]. Then the contribution of each sea-quark loop to a mesonic correlation function is down by a factor $1/n_c$. For mesons the large-n_c limit corresponds to the quenched approximation. Baryons, however, have n_c valence quarks and the baryon mass becomes proportional to n_c as $n_c \to \infty$ [96]. Yet, as we have seen in section 5.6 for the glueballs, ordering various non-baryonic quantities according to powers of $1/n_c$ is quite illuminating even for values of n_c as low as 2 and 3.

Simulations with dynamical fermions ('unquenched') are very time consuming and for illustration we shall now describe the results of a computation with only two dynamical fermion species [97]. An improved action is used, for which larger lattice spacings can be used without discretization errors blowing up. The dynamical fermions are assumed to be the lightest sea quarks, u and d, and their masses are taken to be equal. This is not the actual situation, m_d is roughly twice m_u, both being of the order of 5 MeV.† However, the hadron masses are generally much larger and, neglecting such small $O(5\,\mathrm{MeV})$ effects, one may as well take $m_{\mathrm{sea}}^{(u)} = m_{\mathrm{sea}}^{(d)} = m_{\mathrm{sea}}$ (recall that $m \equiv M - M_c$). The pseudoscalar mesons require special attention in this respect, as will be discussed in the next chapter, but even these depend primarily on the average quark mass $(m_u + m_d)/2$. The other sea quarks in the simulation have effectively infinite mass. The masses in the valence-quark propagators can still be chosen at will; they do not have to be equal to the masses of the sea quarks, so we have $m_{\mathrm{val}}^{(ud)}$, and $m_{\mathrm{val}}^{(s)}$ as valence mass parameters for the hadrons composed of u, d and s. Such computations in which the sea-quark masses differ from the valence-quark masses are called 'partially quenched'.

In the simulations one first produces gauge-field configurations and

---

† This is the reason, for example, why the neutron is 1.3 MeV heavier than the proton, despite the Coulomb self-energy of the proton.

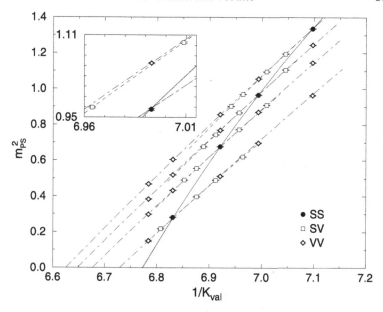

Fig. 7.6. $m_{\rm PS}^2$ as a function of $1/\kappa_{\rm val}$ for $\beta = 2.1$ and four values of $\kappa_{\rm sea}$. From [97].

then computes the average of the hadron-field correlators built from valence-quark propagators. Only valence diagrams of the type (a) in figure 7.2 are computed in this numerical study, since quark propagators corresponding to type (b) are much harder to evaluate. This means an approximation for the masses of mesons with quark–antiquarks of the same flavor (those below the second double line in table 7.1), which makes sense only if one sets $m_u = m_d$ (this follows from the discussion to be given in section 8.2). Diagrams of type (b) cause mixing of the flavor content of the mesons, which is expected to affect the vector mesons less than it does the pseudoscalars. For the η' mass diagrams of type (b) are essential.

Each choice of sea-quark mass implies a separate costly generation of gauge-field configurations, whereas the computation of valence-quark propagators is less expensive, so typically one has many more valence-quark masses than sea-quark masses available for analysis. However, by fitting suitable functions of all the masses involved, it is possible to obtain the desired mass combinations by interpolation and extrapolation. The latter is needed because the simulations need more time as the

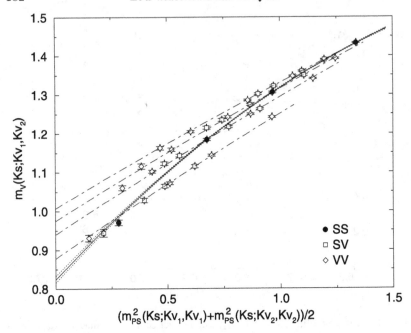

Fig. 7.7. m_V versus m_{PS}^2 for $\beta = 1.8$. From [97].

sea-quark masses are reduced and their small physical values cannot be simulated yet. So this introduces some uncertainty.

It turns out that the dependence of the squared pseudoscalar masses on the quark masses is almost linear, which can be understood as the result of chiral-symmetry breaking (see chapter 8), and the data can be fitted well by a quadratic polynomial in the quark masses. This is done in [97] as follows. For mesons composed of valence quarks 1 and 2 the average valence-quark mass is given by

$$m_{\text{val}} = \tfrac{1}{2}\left(m_{\text{val}}^{(1)} + m_{\text{val}}^{(2)}\right), \quad m = M - M_c = \frac{1}{2\kappa} - \frac{1}{2\kappa_c}, \qquad (7.45)$$

where κ is Wilson's hopping parameter ($r = 1$). In terms of these the pseudoscalar masses are parameterized as

$$m_{PS}^2\left(\kappa_{\text{sea}}; \kappa_{\text{val}}^{(1)}, \kappa_{\text{val}}^{(2)}\right) = b_s m_{\text{sea}} + b_v m_{\text{val}} + c_s m_{\text{sea}}^2 + c_v m_{\text{val}}^2$$
$$+ c_{sv} m_{\text{sea}} m_{\text{val}} + c_{vv} m_{\text{val}}^{(1)} m_{\text{val}}^{(2)}. \qquad (7.46)$$

In figure 7.6 results for the squared pseudoscalar masses are shown as

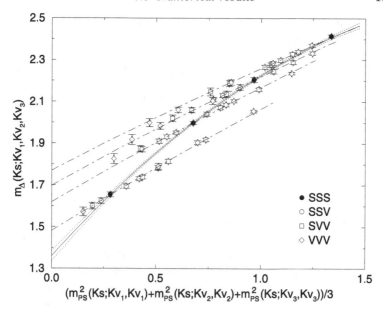

Fig. 7.8. Baryon decuplet masses versus m_{PS}^2 for $\beta = 1.8$. From [97].

a function of the average valence-quark mass, for one of the four values $\{1.8,\ 1.95,\ 2.1,\ 2.2\}$ of the gauge coupling β used in the simulation. Results at the other β values look similar except for a change of vertical scale (the mass in lattice units being smaller at larger β). The labels 'SS', 'SV' and 'VV' mean the following.

VV: $m_{val}^{(1)} = m_{val}^{(2)} = m_{val}$;

SV: $m_{val}^{(2)} = m_{sea}$; then $m_{val}^{(1)}$ can be written as $m_{val}^{(1)} = 2m_{val} - m_{sea}$;

SS: $m_{val}^{(1)} = m_{val}^{(2)} = m_{sea}$.

The lines VV and SV almost coincide and they are almost parallel for different κ_{sea}, so the line SS crosses all the others.

Note that $M_c = 1/2\kappa_c$ is also a free parameter in the fitting formula (7.46). If we read the right-hand side of (7.46) as a function of the inverse κ's, changing κ_c merely shifts all curves in figure 7.6 horizontally; $1/\kappa_c$ is then the value at which $m_{PS}^2(\kappa_{crit}; \kappa_{crit}, \kappa_{crit}) = 0$. Knowing the parameters $b_s,\ \dots,\ c_{sv}$ and κ_c from the fit determines m_{PS}^2 for every combination of the κ's and quark masses.

A similar procedure could be followed for the other hadron masses. Alternatively one can plot them as a function of m_{PS}^2, the procedure followed in [97]. The vector meson masses can be fitted to a quadratic polynomial in $m_{\rm PS}^2$,

$$m_{\rm V}\left(\kappa_{\rm sea};\kappa_{\rm val}^{(1)},\kappa_{\rm val}^{(2)}\right) = A^{\rm V} + B_{\rm s}^{\rm V}\mu_{\rm sea} + B_{\rm v}^{\rm V}\mu_{\rm val}$$
$$+ C_{\rm s}^{\rm V}\mu_{\rm sea}^2 + C_{\rm v}^{\rm V}\mu_{\rm val}^2 + C_{\rm sv}^{\rm V}\mu_{\rm sea}\mu_{\rm val}, \quad (7.47)$$

with

$$\mu_i = m_{\rm PS}^2\left(\kappa_{\rm sea};\kappa_{\rm val}^{(i)},\kappa_{\rm val}^{(i)}\right), \quad \mu_{\rm val} = \tfrac{1}{2}(\mu_1+\mu_2), \quad (7.48)$$

$$\mu_{\rm sea} = m_{\rm PS}^2\left(\kappa_{\rm sea};\kappa_{\rm sea},\kappa_{\rm sea}\right) \quad (7.49)$$

(the data show no need for a term $C_{\rm vv}^{\rm V}\mu_1\mu_2$). A corresponding plot is shown in figure 7.7. Note the shift in vertical scale relative to figure 7.6.

Next the baryon masses are analyzed. The simplest are the decuplet states which are symmetric in the flavor indices. Writing $\mu_{\rm val} = (\mu_1+\mu_2+\mu_3)/3$, the decuplet masses can be fitted by a formula similar to (7.47), see figure 7.8. The octet baryons have a more complicated quark-mass dependence because they have a mixed flavor symmetry; we shall not go into details here (see [97]), but the corresponding figures look roughly similar to figure 7.8.

The gross features of the mass spectrum are that, for the pseudoscalars, the squared mass is approximately linear in the quark masses (and vanishing at $m_{\rm val} = m_{\rm sea} = 0$), whereas for the other hadrons the mass itself is approximately linear.

Having obtained the coefficients from the fits, the physical value of the sea-quark mass can be determined for each β. Ideally this could be done by fixing the computed pion–nucleon mass ratio at the physical value, but in this case there are good reasons to believe that the nucleon mass suffers from finite-volume effects (based on experience in previous computations). A good alternative is to use the pion–rho mass ratio. Setting $\mu_{\rm val} = \mu_{\rm sea} = m_\pi^2$ in (7.47) the equation

$$\frac{m_\pi}{A^{\rm V} + (B_{\rm s}^{\rm V}+B_{\rm v}^{\rm V})m_\pi^2 + (C_{\rm s}^{\rm V}+C_{\rm v}^{\rm V}+C_{\rm sv}^{\rm V})m_\pi^4} = \left[\frac{m_\pi}{m_\rho}\right]_{\rm phys} = 0.176 \quad (7.50)$$

can be solved for m_π.

Using $m_\rho = 768$ MeV, one can then introduce the lattice spacing a by putting $am_\rho =$ denominator in (7.50), and find the value of $1/a$ in MeV units at each β. Knowing the physical values of m_{ud} and the value of $1/a$, the masses of the nucleon and delta can be evaluated from the fits and expressed in MeV units. Linear extrapolation to zero lattice spacing

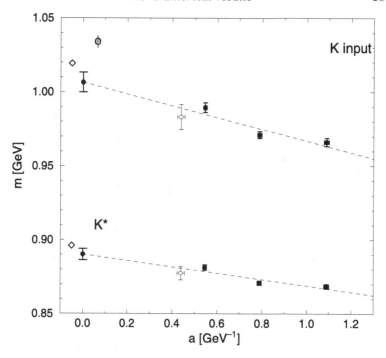

Fig. 7.9. Meson masses as a function of lattice spacing. The linear fit to $a = 0$ uses only the data at the three largest lattice spacings. Experimental values are indicated with diamonds. From [97].

(cf. figures 7.10 and 7.11) gives the results $m_N = 1034(36)$ MeV and $m_\Delta = 1392(58)$ MeV. These 'predictions' are to be compared with the experimental values of 940 and 1232 MeV (recall that in this simulation the physical volume is assumed to be somewhat small for these baryons).

Next the mass of the strange quark can be determined by fitting the kaon–rho mass ratio to the experimental value, $m_{\mathrm{PS}}^2(\kappa_{ud}; \kappa_{ud}, \kappa_s)/m_\rho^2 = m_K^2/m_\rho^2 = (498/768)^2$ (note that m_K is of the type SV). The masses of other hadrons containing strange valence quarks are then 'predictions'. Alternatively, the ϕ–ρ mass ratio was used in [97], the ϕ being of type VV, $m_V(\kappa_{ud}; \kappa_s, \kappa_s)/m_\rho = 1019/768$. The two ways of determining the valence mass of the strange quark are denoted by 'K input' and 'ϕ input'. Figure 7.9 shows such a 'prediction' for mesons as a function of the lattice spacing together with the continuum extrapolation. Examples for the baryon masses are shown in figures 7.10 and 7.11. The improved action allows rather large lattice spacings to be used. It can be seen that

Low-mass hadrons in QCD

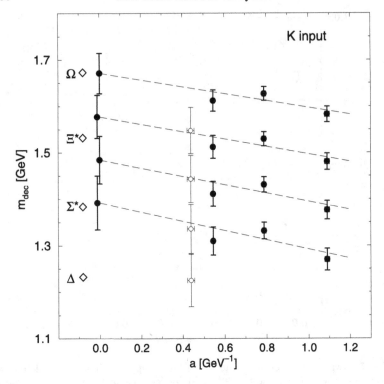

Fig. 7.10. Baryon decuplet masses as functions of a. From [97].

the a-dependence is consistent with 'linear', for the three larger lattice spacings, despite the fact that the baryon masses in lattice units are above 1 for the largest lattice spacing (cf. figure 7.8), which reduces by a factor of about two for the smallest lattice spacing.

The meson masses in the continuum limit are close to experiment at the level of 1%. The masses of baryons with three or two strange valence quarks are also close to experiment, but the discrepancy increases with only one or zero strange valence quarks. This is interpreted as finite-size effects being smaller for the hadrons involving the heavier strange valence quark (the lattice size in physical units is about 2.5 fm).

It is also of considerable phenomenological interest to determine the quark masses in physical units. In QCD the renormalized mass parameters are 'running' with the renormalization scale, similarly to the gauge coupling. An analysis of the quark masses in this simulation leads to the result $m_{ud}^{\overline{MS}} \approx 3.4$ MeV and $m_s^{\overline{MS}} \approx 90$ MeV, at the scale 2 GeV.

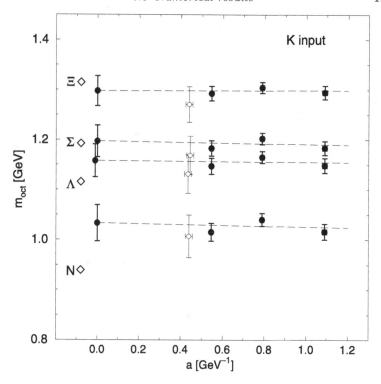

Fig. 7.11. Baryon octet masses as functions of a. From [97].

Comparing with the results of simulations in the quenched approximation, [97] finds that the inclusion of dynamical u and d quarks has improved agreement with experiment. Figure 7.12 shows a comparison; $N_f = 2$ indicates the simulation discussed above while '$N_f = 0$ Improved' denotes a quenched simulation using the same gauge-field action; '$N_f = 0$ Standard' shows the results of an earlier simulation [98] using the standard Wilson action. It is surprising how good the quenched approximation actually is for the hadron spectrum. The effect of dynamical fermions on various physical quantities is not easily established, see e.g. [99]. One may expect that results will further improve with simulations including also a dynamical strange quark, as well as including larger volumes.

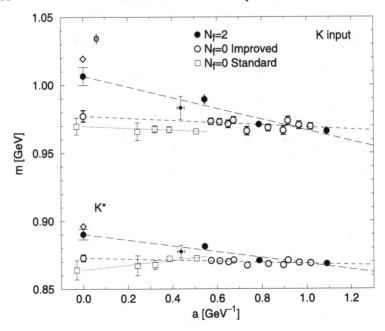

Fig. 7.12. A comparison with the quenched approximation. From [97].

7.6 The parameters of QCD

The parameters in the Wilson action ($r = 1$) are g^2 and $m_f = M_f - M_c$, the critical value $M_c = 1/2\kappa_c$ being determined completely by the gauge coupling. We have seen in the previous sections how these may be determined by the hadron spectrum. In particular, g^2 determines the overall scale, say the proton mass m_p at $m_u = m_d = m_s = 0$, while the quark masses determine the ratios m_{PS}^2/m_p^2. Roughly speaking, m_p, m_{π^+}, m_{K^+}, and m_K^0 are the free parameters of three-flavor QCD.

The renormalized masses and coupling depend in general on the renormalization scheme. In a mass-independent scheme such as minimal subtraction (cf. problem (iii) for a perturbative lattice definition), we get renormalized running coupling and masses at momentum scale μ, $\bar{g}(\mu)$ and $\bar{m}_f(\mu)$. They satisfy the renormalization-group equations

$$\mu \frac{d\bar{g}}{d\mu} = \beta(\bar{g}), \quad \mu \frac{d\bar{m}_f}{d\mu} = \gamma(\bar{g})\,\bar{m}, \quad (7.51)$$

with†

$$\beta(\bar{g}) = -\beta_1 \bar{g}^3 - \beta_2 \bar{g}^5 - \cdots, \tag{7.52}$$

$$\gamma(\bar{g}) = -\gamma_1 \bar{g}^2 - \gamma_2 \bar{g}^4 - \cdots. \tag{7.53}$$

Here β_1, β_2 and γ_1 are universal and given by

$$\beta_1 = \frac{1}{16\pi^2}\left(\frac{11}{3}n_c - \frac{2}{3}n_f\right), \tag{7.54}$$

$$\beta_2 = \frac{1}{(16\pi)^2}\left[\frac{34}{3}n_c^2 - \frac{10}{3}n_c n_f - \frac{(n_c^2-1)n_f}{n_c}\right], \tag{7.55}$$

$$\gamma_1 = \frac{1}{16\pi^2}\frac{3(n_c^2-1)}{n_c}, \tag{7.56}$$

where n_c is the number of colors and n_f the number of dynamical flavors.

We have already seen in section 5.2 how an overall scale Λ may be defined in terms of the gauge coupling,

$$\Lambda = \mu(\beta_1\bar{g}^2)^{-\beta_2/2\beta_1^2}e^{-1/2\beta_1\bar{g}^2}$$
$$\times \exp\left\{-\int_0^{\bar{g}} dg\left[\frac{1}{\beta(g)} + \frac{1}{\beta_1 g^3} - \frac{\beta_2}{\beta_1^2 g}\right]\right\}. \tag{7.57}$$

This scale is *renormalization-group invariant*, i.e. $d\Lambda(\mu,\bar{g}(\mu))/d\ln\mu = 0$. Similarly one defines renormalization-group-invariant quark masses

$$m_f^{\text{rgi}} = \bar{m}_f(\mu)(2\beta_1\bar{g}^2)^{-\gamma_1/2\beta_1}$$
$$\times \exp\left\{-\int_0^{\bar{g}} dg\left[\frac{\gamma(g)}{\beta(g)} - \frac{\gamma_1}{\beta_1 g}\right]\right\}, \tag{7.58}$$

which satisfy $dm_f^{\text{rgi}}(\mu,\bar{g}(\mu))/d\ln\mu = 0$. As we have seen in section 5.2, the scale Λ depends on the renormalization scheme; however, using similar arguments it follows that the m_f^{rgi} are scheme-independent. In [63] special techniques are used to compute the renormalization-group functions $\beta(g)$ and $\gamma(g)$ non-perturbatively.

Asymptotic freedom ($\beta_1 > 0$) is guaranteed for $n_f < 11n_c/2$, or $n_f \leq 16$ for QCD. We also see from the μ-independence of m_f^{rgi} that the running mass $\bar{m}_f(\mu)$ goes to zero $\propto (\bar{g}^2)^{\gamma_1/2\beta_1}$ as $\mu \to \infty$. The same is true for the bare quark mass m_f as the lattice spacing $a \to 0$ (in minimal subtraction the bare parameters run in the same way the renormalized ones, cf. problem (iv)).

† The subscript k of β_k and γ_k indicates that the coefficient corresponds to diagrams with k loops. Another notation, often used, is $\beta_k \to \beta_{k-1}$, $\gamma_k \to \gamma_{k-1}$, and [63] $\beta_k \to b_{k-1}$, $k = 1, 2, \ldots$, $\gamma(\bar{g}) \to \tau(\bar{g}) = -d_0\bar{g}^2 - d_1\bar{g}^4 - \cdots$.

7.7 Computing the gauge coupling from the masses

At long distances the non-perturbative methods of lattice gauge theory allow us to compute the properties of hadrons. At short distances we know that weak-coupling perturbation theory works well. Many physical properties have been successfully related with the methods of perturbative QCD. The essential parameter in these calculations is the renormalized coupling constant g_R. A useful characterization of the coupling strength is the value of the running coupling $g_{\overline{MS}}(\mu)$ in the $\overline{MS}$ scheme, which is customarily taken at the scale set by the mass of the Z-boson, $\mu = m_Z$, or rather the value of the 'strong fine-structure constant' $\alpha_s(m_Z) = g_{\overline{MS}}^2(m_Z)/4\pi$. It is not a free parameter; its value can be predicted just like other physical quantities such as mass ratios. Let us see in more detail how this can be done.

Suppose that we compute the static quark–antiquark potential V at short distances. From the force

$$F(r) = \frac{\partial V}{\partial r} = C_2 \frac{g_V^2(1/r)}{4\pi r^2}, \qquad (7.59)$$

we know $g_V^2(1/r)$ at some distance r/a in lattice units, for some bare g and quark-mass parameters aM, chosen such that m_π/m_p, m_K/m_p, $\ldots$ have the experimental values to reasonable accuracy. From the value of the proton mass in lattice units, am_p, we then also know the distance r in units of m_p, rm_p. Provided that rm_p is small enough, we can then use the perturbative renormalization group

$$\frac{\mu \, dg_V}{d\mu} = \beta(g_V), \qquad \beta(g_V) = -\beta_1 g_V^3 - \beta_2 g_V^5 + \cdots, \qquad \mu = 1/r, \quad (7.60)$$

to relate the computed $g_V^2(1/r)$ to g_V^2 at higher μ. At sufficiently large μ we can use the perturbative connection between g_V^2 and $g_{\overline{MS}}^2$ parameterized by the ratio of the scales $\Lambda_{\overline{MS}}/\Lambda_V$.

This program is difficult to implement because the lattices for simulations with dynamical fermions in spectrum computations tend to be small. Other renormalized coupling constants have been proposed in place of g_V, which are useful for numerical computations, e.g. the 'Schrödinger functional method' [94].

7.8 Problems

(i) *Effective action*

The exponent in (7.17) can be interpreted as an effective action for the gauge field. Calculate the contribution of the smallest closed loop (around a plaquette). Show that it corresponds to an *decrease* (i.e. increase of the effective $\beta = 6/g^2$) of the effective gauge coupling.

(ii) *Three flavors*

Devise a method for analyzing numerical hadron-mass data with dynamical up, down and strange quarks, with $m_u = m_d \neq m_s$.

(iii) *Minimal subtraction revisited*

In the following g_0 and m_0 denote the bare gauge coupling and quark masses, and g and m the renormalized ones (we suppress the flavor label f). For Wilson fermions $m_0 \equiv M - M_c(g_0, r)$. For staggered fermions we may think of m_0 simply being the parameter appearing in the action (see [73] for more details). The critical mass M_c is linearly divergent and the bare m_0 has to absorb the remaining logarithmic divergences as the lattice spacing $a \to 0$. The coupling g_0 is logarithmically divergent. We shall now follow similar steps to those in problem 3(iv) for the QCD case.

Both g_0 and m_0 are multiplicatively renormalized,

$$g_0 = g Z_g(g, \ln a\mu), \tag{7.61}$$

$$m_0 = m Z_m(g, \ln a\mu), \tag{7.62}$$

$$Z_g(g, \ln a\mu) = 1 + \sum_{n=1}^{\infty} \sum_{k=0}^{n} Z_{nk}^g g^{2n} (\ln a\mu)^k$$

$$= \sum_{k=0}^{\infty} Z_k^g(g)(\ln a\mu)^k, \tag{7.63}$$

and similarly for Z_m,

$$Z_m(g, \ln a\mu) = \sum_{k=0}^{\infty} Z_k^m(g)(\ln a\mu)^k. \tag{7.64}$$

Terms vanishing as $a \to 0$ have been neglected, order by order in perturbation theory. In principle we can allow any choice of the coefficients $Z_{nk}^{g,m}$ which lead to a series in g^2 for the renormalized vertex functions in which the dependence on a cancels out. In minimal subtraction one chooses

$$Z_0^g(g) \equiv 1, \quad Z_0^m(g) \equiv 1. \tag{7.65}$$

The renormalized g and m depend on the physical scale μ but not on a whereas g_0 and m_0 are supposed to depend on a but not on μ. Then

$$0 = \left[\mu \frac{\partial}{\partial \mu} + \beta(g) \frac{\partial}{\partial g}\right] g Z_g(g, \ln a\mu), \qquad (7.66)$$

$$0 = \left[\mu \frac{\partial}{\partial \mu} + \beta(g) \frac{\partial}{\partial g} + \gamma(g)\right] Z_m(g, \ln a\mu), \qquad (7.67)$$

where

$$\beta(g) = \mu \frac{dg}{d\mu}, \quad \gamma(g) = \frac{\mu}{m} \frac{dm}{d\mu}. \qquad (7.68)$$

By going through similar arguments to those in problem 3(iv), show that in minimal subtraction

$$\beta(g) = -Z_1^g(g), \quad \gamma(g) = -Z_1^m(g). \qquad (7.69)$$

Verify that, in minimal subtraction, the renormalization-group functions for the bare parameters are identical to those for the rernormalized ones, $\beta_0(g_0) = \beta(g_0)$ and $\gamma_0(g_0) = \gamma(g_0)$. Verify the RG-independence of Λ and m^{rgi} in (7.57) and (7.58).

8

Chiral symmetry

In this chapter we pay attention to a very important aspect of QCD and the Standard Model: chiral symmetry. It is a symmetry that is natural in the continuum but it poses special problems for regularizations, including the lattice. We review first some aspects of chiral symmetry in QCD, then discuss chiral aspects of QCD on a lattice, and finally give a brief introduction to chiral gauge theories, of which the Standard Model is an example.

8.1 Chiral symmetry and effective action in QCD

Consider the mass terms in the QCD action,

$$S_{\text{mass}} = -\int d^4x \, \bar{\psi} m \psi, \tag{8.1}$$

where $m = \text{diag}\,(m_u, m_d, m_s, \cdots)$ is the diagonal matrix of mass parameters. We know that the first three quarks u, d and s have relatively small masses compared with a typical hadronic scale such as the Regge slope $(\alpha')^{-1/2} \approx 1100$ MeV or the string tension $\sqrt{\sigma} \approx 400$ MeV. (Recall that $m_{ud} = 4.4$ MeV and $m_s \approx 90$ MeV in section 7.5.) Suppose we set the first n_{f} quark-mass parameters to zero. In that case the QCD action has $U(n_{\text{f}}) \times U(n_{\text{f}})$ symmetry, loosely called chiral symmetry, in which the left- and right-handed components of the Dirac field are subjected to independent flavor transformations $V_{\text{L,R}} \in U(n_{\text{f}})$,

$$
\begin{aligned}
\psi &\to V\psi, \quad V = V_{\text{L}} P_{\text{L}} + V_{\text{R}} P_{\text{R}}, \\
\bar{\psi} &\to \bar{\psi}\bar{V}, \quad \bar{V} = V_{\text{L}}^{\dagger} P_{\text{R}} + V_{\text{R}}^{\dagger} P_{\text{L}} = \beta V^{\dagger}\beta, \\
P_{\text{L}} &= \tfrac{1}{2}(1 - \gamma_5), \quad P_{\text{R}} = \tfrac{1}{2}(1 + \gamma_5).
\end{aligned}
\tag{8.2}
$$

Here $V_{L,R}$ act only on the first n_f flavor indices of the quark field. The matrix $\gamma_5 \equiv i\gamma^0\gamma^1\gamma^2\gamma^3 = -\gamma_1\gamma_2\gamma_3\gamma_4$ has the properties $\gamma_5^\dagger = \gamma_5$, $\gamma_5^2 = 1$ and it anticommutes with the γ_μ, i.e. $\gamma_5\gamma_\mu = -\gamma_\mu\gamma_5$. The $P_{L,R}$ are orthogonal projectors, $P_L^2 = P_L$, $P_R^2 = P_R$, $P_L P_R = 0$, $P_L + P_R = 1$. Because of these properties the derivative terms in the action are invariant,

$$\bar{\psi}\gamma^\mu D_\mu\psi \to \bar{\psi}\bar{V}\gamma^\mu V D_\mu\psi = \bar{\psi}\gamma^\mu(V_L^\dagger V_L + V_R^\dagger V_R)D_\mu\psi = \bar{\psi}\gamma^\mu D_\mu\psi. \quad (8.3)$$

The mass terms transform as

$$\bar{\psi}m\psi \to \psi\bar{V}mV\psi = \bar{\psi}(V_R^\dagger m V_L P_L + V_L^\dagger m V_R P_R)\psi, \quad (8.4)$$

so they break the symmetry. A flavor-symmetric mass term has $m \propto 1$. Such a mass term is invariant under *flavor transformations*, for which $V_L = V_R$. However, it is not invariant under transformations with $V_L \neq V_R$. A special case of these are *chiral transformations* in the narrow sense,[1] for which $V_L = V_R^\dagger$. For $m = 0$ in the $n_f \times n_f$ subspace the action is invariant under chiral $U(n_f) \times U(n_f)$ transformations.

In the quantum theory the $U(n_f) \times U(n_f)$ symmetry is reduced to $SU(n_f) \times SU(n_f) \times U(1)$ by so-called *anomalies* (this will be reviewed in section 8.4). Here $U(1)$ is the group of ordinary (Abelian) phase transformations $\psi \to e^{i\omega}\psi$, $\bar{\psi} \to e^{-i\omega}\bar{\psi}$. Furthermore, the dynamics is such that the $SU(n_f) \times SU(n_f)$ symmetry is spontaneously broken.

An informative way to exhibit the physics of this situation is by using an effective action. We have met already in chapter 3 the $O(4)$ model for pions (which can be extended to include nucleons, cf. problem (i)). This illustrates the case $n_f = 2$ (the group $SO(4)$ is equivalent to $SU(2) \times SU(2)/Z_2$, cf. (D.19) in appendix D). One introduces effective fields ϕ which transform in the same way as the quark bilinear scalar fields $\bar{\psi}_g\psi_f$ and pseudoscalar fields $\bar{\psi}_g i\gamma_5\psi_f$, $f,g = 1,\ldots,n_f$. We start with

$$\Phi_{fg} \equiv \bar{\psi}_g P_L \psi_f, \quad (8.5)$$

which transforms as

$$\Phi_{fg} \to (V_L)_{ff'}(V_R^\dagger)_{g'g} \Phi_{f'g'}, \quad (8.6)$$

or, in matrix notation,

$$\Phi \to V_L \Phi V_R^\dagger. \quad (8.7)$$

The other possibility leads to $\Phi^\dagger$:

$$\bar{\psi}_g P_R \psi_f = (\psi_f^\dagger P_R \,\beta\psi_g)^* = (\bar{\psi}_f P_L \psi_g)^* = (\Phi_{gf})^* \quad (8.8)$$

$$\equiv (\Phi^\dagger)_{fg}. \quad (8.9)$$

Under parity Φ and $\Phi^\dagger$ are interchanged,

$$\Phi(x^0, \mathbf{x}) \xrightarrow{P} \Phi^\dagger(x^0, -\mathbf{x}). \tag{8.10}$$

Ignoring the symmetry breaking due to anomalies, the effective action for the *effective field* $\phi \leftrightarrow \Phi$ has the same chiral transformation properties as the QCD action. We shall first examine the form of this effective action and derive some consequences, and later take into account that anomalies reduce the $U(n_f) \times U(n_f)$ symmetry to $SU(n_f) \times SU(n_f) \times U(1)$.

For $m = 0$ we want an invariant action. The combination $\mathrm{Tr}[(\phi\phi^\dagger)^k]$ is invariant under (8.7). An invariant action is given by

$$S = -\int d^4x\, \mathrm{Tr}(F_2 \partial_\mu \phi^\dagger F_1 \partial^\mu \phi + G), \tag{8.11}$$

where $F_{1,2}$ and G have the forms

$$F_1 = \sum_k f_{1k}(\phi\phi^\dagger)^k, \quad F_2 = \sum_k f_{2k}(\phi^\dagger\phi)^k, \tag{8.12}$$

$$G = \sum_k g_k(\phi\phi^\dagger)^k. \tag{8.13}$$

Reality of the action requires the coefficients f_{1k}, f_{2k} and g_k to be real. Invariance under parity requires

$$f_{1k} = f_{2k}. \tag{8.14}$$

The action might also contain terms of the type

$$\mathrm{Tr}[(\phi\phi^\dagger)^k]\,\mathrm{Tr}[(\phi\phi^\dagger)^l]. \tag{8.15}$$

At this point we assume such terms to be absent and come back to them later.

There may also be higher derivative terms. Their systematic inclusion is part of *chiral perturbation theory*, see e.g. [19]. For slowly varying fields, which is all we need for describing physics on the low-energy–momentum scale, we may assume such higher derivative terms to be negligible.

The classical ground state will be characterized by $\partial_\mu \phi = 0$ and correspond to a minimum of $\mathrm{Tr}\,G$. Let $\lambda_1, \ldots, \lambda_{n_f}$ be the eigenvalues of the Hermitian matrix $\phi\phi^\dagger$. Then

$$\mathrm{Tr}\,G = \sum_k g_k(\lambda_1^k + \cdots + \lambda_{n_f}^k). \tag{8.16}$$

A stationary point of $\mathrm{Tr}\,G$ has to satisfy

$$0 = \frac{\partial}{\partial \lambda_j}\mathrm{Tr}\,G = \sum_k g_k k \lambda_j^{k-1}, \tag{8.17}$$

which is the same equation for each j. Hence, the solution is

$$\lambda_1 = \cdots = \lambda_{n_f} \equiv \lambda, \quad \phi\phi^\dagger = \lambda\mathbb{1}. \tag{8.18}$$

Since $\phi\phi^\dagger \geq 0$ (i.e. all eigenvalues are ≥ 0), $\lambda \geq 0$.

We shall now assume that $\lambda \neq 0$ at the minimum of $\mathrm{Tr}\,G$. The symmetry is then spontaneously broken, because a non-zero ϕ in the ground state is not invariant under $U(n_f) \times U(n_f)$ transformations. It is helpful to use a generalized polar decomposition for ϕ,

$$\phi = HU, \quad H = H^\dagger, \quad U^\dagger = U^{-1}. \tag{8.19}$$

The H and U can be found as follows: H can be calculated from $H = \pm\sqrt{\phi\phi^\dagger}$ and then $U = H^{-1}\phi$. In the ground state $H = \pm\sqrt{\lambda}\mathbb{1}$. The degeneracy of the ground state is described by U, which is an element of $U(n_f)$. It transforms as $U \to V_L U V_R^\dagger$. Without loss of generality we may assume that $U = \mathbb{1}$ and $H = -\sqrt{\lambda}\mathbb{1}$ (the minus sign becomes natural on taking into account the explicit symmetry breaking due to the quark masses). This exhibits clearly the residual degeneracy of the ground state: it is invariant under the diagonal $U(n)$ subgroup, for which $V_L = V_R$. The pattern of spontaneous symmetry breaking is

$$U(n_f) \times U(n_f) \to U(n_f). \tag{8.20}$$

The variables of U (e.g. using the exponential parameterization) are analogous to angular variables for the $O(n)$-vector field φ_α in the $O(n)$ model. We expect these to correspond to Nambu–Goldstone bosons.

Let us linearize the effective action about the ground state, writing $H = -v + h$, $U = \exp(i\alpha)$,

$$\phi = (-v + h)\left(1 + i\alpha - \tfrac{1}{2}\alpha^2 + \cdots\right), \quad v = \sqrt{\lambda} > 0 \tag{8.21}$$

(from now on we no longer distinguish between 1 and $\mathbb{1}$). We keep only terms up to second order in h and α. Since $\partial_\mu\phi$ is of first order, we may replace ϕ by v in $F_{1,2}$ in (8.12),

$$F_1 = F_2 \equiv F, \quad \text{for } \phi = -v, \tag{8.22}$$

and obtain

$$S = -\int d^4x\, \mathrm{Tr}\left(F^2 v^2\, \partial_\mu\alpha\partial^\mu\alpha + F^2 \partial_\mu h\partial^\mu h + r\, h^2\right) + \cdots. \tag{8.23}$$

The $\cdots$ also include the ground-state value of S. The coefficient r follows from

$$\text{Tr}\, G = \sum_k g_k v^{2k}\, \text{Tr}\, [(-1 + h/v)^{2k}] \tag{8.24}$$

$$= \sum_k g_k v^{2k}[n_{\text{f}} + k(2k-1)v^{-2}\,\text{Tr}\, h^2 + \cdots], \tag{8.25}$$

where the term linear in h vanishes because $\text{Tr}\, G$ is stationary at $h = 0$. Since we expand around a minimum of $\text{Tr}\, G$, the coefficient

$$r = \sum_k g_k k(2k-1)v^{2k-2} \tag{8.26}$$

is positive. The form (8.23) shows that the α fields have zero mass parameter – they correspond to n_{f}^2 Nambu–Goldstone bosons. The h fields have a mass given by

$$m_h^2 = r/F^2. \tag{8.27}$$

At this point we shall make the useful approximation of 'freezing' the 'radial' degrees of freedom H to their ground-state value $H = -v$, or $h = 0$. This approximation is justified when m_h is sufficiently large compared with the momenta of interest (cf. problem (i) for numbers) and it simplifies the derivations to follow. Thus we get

$$\phi(x) = -v\, U(x), \tag{8.28}$$

$$S = -\int d^4x\, \frac{f^2}{4}\, \text{Tr}\, (\partial_\mu U^\dagger \partial^\mu U), \quad f^2 = 4F^2v^2, \tag{8.29}$$

where we omitted a constant term.

We now comment on the terms of the form (8.15). When these are included the uniqueness of the form (8.18) is no longer compelling and other solutions with $\lambda_j \neq \lambda_k$ are also possible. This depends on the details of the action. However, arguments based on the large-n_{c} behavior of the generalization of QCD to an $SU(n_{\text{c}})$ gauge theory suggest that terms of the form (8.15) are subdominant [100]. The ground-state solution of the complete effective action including terms of the form (8.15) is still expected to have the symmetric form (8.18), and the symmetry-breaking pattern is still expected to be $U(n_{\text{f}}) \times U(n_{\text{f}}) \to U(n_{\text{f}})$.

The quark-mass terms in the QCD action explicitly break chiral symmetry. They have the form of an external source coupled to quark bilinears,

$$S_{\text{mass}} = -\int d^4x \, \bar\psi m(P_{\text{L}} + P_{\text{R}})\psi \tag{8.30}$$

$$= \int d^4x \, \text{Tr}\,(J^\dagger \Phi + \Phi^\dagger J), \quad J = -m. \tag{8.31}$$

Hence, we can absorb the quark-mass terms in the coupling to such an external source. The total effective action including this source has the form $\ln Z(J) = S(\phi) + \int d^4x \, \text{Tr}\,(J^\dagger \phi + \phi^\dagger J)$, where ϕ is again the effective field. Setting $J = -m$ thus leads to an addition ΔS in the effective action

$$\Delta S = -\int d^4x \, \text{Tr}\,[m(\phi + \phi^\dagger)]. \tag{8.32}$$

Expanding† $\phi = -vU = -v\exp(i\alpha)$ to second order in α gives

$$\Delta S = -\int d^4x \, v\,\text{Tr}\,(m\alpha^2) + \cdots = -\int d^4x \, v \sum_{fg} m_f \alpha_{fg}\alpha_{gf} + \cdots . \tag{8.33}$$

Since U is unitary, $\alpha_{gf} = \alpha_{fg}^*$. Taking α_{fg} with $f \le g$ as independent variables leads to

$$\Delta S = -\int d^4x \, v \left[\sum_{f<g}(m_f + m_g)\alpha_{fg}\alpha_{fg}^* + \sum_f m_f \alpha_{ff}^2 + \cdots \right]. \tag{8.34}$$

Similarly, expanding the gradient term (8.29) gives

$$S = -\int d^4x \, \frac{f^2}{4}\left(2\sum_{f<g}\partial_\mu \alpha_{fg}^* \, \partial^\mu \alpha_{fg} + \sum_f \partial_\mu \alpha_{ff} \, \partial^\mu \alpha_{ff} + \cdots \right). \tag{8.35}$$

As expected from the $O(4)$ model, ΔS gives a mass to the Goldstone bosons, which for small m is linear in m,

$$m_{fg}^2 = B(m_f + m_g), \quad B = 2v/f^2. \tag{8.36}$$

In the next section we shall confront these mass relations with experiment.

By coupling the effective action to the electroweak gauge fields it can be shown that the constant f determines the leptonic decays

† We neglect here the effect of the quark masses on the ground-state value of ϕ.

of the pseudoscalar mesons. It is known as the pion decay constant, $f = f_\pi \approx 93$ MeV. This constant also determines the size of the s-wave pi–pi scattering lengths in good agreement with experiment.

To end this section we note a relation between the pion decay constant f, the unrenormalized *chiral condensate* $\langle \bar{\psi}\psi \rangle = \sum_f \langle \bar{\psi}_f \psi_f \rangle = 2\sum_f \langle \phi_{ff} \rangle = -2n_f v$, and the wave-function renormalization constant Z of the pseudoscalar fields $P_{fg} \equiv i(\phi^\dagger - \phi)_{fg} \leftrightarrow \bar{\psi}_g i\gamma_5 \psi_f$. The constant Z can be read off from $S = -\int d^4x\, Z^{-1} \sum_{f<g} \partial_\mu P^*_{fg} \partial^\mu P_{fg} + \cdots$, using $U = -\phi/v$ and (8.29) and (8.35): $Z^{-1} = f^2/8v^2$. Hence, f is given by the renormalized chiral condensate $\langle \bar{\psi}\psi \rangle/\sqrt{Z}$,

$$f = \frac{2\sqrt{2}v}{\sqrt{Z}} = \frac{-\sqrt{2}\langle \bar{\psi}\psi \rangle}{n_f \sqrt{Z}}. \tag{8.37}$$

8.2 Pseudoscalar masses and the $U(1)$ problem

The candidate Nambu–Goldstone (NG) bosons and their masses are

$$\pi^\pm: \quad m^2_{\pi^+} = m^2_{ud} = 0.0195 \text{ GeV}^2$$
$$K^\pm: \quad m^2_{K^+} = m^2_{us} = 0.244 \text{ GeV}^2$$
$$K^0, \bar{K}^0: \quad m^2_{K^0} = m^2_{ds} = 0.248 \text{ GeV}^2$$
$$\pi^0: \quad m^2_{\pi^0} = 0.0182 \text{ GeV}^2$$
$$\eta: \quad m^2_\eta = 0.301 \text{ GeV}^2$$
$$\eta': \quad m^2_{\eta'} = 0.917 \text{ GeV}^2 \tag{8.38}$$

For the unequal-flavor particles ($f \neq g$) we have indicated the quark labels. For the neutral π^0, η and η' the quark assignment turns out to be less straightforward.

Consider two light flavors, $n_f = 2$. The mass formula (8.36) with $f = u, d$ and $g = u, d$ predicts four NG bosons in this case. The obvious candidates are $\pi^\pm$, π^0 and η, with

$$m^2_{\pi^+} = m^2_{ud} = B(m_u + m_d). \tag{8.39}$$

According to (8.36), the other eigenstates are $\bar{u}u$ and $\bar{d}d$. If we try to assign $\pi^0 \leftrightarrow \bar{u}u$, $\eta \leftrightarrow \bar{d}d$, the relation

$$m^2_{ud} = \tfrac{1}{2}(m^2_{uu} + m^2_{dd}) \tag{8.40}$$

cannot be fulfilled at all. If we assume that $m_u \approx m_d$ and π^0 is an equal mixture of $\bar{u}u$ and $\bar{d}d$ to get $m^2_{\pi^0} \approx m^2_{\pi^+}$, the orthogonal combination of

$\bar{u}u$ and $\bar{d}d$ should have approximately the same mass as π^0: the η does not fit in.

Consider next three light flavors, $n = 3$. The mass formulas now predict nine NG bosons. We find

$$\frac{m_u + m_d}{m_u + m_s} = \frac{m_{\pi^+}^2}{m_{K^+}^2} \equiv R_1, \quad \frac{m_u + m_s}{m_d + m_s} = \frac{m_{K^+}^2}{m_{K^0}^2} \equiv R_2, \qquad (8.41)$$

and from this

$$\frac{m_s}{m_u} = \frac{R_2(R_1 - 1)}{1 - R_2 - R_1 R_2} = 31, \quad \frac{m_s}{m_d} = \frac{R_2}{1 - R_2 + m_u/m_s} = 20. \quad (8.42)$$

Hence $m_u : m_d : m_s \approx 1 : 1.5 : 30$. The effective action furthermore predicts particles with masses

$$m_{uu}^2 = \frac{2m_u}{m_u + m_d} m_{\pi^+}^2 = 0.0155 \text{ GeV}^2, \qquad (8.43)$$

$$m_{dd}^2 = \frac{2m_d}{m_u + m_d} m_{\pi^+}^2 = 0.0235 \text{ GeV}^2, \qquad (8.44)$$

$$m_{ss}^2 = \frac{2m_s}{m_u + m_s} m_{K^+}^2 = 0.473 \text{ GeV}^2. \qquad (8.45)$$

The candidates π^0, η and η' do not fit into the $n = 3$ formulas either. The effective action obtained so far must be wrong.

This is an aspect of the notorious $U(1)$ problem. The problem is the chiral $U(1)$ invariance contained in $U(n_f) \times U(n_f)$. These are the transformations of the type $V_L = V_R^\dagger = \exp(i\omega)\, \mathbb{1}$, or more generally, transformations $V_L = V_R^\dagger$ with $\det V_L \neq 1$. We know that this invariance of the classical QCD action is broken in the quantum theory by 'anomalies': QCD has only approximate $SU(n_f) \times SU(n_f)$ chiral symmetry, plus the flavor $U(1)$ symmetry $V_L = V_R = \exp(i\omega)\, \mathbb{1}$ corresponding to quark-number conservation.

The resolution of the $U(1)$ problem through 'anomalies' turned out to be a difficult but very interesting task. Here we shall simply add terms to the action that break the chiral $U(1)$ symmetry and see what this implies for the mass formulas. We need to introduce terms of the type $\det U$, which is invariant under $SU(n_f) \times SU(n_f)$ but not under $U(n_f) \times U(n_f)$:

$$\det U \rightarrow \det(V_L U V_R^\dagger) = \det(U) \det(V_L V_R^\dagger) = \det U, \qquad (8.46)$$

for $V_{L,R} \in SU(n)$. A term like

$$\Delta' S = \int d^4x \, c \, (\det U + \det U^\dagger) \qquad (8.47)$$

would do, but considerations of the large-n_c behavior of 'n_c-color QCD' suggest using instead the form

$$\Delta'S = \int d^4x\, c\, (\mathrm{Tr}\, \ln U - \mathrm{Tr}\, \ln U^\dagger)^2. \tag{8.48}$$

In fact, $c \propto 1/n_c$. (Both choices for $\Delta'S$ lead to the same form of the mass matrix for the neutral pseudoscalar mesons to be derived below.) Writing $U = \exp(i\alpha)$ gives

$$\Delta'S = -\int dx\, 4c \left(\sum_f \alpha_{ff}\right)^2. \tag{8.49}$$

Hence, the masses of α_{fg}, $f < g$, are unaffected, but the α_{ff} modes are now coupled by a mass matrix of the form

$$m^2_{ff,gg} = 2Bm_f\delta_{fg} + \lambda, \quad \lambda = 16c/f^2, \tag{8.50}$$

or

$$m^2 = 2B\begin{pmatrix} m_u & 0 & 0 \\ 0 & m_d & 0 \\ 0 & 0 & m_s \end{pmatrix} + \lambda\begin{pmatrix} 1 & 1 & 1 \\ 1 & 1 & 1 \\ 1 & 1 & 1 \end{pmatrix}. \tag{8.51}$$

We shall treat the quark-mass term as a perturbation to the λ term. For $m_f = 0$ we have the eigenvectors and eigenvalues

$$\phi_0 = \frac{1}{\sqrt{3}}(1,1,1), \quad m^2 = 3\lambda, \tag{8.52}$$

$$\phi_3 = \frac{1}{\sqrt{2}}(1,-1,0), \quad m^2 = 0, \tag{8.53}$$

$$\phi_8 = \frac{1}{\sqrt{6}}(1,1,-2), \quad m^2 = 0. \tag{8.54}$$

Using $m_{u,d,s}$ as a perturbation (in the way familiar from quantum mechanics) leads to the following mass formulas:

$$m^2_{\eta'} = 3\lambda + B(\tfrac{2}{3}m_u + \tfrac{2}{3}m_d + \tfrac{2}{3}m_s), \tag{8.55}$$

$$m^2_{\pi^0} = B(m_u + m_d), \tag{8.56}$$

$$m^2_{\eta} = B(\tfrac{1}{3}m_u + \tfrac{1}{3}m_d + \tfrac{4}{3}m_s), \tag{8.57}$$

which hold for the mass ratios (8.42) up to tiny corrections. The eigenvectors are also interesting, but here we merely mention that π^0 and η are mainly ϕ_3 and ϕ_8, whereas the η' is predominantly ϕ_0. From (8.55) we can determine the chiral $U(1)$ breaking strength λ,

$$3\lambda = m^2_{\eta'} - \tfrac{1}{2}(m^2_{\pi^0} + m^2_\eta) = 3(0.252)\, \mathrm{GeV}^2. \tag{8.58}$$

The mass terms in the effective action depend on four parameters, Bm_u, Bm_d, Bm_s and λ. Hence we have two predictions for the five pseudoscalar masses:

$$m_{\pi^0}^2 = m_{\pi^+}^2, \tag{8.59}$$
$$m_\eta^2 = \tfrac{1}{6}(m_{uu}^2 + m_{dd}^2) + \tfrac{2}{3}m_{ss}^2 = 0.322 \text{ GeV}^2, \tag{8.60}$$

which agree reasonably well with experiment. It should be kept in mind that electromagnetic corrections, which affect in particular the electrically charged particles, are neglected.

In the early days the near equality of m_{π^0} and m_{π^+} was interpreted as an aspect of approximate flavor symmetry, $m_u \approx m_d$. Now we know that m_d is substantially larger than m_u and that the approximate flavor symmetry is due to approximate chiral symmetry, $m_{u,d} \ll \sqrt{\sigma}$, the spontaneous-symmetry-breaking pattern $U(n_\mathrm{f}) \times U(n_\mathrm{f}) \rightarrow U(n_\mathrm{f})_\mathrm{flavor}$, and the flavor-singlet character of the chiral-anomaly term $\Delta' S$.

8.3 Chiral anomalies

The Noether argument tells us that to each continuous symmetry of the action corresponds a 'conserved current' j^μ, $\partial_\mu j^\mu = 0$, and a conserved 'charge' $Q = \int d^3x \, j^0(x)$, $\partial_0 Q = 0$. This is true in the classical theory but not necessarily in the quantum theory, which needs more specification than merely giving the action, such as the precise definition of the path integral. In case the quantum analog of j^μ is not conserved, one speaks of an *anomaly* $\mathcal{A} \equiv \partial_\mu j^\mu$. In four space–time dimensions $\mathcal{A}$ is typically $\propto \epsilon^{\kappa\lambda\mu\nu} \operatorname{Tr}(G_{\kappa\lambda} G_{\mu\nu})$, where $G_{\mu\nu}$ is a gauge-field tensor. Relations like $\partial_\mu j^\mu = \mathcal{A}$ can be found in perturbation theory by studying correlation functions of j^μ and $\mathcal{A}$ with other fields.

Chiral anomalies correspond to diagrams of the type shown in figure 8.1, and related diagrams, in which one vertex corresponds to a (polar) vector current, $\bar\psi i\gamma^\mu\psi$, or an axial vector current, $\bar\psi i\gamma^\mu\gamma_5\psi$, and the other two vertices to gauge fields. There must be an odd number of γ_5's in the trace over the Dirac indices ($\operatorname{Tr}(\gamma_5\gamma_\kappa\gamma_\lambda\gamma_\mu\gamma_\nu) = 4i\epsilon_{\kappa\lambda\mu\nu}$), hence the name 'chiral anomalies'. These γ_5 may come from the gauge-field vertices or from the current.

In QCD there is no γ_5 associated with the gauge-field vertices and only axial vector currents can have an anomaly. In the Euclidean formulation

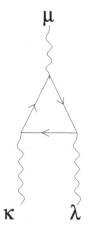

Fig. 8.1. Triangle diagram in which chiral anomalies show up.

their divergence reads†

$$\partial_\mu(\bar\psi_f i\gamma_\mu\gamma_5\psi_g) = (m_f + m_g)\bar\psi_f i\gamma_5\psi_g + \delta_{fg}\, 2iq, \qquad (8.61)$$

$$q = \frac{g^2}{32\pi^2}\,\epsilon_{\kappa\lambda\mu\nu}\,\mathrm{Tr}\,(G_{\kappa\lambda}G_{\mu\nu}). \qquad (8.62)$$

For zero quark masses the right-hand side of (8.61) is the anomaly. The vector currents have no such anomaly. Their divergence reads

$$\partial_\mu(\bar\psi_f i\gamma_\mu\psi_g) = i(m_f - m_g)\bar\psi_f\psi_g, \qquad (8.63)$$

which is zero in the symmetry limit $m_f = m_g$, hence also in the chiral limit $m_f = m_g = 0$. The right-hand sides of the divergence equations (8.61) and (8.63) are zero for the currents corresponding to $SU(n_{\mathrm{f}}) \times SU(n_{\mathrm{f}})$ symmetry, obtained by contraction of $\bar\psi_f i\gamma^\mu P_{\mathrm{L,R}}\psi_g$ with the $n_{\mathrm{f}}^2 - 1$ flavor $SU(n_{\mathrm{f}})$ generators $(\lambda_k)_{fg}/2$, $\mathrm{Tr}\,\lambda_k = 0$. Hence, the anomaly in (8.61) breaks only chiral $U(1)$ invariance corresponding to $\lambda_0 \propto \mathbb{1}$ with $\partial_\mu \sum_f \bar\psi_f i\gamma_\mu\gamma_5\psi_f = 2n_{\mathrm{f}}iq$.

The quantity q is called the topological charge density. Continuum gauge fields on topologically non-trivial manifolds (such as the torus T^4 which corresponds to periodic boundary conditions) fall into so-called Chern classes characterized by an integer, the Pontryagin index or

† The gauge fields are normalized here according to $S = -\int d^4x\, G^k_{\mu\nu}G^k_{\mu\nu}/4 + \cdots$ with $G^k_{\mu\nu} = \partial_\mu G^k_\nu - \partial_\nu G^k_\mu + g f_{klm}G^l_\mu G^m_\nu$.

topological charge Q_{top}:

$$Q_{\text{top}} = \int d^4x \, q(x). \tag{8.64}$$

An important example of configurations with topological charge is given by superpositions of (anti)instantons. The latter are solutions of the Euclidean field equations (hence they are saddle points in the path integral) with localized action density, non-perturbative action $S = 8\pi^2/g^2$ and topological charge ± 1. In this context we mention also the Atiyah–Singer index theorem:

$$Q_{\text{top}} = n_+ - n_-, \tag{8.65}$$

where $n_\pm$ are the numbers of zero modes (eigenvectors with zero eigenvalue) of the Dirac operator $\gamma_\mu D_\mu$ with chirality $\gamma_5 = \pm 1$ (cf. problem (iii)).

The significance of all this for our pseudoscalar particle mass spectrum is that the phenomenologically required chiral $U(1)$ breaking is present indeed in *quantum* chromodynamics, provided that gauge-field configurations with topological charge density give sufficiently important contributions to the path integral. The analysis of this is complicated [101] but fortunately there is a simple approximate formula which expresses the effect of the chiral anomaly on the neutral pseudoscalar masses, the Witten–Veneziano formula [102, 103]:

$$\lambda \approx \frac{1}{2f_\pi^2} \chi_{\text{top}}, \quad \text{no quarks.} \tag{8.66}$$

Here λ is the $U(1)$-breaking mass term introduced in (8.50) and χ_{top} is the *topological susceptibility*,

$$\chi_{\text{top}} = \int d^4x \, \langle q(x)q(0) \rangle. \tag{8.67}$$

Note that in (8.66) χ_{top} is to be computed in the pure gauge theory without quarks, although it can of course also be evaluated in the full theory with dynamical fermions. From (8.58) we have $\chi_{\text{top}} \approx (180 \, \text{MeV})^4$.

8.4 Chiral symmetry and the lattice

With Wilson's fermion method chiral symmetry is explicitly broken by two large mass terms $\propto M$ and r/a in the action. With staggered fermions there are not even any flavor indices to act on with chiral transformations (cf. (6.67)). So we have a problem translating the continuum lore in the previous section to the lattice using these fermion formulations. As will be mentioned at the end of this section, this problem can be avoided or at least ameliorated with formulations of the 'Ginsparg–Wilson variety', but an introduction in terms of Wilson fermions is instructive and this will be the focus of our immediate attention.

Let us first derive the Noether currents of chiral symmetry. Consider the fermion part of the action,

$$S_{\mathrm{F}} = \sum_{x\mu f} \tfrac{1}{2}\big[\bar{\psi}_{fx}(r - \gamma_\mu)U_{\mu x}\psi_{fx+\hat{\mu}} + \bar{\psi}_{fx+\hat{\mu}}(r + \gamma_\mu)U_{\mu x}^\dagger\psi_{fx}\big]$$
$$- \sum_{xf} M_f\bar{\psi}_{fx}\psi_{fx}, \tag{8.68}$$

where we have explicitly indicated the flavor index f in addition to x. We make a variation of ψ and $\bar{\psi}$ that looks like a chiral transformation,

$$\psi'_{fx} = V_{fgx}\,\psi_{gx}, \quad \bar{\psi}'_{fx} = \bar{\psi}_{gx}\,\bar{V}_{gxf}, \tag{8.69}$$

in which V has been generalized to depend on the space–time point x:

$$V_{fgx} = \delta_{fg} + i\omega^{\mathrm{L}}_{fgx}P_{\mathrm{L}} + i\omega^{\mathrm{R}}_{fgx}P_{\mathrm{R}} + O(\omega^2) \tag{8.70}$$
$$\equiv \delta_{fg} + i\omega^{\mathrm{V}}_{fgx} + i\omega^{\mathrm{A}}_{fgx}\gamma_5 + \cdots, \tag{8.71}$$
$$\bar{V}_{fgx} = \delta_{fg} - i\omega^{\mathrm{V}}_{fgx} + i\omega^{\mathrm{A}}_{fgx}\gamma_5 + \cdots, \tag{8.72}$$

where $\omega_{fg} = \omega^*_{gf}$ for L, R, V and A. The variation of the action can be written for infinitesimal ω's as

$$\delta S_{\mathrm{F}} = S_{\mathrm{F}}(\psi',\bar{\psi}') - S_{\mathrm{F}}(\psi,\bar{\psi})$$
$$= -\sum_x \big[V^\mu_{fgx}\partial_\mu\omega^{\mathrm{V}}_{fgx} + A^\mu_{fgx}\partial_\mu\omega^{\mathrm{A}}_{fgx}$$
$$+ D^{\mathrm{V}}_{fgx}\omega^{\mathrm{V}}_{fgx} + D^{\mathrm{A}}_{fgx}\omega^{\mathrm{A}}_{fgx} + O(\omega^2)\big] \tag{8.73}$$
$$= \sum_x \big[(\partial'_\mu V^\mu_{fgx} - D^{\mathrm{V}}_{fgx})\omega^{\mathrm{V}}_{fgx} + (\partial'_\mu A^\mu_{fgx} - D^{\mathrm{A}}_{fgx})\omega^{\mathrm{A}}_{fgx}\big].$$
$$\tag{8.74}$$

We recall that ∂_μ and ∂'_μ denote the forward and backward lattice derivatives, $\partial_\mu\omega_x = \omega_{x+\hat{\mu}} - \omega_x$ and $\partial'_\mu\omega_x = \omega_x - \omega_{x-\hat{\mu}}$. In (8.73), the terms without derivatives of ω are due to symmetry breaking, while the terms containing $\partial_\mu\omega$ are a consequence of the fact that ω depends on x – they serve to identify the vector (V^μ) and axial-vector (A^μ) currents. The classical Noether argument can be given as follows: if ψ and $\bar{\psi}$

satisfy the equations of motion, the action is stationary, $\delta S_F = 0$, and consequently

$$\partial'_\mu V^\mu_{fg} = D^V_{fg}, \quad \partial'_\mu A^\mu_{fg} = D^A_{fg}. \tag{8.75}$$

Explicitly we have

$$V^\mu_{fgx} = \frac{1}{2}[\bar{\psi}_{fx}i(\gamma_\mu - r)U_{\mu x}\psi_{gx+\hat{\mu}} + \bar{\psi}_{fx+\hat{\mu}}i(\gamma_\mu + r)U^\dagger_{\mu x}\psi_{gx}], \tag{8.76}$$

$$A^\mu_{fgx} = \frac{1}{2}[\bar{\psi}_{fx}i\gamma_\mu\gamma_5 U_{\mu x}\psi_{gx+\hat{\mu}} + \bar{\psi}_{fx+\hat{\mu}}i\gamma_\mu\gamma_5 U^\dagger_{\mu x}\psi_{gx}], \tag{8.77}$$

$$D^V_{fgx} = i(M_f - M_g)\bar{\psi}_{fx}\psi_{gx}, \tag{8.78}$$

$$D^A_{fgx} = (M_f + M_g)\bar{\psi}_{fx}i\gamma_5\psi_{gx}$$
$$- \frac{r}{2}\sum_\mu [\bar{\psi}_{fx}i\gamma_5(U_{\mu x}\psi_{gx+\hat{\mu}} + U^\dagger_{\mu x-\hat{\mu}}\psi_{gx-\hat{\mu}})$$
$$+ (\bar{\psi}_{fx+\hat{\mu}}U^\dagger_{\mu x} + \bar{\psi}_{fx-\hat{\mu}}U_{\mu x-\hat{\mu}})i\gamma_5\psi_{gx}]. \tag{8.79}$$

We see that, in the flavor-symmetry limit $M_f = M_g = M$, the vector-current divergence $D^V = 0$. For the axial-vector divergence the story is more subtle: we can set all mass parameters M_f and r to zero, in which case $D^A = 0$, but then we get back the species doublers, which is not Wilson's method. To get chiral symmetry without fermion doubling, we have to take the continuum limit. In the classical continuum limit we expect D^A_{fg} to be proportional to the quark masses because then the mass terms in the action reduce to $\int d^4x\, \bar{\psi}m\psi$, by construction (recall (6.58)):[2]

$$D^A_{fg}(x) = (m_f + m_g)\bar{\psi}_f(x)i\gamma_5\psi_g(x) + O(a). \tag{8.80}$$

Hence, the *classical* D^A vanishes in the chiral limit, which is 'Noether's theorem' for Wilson fermions.

In the quantum theory the fields become operators. Their correlation functions can be obtained with the path integral. Consider the expectation value of an arbitrary set of fields $\phi_1 \cdots \phi_n \equiv F$, composed of the fermion fields and/or gauge fields,

$$\langle F \rangle = \frac{1}{Z}\int D\bar{\psi}D\psi\, DU\, e^S\, F, \quad Z = \int D\bar{\psi}D\psi\, DU\, e^S, \tag{8.81}$$

and let us make the transformation of variables (8.69). A transformation of variables cannot change the integrals, so $Z' = Z$ and $\langle F \rangle' = \langle F \rangle$. However, by following how the path-integral measure and the integrant transform we can derive useful relations, called Ward–Takahashi identities. The path-integral measure is *invariant*,

$$D\bar{\psi}' D\psi' \equiv \prod_{xa\alpha f} d\bar{\psi}'_{xa\alpha f}\, d\psi'_{xa\alpha f} = \prod_{xa\alpha f} d\bar{\psi}_{xa\alpha f}\, d\psi_{xa\alpha f} (\det V_x \det \bar{V}_x)^{-1}$$

$$= D\bar{\psi} D\psi, \tag{8.82}$$

because $\det V \det \bar{V} = \det(V\bar{V}) = \det(V_L V_R^\dagger P_L + V_R V_L^\dagger P_R) = \det(V_L V_R^\dagger)$ $\times \det(V_R V_L^\dagger) = \det(V_L V_R^\dagger V_R V_L^\dagger) = \det \mathbb{1} = 1$. On the other hand, the change in the action is given in (8.74) and the fields in F may also change, $F' = F + \sum_{fgx} \omega_{fgx}^A \partial F / \partial_{fgx}^A + \cdots + A \to V$. So we get the identity, e.g. for a chiral transformation,

$$0 = \frac{\partial}{\partial \omega_{fgx}^A} \langle F \rangle' = \frac{\partial}{\partial \omega_{fgx}^A} \left(\frac{1}{Z'} \int D\bar{\psi}' D\psi' DU\, e^{S'} F' \right)$$

$$= \frac{1}{Z} \int D\bar{\psi} D\psi\, DU\, \frac{\partial}{\partial \omega_{fgx}^A} \left(e^{S'} F' \right) = \left\langle \frac{\partial S}{\partial \omega_{fgx}^A} F + \frac{\partial F}{\partial \omega_{fgx}^A} \right\rangle$$

$$= \left\langle (\partial'_\mu A^\mu_{fgx} - D^A_{fgx}) F + \frac{\partial F}{\partial \omega_{fgx}^A} \right\rangle. \tag{8.83}$$

The content of such relations may be studied in perturbation theory. To one-loop order this can be done in the way seen in section 3.4 and the problems in section 6.6. A crucial example is the case $F = G_{\kappa x} G_{\lambda y}$, for which $\partial F / \partial \omega_{fg}^A = 0$ since F consists only of gluon fields, which leads to triangle-diagram contributions of the type shown in figure 8.1. A calculation [70] shows that, for this case, $D^A_{fg} \to (m_f + m_g) \bar{\psi}_f i \gamma_5 \psi_g + \delta_{fg} 2iq$ in the continuum limit. The topological-charge-density contribution is due to the Wilson mass term and the coefficient of q is formally $\propto r$, but actually independent of r, provided that it is non-zero.

Another example is $F = \psi_{fx} \bar{\psi}_{gy}$, which leads to the conclusion that, for this case, $D^A_{fg} \to (m_f + m_g) \kappa_P \bar{\psi}_f i \gamma_5 \psi_g - (\kappa_A - 1) \partial_\mu A^\mu_{fg}$, where κ_P and κ_A are finite renormalization constants of order g^2 (cf. [70, 109, 104]). The topological charge density does not contribute here in this order because it is already of order g^2.

At one-loop order we get the same contributions as those found in continuum perturbation theory because the bare vertex functions reduce to the continuum ones (in the balls around the origin of the loop-momentum integration) in the classical continuum limit. There are also differing contributions, which are, however, only contact terms.† These translate into the finite renormalization constants κ.‡ The anomaly is a singlet under flavor transformations (i.e. $\propto \delta_{fg}$) because the r-mass term is a flavor singlet. Note that, by a global finite chiral transformation, we could transform $r\delta_{fg} \to r(\bar{V}V)_{fg}$, implying that $M_{\rm c} \propto \bar{V}V$. However, this is merely a change of reference frame and the physics cannot depend on it. The quark masses have to be identified as the mismatch between M and $M_{\rm c}$.

The above examples show the phenomenon of operator mixing: operators (fields) with the same quantum numbers tend to go over into linear combinations of each other in the continuum limit (the scaling region). Such mixing is restricted by the symmetries of the model and there is more mixing on the lattice than there is in the continuum because there is less symmetry on the lattice. The κ's above are due to the chiral-symmetry breaking of the Wilson mass term at non-zero lattice spacing. On general grounds of scaling and universality one assumes these results to be qualitatively valid also non-perturbatively. One introduces renormalized field combinations that are finite as $a \to 0$ that satisfy some standard normalization conditions. Before writing these down, let us introduce a lattice field that reduces to the topological charge density q in the classical continuum limit. There are many of course, as usual, e.g. the one introduced in [105],

$$
q_x = -\frac{1}{32\pi^2} \left[\sum_{\kappa\lambda\mu\nu} \epsilon_{\kappa\lambda\mu\nu} \mathrm{Tr}\left(U_{\kappa\lambda x} U_{\mu\nu x}\right) \right]_{\mathrm{symmetrized}} , \qquad (8.84)
$$

where the symmetrization is such that q_x transform as a scalar under lattice rotations. Denoting the renormalized fields by a 'bar', they can be written as [104]

$$
\bar{A}^\mu_{fg} = \kappa_{\rm A} A^\mu_{fg} + \delta_{fg}(Z_{\rm A} - 1)\kappa_{\rm A} \frac{1}{2n_{\rm f}} \sum_f \partial'_\mu A^\mu_{ff}, \qquad (8.85)
$$

$$
\bar{D}^{\rm A}_{fg} = D^{\rm A}_{fg} + (\kappa_{\rm A} - 1)\partial'_\mu A^\mu_{fg} \\
+ \delta_{fg}(Z_{\rm A} - 1)\kappa_{\rm A} \frac{1}{2n_{\rm f}} \sum_f \partial'_\mu A^\mu_{ff}, \qquad (8.86)
$$

---

† Recall that contact corresponds in momentum space to polynomials in the momenta, of degree less than or equal to the mass dimension of the vertex function under consideration.

‡ In the literature these κ's are often denoted by Z, which notation we have reserved for renormalizations diverging when $a \to 0$.

$$\partial'_\mu \bar{A}^\mu_{fg} = \bar{D}^A_{fg}, \tag{8.87}$$

$$\bar{q} = \kappa_q q - i(Z_A - 1)\kappa_A \frac{1}{2n_f} \sum_f \partial'_\mu A^\mu_{ff}, \tag{8.88}$$

where Z_A is a diverging renormalization constant of order g^4. The operator subtractions $\propto (Z_A - 1)$ are suggested by analysis at two-loop order in the continuum [106]. In the quenched approximation $Z_A = 1$. In the scaling region

$$\bar{D}^A_{fg} = (m_f + m_g)\kappa_P \bar{\psi}_f i\gamma_5 \psi_g + \delta_{fg} 2i\bar{q} + O(a), \tag{8.89}$$

with $m_f = M_f - M_c$. Similar analysis of Ward–Takahashi identities shows that the vector currents V^μ_{fg} need no finite renormalization, $\bar{V}^\mu_{fg} = V^\mu_{fg}$, $\kappa_V = 1$. The reason is that they are conserved if $m_f = m_g$ even for $a \neq 0$.

The implications of the lattice Ward–Takahashi identities can of course be studied also non-perturbatively. As a first step one can use only external gauge fields with $F = 1$ and test the index theorem (8.65), using topologically non-trivial gauge fields transcribed from the continuum to the lattice [107, 108]. Adding dynamical gauge fields, we can then also use the Ward–Takahashi identities to determine the renormalization constants κ in the quenched approximation [104, 70, 109]. The computation of the topological susceptibility turns out to be complicated by the fact that $\langle \bar{q}_x \bar{q}_y \rangle$ has divergent contact terms that severely influence the value of $\sum_x \langle \bar{q}_x \bar{q}_0 \rangle$. One can try to subtract this contribution,

$$\chi_{\text{top}} = \frac{1}{V} \sum_{xy} \langle \bar{q}_x \bar{q}_y \rangle_U - \text{contact contribution} \tag{8.90}$$

(assuming periodic boundary conditions, space–time volume $V \to \infty$), but it is hard to define it unambiguously [110]. In practice it appears to work well [111]. By 'cooling' the gauge fields after they have been generated by a Monte Carlo process this problem can be reduced further (see e.g. [112] and also [108]).

A different approach to the topological susceptibility is to accept that the configurations in the path integral are inherently not smooth functions of space–time and to avoid defining a topological integer from a collection of wildly fluctuating lattice variables. Instead, one can return to the physical role played by χ_{top} and derive the Witten–Veneziano formula entirely within the lattice formulation. This can be done by studying the pseudoscalar meson contribution in the $\langle \bar{A}^\mu \bar{A}^\nu \rangle$

Chiral symmetry

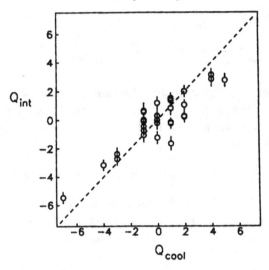

Fig. 8.2. Correlation between 'fermionic' and 'cooling' topological charge assignments for 32 $SU(3)$ gauge-field configurations at $\beta = 6.0$. From [117].

and $\langle \bar{D}^A \bar{D}^A \rangle$ correlators. The analysis is subtle [104] but results in the simple formula

$$\chi_{\text{top}} = \frac{\kappa_{\text{P}}^2 m_f^2}{V} \langle \text{Tr} \, [\gamma_5 S_{ff}(U)] \, \text{Tr} \, [\gamma_5 S_{ff}(U)] \rangle_{\text{U}} . \qquad (8.91)$$

Here $S_{ff}(U)$ is the fermion propagator in the gauge field U and the trace is over all non-flavor indices (x, a and α). The large-n_c limit is not taken in this derivation, only the quenched approximation. From this, the formula in terms of $\bar{q}$ can be understood from (8.87), (8.89), and $\sum_x \partial'_\mu \bar{A}^\mu = 0$ for periodic boundary conditions. A derivation for staggered fermions can also be given [113]. The limit $m_f \to 0$ is needed in order to avoid divergences (this limit must be carefully controlled by taking m_f at the lower end of a scaling window that extends to zero as $a \to 0$).

In the two-dimensional $U(1)$ model the properties of (8.91) have been studied and compared with the index theorem as well as with definitions of χ_{top} in terms of the gauge field only [114, 115]. The staggered form was explored in numerical $SU(3)$ simulations [116, 117]. Figure 8.2 shows that the individual topological charges obtained with this 'fermionic method' are at $\beta = 6/g^2 = 6.0$ already quite correlated to the charges obtained with the cooling method. This is expected to improve at higher β but at lower β the gauge fields are too 'rough' on the lattice scale for notions of topology to make sense (also, the staggered-fermion renormalization factor κ_P becomes uncannily very large [116]). The resulting $\chi_{\text{top}} \approx (154\pm17 \text{ MeV})^4$ seems a bit low compared with the experimental value following from the Witten–Veneziano formula $(180 \text{ MeV})^4$, but this may be due to the somewhat low value $a^{-1} = 1900$ MeV used for conversion to physical units. Using $a^{-1} = 2216$ MeV inferred from the values 1934 MeV ($\beta = 5.9$) and 2540 MeV ($\beta = 6.1$) recorded in [98] would give $\chi_{\text{top}} \approx (180 \pm 20 \text{ MeV})^4$.

By contracting the currents with the $n_{\text{f}}^2 - 1$ $SU(n_{\text{f}})$ generators $\lambda_{fg}^k/2$, we can form the left- and right-handed currents $j_{\mu k}^{\text{L,R}} = (\bar{V}_{fg}^\mu \pm \bar{A}_{fg}^\mu) \times (\lambda_k)_{fg}/4$. According to (8.78) and (8.89), these currents and the $U(1)$ vector current $V_\mu = \sum_f V_{ff}^\mu$ are conserved in the limit $m_f \to 0$. Further Ward–Takahashi identities can be derived to fix renormalization constants and ensure that the currents satisfy 'current algebra' [118]. The corresponding charges would then satisfy the algebra of generators of $SU(n_{\text{f}}) \times SU(n_{\text{f}})$, were it not that the symmetry is supposed to be broken spontaneously. It should also be possible to introduce the QCD theta parameter (cf. problem (iv)).

From the chiral-symmetry point of view there are now much better lattice fermion methods. Ginsparg and Wilson made a renormalization-group 'block-spin' transformation for fermions from the continuum to the lattice, paying special attention to chiral symmetry [124]. More recently such transformations were studied in search of 'perfect actions' [125]. The continuum action is chirally symmetric for zero mass parameters but this symmetry is hidden in the resulting lattice action, because the blocking transformation to the lattice breaks chiral symmetry to avoid fermion doubling. Writing the massless fermion action as $S_{\text{F}} = -\bar{\psi}D\psi$, chiral symmetry in the continuum can be expressed as $\gamma_5 D + D\gamma_5 = 0$. On the lattice there is a remnant of this: the blocked D satisfies the Ginsparg–Wilson relation

$$\gamma_5 D + D\gamma_5 = aD\, 2R\gamma_5\, D, \qquad (8.92)$$

where we used matrix notation also for the space–time indices; R is a matrix commuting with γ_5 that enters in the renormalization-group blocking transformation. It is *local*, which means that R_{xy} falls off exponentially fast as $|x - y| \to \infty$ (on the lattice scale, in physical units it resembles a delta function). So D_{xy} practically anticommutes with γ_5 for physical separations, provided that it is itself local, as it should be (this is a basic requirement for universality). Taking (8.92) as a starting point, one can take $R_{xy} = \frac{1}{2}\delta_{xy}$. Dirac matrices D_{xy} satisfying

(8.92) are complicated, because for given x all y contribute, albeit with exponentially falling magnitude as $|x - y|$ increases. An explicit solution [126], arrived at via the 'overlap' approach to chiral gauge theories (see the next section), has the form

$$aD = 1 - A(A^\dagger A)^{-1/2}, \quad A = 1 - aD_{\rm W}, \quad R = \tfrac{1}{2}, \tag{8.93}$$

where $D_{\rm W}$ is Wilson's lattice Dirac operator with zero bare mass ($r = 1$, $M = 4/a$). Adding mass terms the resulting lattice QCD action has very nice properties with respect to broken chiral symmetry and topology, which can be studied again by deriving Ward–Takahashi identities [130].

Moreover, the resulting action has (for $m = 0$) an exact chiral symmetry [131] under

$$\delta\psi = i\omega\gamma_5\big(1 - \tfrac{1}{2}aD\big)\psi, \quad \delta\bar\psi = i\bar\psi\big(1 - \tfrac{1}{2}aD\big)\omega, \tag{8.94}$$

with infinitesimal ω_{fg}. (Note that such a finite chiral transformation is non-local as it involves arbitrarily high powers of D.) The chiral anomaly in this formulation comes from a non-invariance of the fermion measure [131], similar to continuum derivations [132]. Domain-wall fermions [128, 129] are closely related. At the time of writing the research into these directions is very active; for a review, see [135]. Applications to the topological susceptibility can be found in [136, 137].

8.5 Spontaneous breaking of chiral symmetry

We now turn to the question of *spontaneous* chiral-symmetry breaking. One would like to compute the expectation value of the order field $\bar\psi_f P_{\rm L}\psi_g$ at vanishing quark masses and verify that $SU(n_{\rm f}) \times SU(n_{\rm f})$ symmetry is broken spontaneously to $SU(n_{\rm f})$. As for the $O(n)$ model (cf. (3.157)), this could be done by introducing explicit symmetry-breaking quark masses and studying the infinite-volume limit.

However, with Wilson fermions we cannot simply use $\bar\psi_f P_{\rm L}\psi_g$ as an order field because the cancellation of the chiral-symmetry breaking by the M and r terms is a subtle issue. Even for free fermions $\langle\bar\psi_f P_{\rm L}\psi_g\rangle \neq 0$ at $m_f = M_f - 4r/a = 0$: it diverges in the continuum limit (cf. problem (ii)). The chiral-symmetry breaking causes $\bar\psi_f P_{\rm L}\psi_g$ to mix with the unit operator, with a coefficient $c(g^2, m)\,\delta_{fg} = [c_0(g^2)a^{-3} + c_1(g^2)ma^{-2} + c_2(g^2)m^2a^{-1}]\,\delta_{fg}$ that diverges in the limit $a \to 0$ (for simplicity we assume here all quark masses to be equal). The identification of $c(g^2, m)$, and a computation of the subtracted expectation value $\langle\bar\psi_f P_{\rm L}\psi_g\rangle - c(g^2, m)\delta_{fg}$ in the limit of zero quark mass is a hazardous endeavor, because several powers of a^{-1} have to cancel out, and, moreover, because the gauge coupling g^2 also depends on a.

On the other hand, we have seen (section 7.5) that the relation (8.36), i.e. $m_{fg}^2 \approx B(m_f + m_g)$, $B = 2v/f_\pi^2$, is borne out by the numerical results when $f \neq g$. So, using this relation, we could *define* $\langle \bar{\psi}\psi \rangle$ by $\sum_f \langle \bar{\psi}_f \psi_f \rangle = -n_f B f_\pi^2$, with $m_u = m_d \equiv m_{ud} \to 0$, or $\langle \bar{\psi}_u \psi_u \rangle = -f_\pi^2 m_\pi^2/2m_{ud}$. Using renormalized quark masses instead of the bare m_{ud} would give a renormalized B and a correspondingly renormalized $\langle \bar{\psi}_f \psi_f \rangle$. Note that $m_{ud} \langle \bar{\psi}_u \psi_u \rangle$ should be renormalization-group invariant. Using e.g. $m_{ud} = 3.4$ MeV (the result of [97]), the value of $\langle \bar{\psi}_u \psi_u \rangle$ is about $(290\,\mathrm{MeV})^3$ in the MS-bar scheme on the scale $\mu = 2$ GeV.

We may appeal to continuity at any fixed gauge coupling $0 < g < \infty$ by sending the symmetry-breaking parameters M and r to zero and studying spontaneous breaking of chiral symmetry there. Actually, at $r = M = 0$ the staggered-fermion form (6.66) of the action is more appropriate and it shows that the Dirac labels are to be interpreted as flavor indices. At $M = r = 0$ the symmetry of the action enlarges to $U(4n_f) \times U(4n_f)$. Combining the Dirac (α) and flavor (f) indices into one label $A = (\alpha, f)$ the transformation is

$$\chi_{Ax} \to \left(\mathcal{V}_{AB}^{\mathrm{L}} \frac{1 - \epsilon_x}{2} + \mathcal{V}_{AB}^{\mathrm{R}} \frac{1 + \epsilon_x}{2} \right) \chi_{Bx},$$

$$\bar{\chi}_{Ax} \to \bar{\chi}_{Bx} \left(\mathcal{V}_{BA}^{\mathrm{R}\dagger} \frac{1 - \epsilon_x}{2} + \mathcal{V}_{BA}^{\mathrm{L}\dagger} \frac{1 + \epsilon_x}{2} \right) \tag{8.95}$$

with $\epsilon_x = (-1)^{x_1 + \cdots x_4}$. Moreover, in the scaling region at weak coupling the *staggered-fermion flavors* also emerge, implying a further multiplication of the number of flavors by four. With such a large number of flavors (i.e. $16n_f$) and only three colors, asymptotic freedom is lost as soon as $n_f > 1$ (recall (7.54)) and we can expect continuity in $M, r \to 0$ only if we consider a sufficiently large number of colors n_c. Assuming this to be the case, we can get analytic insight at strong coupling [119, 120, 83, 84, 82, 121].

At strong gauge coupling and for a large number of colors the exact continuous symmetry breaks spontaneously as $U(4n_f) \times U(4n_f) \to U(4n_f)$, resulting in $16n_f^2$ NG bosons. The baryons acquire a mass $\propto n_c$ from the spontaneous symmetry breaking. Suppose now $n_f = 3$. Turning on the symmetry-breaking parameters M and r, it is possible to keep the pions, kaons and eta massless by choosing $M = M_c(g, r)$ and in the process all other NG bosons become massive. We need to keep $M - M_c(g, r)$ infinitesimally positive to let the order field $\bar{\psi}_f P_L \psi_g$ acquire its expectation value in the direction $-v\delta_{fg}$, with real positive v. Otherwise we might induce complex v, which corresponds to non-zero $\langle \bar{\psi}_f i\gamma_5 \psi_g \rangle$ and spontaneous breaking of parity [119, 122]. The situation is similar in the model using continuous time (the Hamiltonian method), in which the symmetry breaking at strong coupling is actually $U(4n_f) \to U(2n_f) \times U(2n_f)$. At non-zero r the $U(1)$ problem is also qualitatively resolved by giving the flavor-singlet boson a (small) non-zero mass [119, 122, 123].

So in this way, connecting with $M = r = 0$, we can understand spontaneous breaking of chiral symmetry in multicolor QCD with Wilson fermions. However, it is conceptually simpler to study the corresponding order field for staggered fermions.

The staggered-fermion action (6.67) has for $m = 0$ a chiral $U(1) \times U(1)$ symmetry, which is (8.95) with phase factors $\mathcal{V}^{L,R}$ (since there is no spin–flavor index A to act on). The axial $U(1)$ transformation contained in this $U(1) \times U(1)$, i.e. $\mathcal{V}^L = \mathcal{V}^{R*} = \exp(i\omega^A)$, is in the staggered-fermion interpretation [74] a flavor-*non-singlet* transformation, of the form $\exp(i\omega^A \xi_5)$ with $\text{Tr}\,\xi_5 \neq 0$. In the scaling region, where the symmetry enlarges to $SU(4) \times SU(4) \times U(1)_V$, this ξ_5 is a linear combination of the generators of $SU(4)$. So it is natural to study spontaneous breaking of this $U(1)$ remnant of $SU(4) \times SU(4)$ chiral symmetry. A suitable order field for this symmetry is the coefficient of the quark mass m in the action, i.e. $\bar{\chi}_x \chi_x$, which together with $\epsilon_x \bar{\chi}_x \chi_x$ forms a doublet under the chiral $U(1)$. In the scaling region $\bar{\chi}_x \chi_x \to \sum_{f=1}^4 \bar{\psi}_f(x)\psi_f(x)$ and $\epsilon_x \bar{\chi}_x \chi_x \to \sum_{fg} \bar{\psi}_f(x)\xi_{5fg}\gamma_5\psi_g(x)$.

A definition of $\Sigma \equiv -\langle \bar{\chi}\chi \rangle$ in which the quark mass is introduced as a symmetry breaker, which is to be taken to zero after taking the infinite-volume limit, as in (3.157) for the $O(n)$ model, is hard to implement in practice. This can be circumvented by using a method based on the eigenvalues of the Dirac operator, which we shall denote by $D(U)$, where U is a given gauge-field configuration. In the continuum D is anti-Hermitian, $D(U) = -D(U)^\dagger$, and therefore its eigenvalues are purely imaginary. On the lattice the staggered-fermion Dirac matrix

$$D(U)_{xa,yb} = \sum_\mu \eta_{\mu x}[(U_{xy})_{ab}\delta_{x+\hat{\mu},y} - (U_{yx})_{ab}\delta_{y+\hat{\mu},x}] \tag{8.96}$$

has the same property (unlike the Wilson–Dirac operator $D(U) = \slashed{D}(U) + M - W(U)$ which is the sum of an anti-Hermitian and a Hermitian matrix). Let u_r denote the complete orthonormal set of eigenvectors with eigenvalues $i\lambda_r$,

$$D\,u_r = i\lambda_r\,u_r, \qquad u_r^\dagger u_s = \delta_{rs}, \qquad \sum_r u_r u_r^\dagger = \mathbb{1}. \tag{8.97}$$

The matrix $\epsilon_{xy} = \epsilon_x \delta_{xy}$ anticommutes with D, $D\epsilon = -\epsilon D$, so

$$D\,\epsilon u_r = -i\lambda_r\,\epsilon u_r, \tag{8.98}$$

and for every eigenvalue λ_r there is also an eigenvalue $-\lambda_r$. The expectation value of the order field at finite quark mass, $\Sigma \equiv -\langle \bar{\chi}_x \chi_x \rangle$, can be written as

$$\Sigma = \frac{1}{V} \sum_x \langle \chi_x \bar{\chi}_x \rangle = \frac{1}{V} \langle \mathrm{Tr}\left[(D+m)^{-1}\right] \rangle_U$$

$$= \frac{1}{V} \left\langle \sum_r \frac{1}{i\lambda_r + m} \, \mathrm{Tr}\left(u_r u_r^\dagger\right) \right\rangle_U = \frac{1}{V} \left\langle \sum_r \frac{1}{i\lambda_r + m} \right\rangle_U$$

$$= \frac{1}{V} \left\langle \sum_r \frac{m}{\lambda_r^2 + m^2} \right\rangle_U. \tag{8.99}$$

In terms of the spectral density $\rho(\lambda)$,

$$\rho(\lambda) = \frac{1}{V\,\Delta\lambda} \langle n(\lambda + \Delta\lambda, \lambda) \rangle_U, \qquad \Delta\lambda \to 0, \tag{8.100}$$

where $n(\lambda + \Delta\lambda, \lambda) = \sum_r \theta((\lambda + \Delta\lambda - \lambda_r)\theta(\lambda_r - \lambda)$ is the number of eigenvalues of $D(U)$ in the interval $(\lambda, \lambda + \Delta\lambda)$, this can be written as [138]

$$\Sigma = \lim_{m \to 0} \lim_{V \to \infty} \int d\lambda\, \rho(\lambda)\, \frac{m}{\lambda^2 + m^2} \tag{8.101}$$

$$= \pi\rho(0). \tag{8.102}$$

Here we used the identity $\lim_{\epsilon \to 0} \epsilon/(x^2 + \epsilon^2) = \pi\delta(x)$. Note that $\rho(\lambda)$ depends on the gauge coupling, the dynamical quark mass and the volume V; it furthermore satisfies $\rho(\lambda) = \rho(-\lambda)$ because of (8.98).

The spectral density can be computed numerically by counting the number of eigenvalues in small bins and figure 8.3 shows an example for the gauge group $SU(2)$ in the quenched approximation. The quantity $\rho(\lambda)/V$ in the plot is our density $\rho(\lambda)$ in lattice units, i.e. $a^3\rho(\lambda)$. The value $\rho(0)$ may be determined by extrapolating $\lambda \to 0$, it is nearly equal to the value in the first bin. The resulting $a^3\Sigma$ drops rapidly from the value $0.1247(22)$ to $0.00863(48)$ as β is increased from 2.0 to 2.4 and the lattice spacing decreases accordingly. Actually, the value $\beta = 2.0$ is near

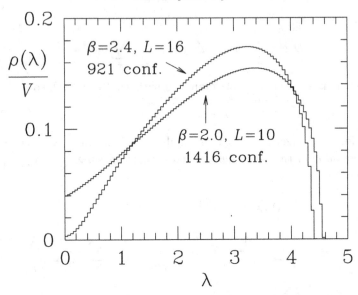

Fig. 8.3. The quenched spectral density in lattice units of the $SU(2)$ staggered-fermion matrix for two values of $\beta = 4/g^2$ and lattice volumes $V = L^4$. The number of gauge-field configurations used is also indicated. From [142].

the edge of the scaling region on the strong-coupling side, while $\beta = 2.4$ is more properly in the scaling region[3] (see e.g. figure 8 in [69]).

The volume dependence of Σ obtained this way is expected to be small. This can be made more precise by using scaling arguments based on a remarkable connection with random-matrix theory (for a review see [140]). From (8.100) and (8.102) we see that, in the neighborhood of the origin, $\rho(\lambda)$ behaves like $1/[V\Sigma d(\lambda)]$, with $d(\lambda)$ the average distance between two eigenvalues. This observation leads one to define the *microscopic spectral density* [141]

$$\rho_{\rm s}(\zeta) = \frac{1}{\Sigma} \rho\left(\frac{\zeta}{\Sigma V}\right), \qquad (8.103)$$

in which the region around the origin is blown up by the factor ΣV. The function $\rho_{\rm s}(\zeta)$ is predicted to be a universal function in random-matrix theory depending only on the gauge group and the representation carried by the fermions, provided that it is evaluated for gauge fields with fixed topological charge $Q_{\rm top} = \nu$. For example for $SU(n_{\rm c} > 2)$ and $n_{\rm f}$

dynamical fermions it is given by

$$\rho_s^{(\nu)}(\zeta) = \frac{\zeta}{2}\left[J_{n_f+\nu}(\zeta)^2 - J_{n_f+\nu-1}(\zeta)J_{n_f+\nu+1}(\zeta)\right], \tag{8.104}$$

where J is the Bessel function. So, by fitting $\rho^{(\nu)}(\lambda)$ according to (8.103) and (8.104) with only one free parameter (Σ), one obtains the *infinite-volume* value of Σ.

Zero modes corresponding to the index theorem should be ignored here. This is not easy with staggered fermions as the would-be zero modes fluctuate away from zero and can be identified only by the expectation value of the 'staggered γ_5' (cf. (8.134) and (8.135)) [107, 113, 114, 116].

The prediction (8.104) works well using staggered fermions and $SU(2)$ [142] or $SU(3)$ [143] quenched ($n_f = 0$) gauge-field configurations at relatively strong gauge coupling and selecting[4] $\nu = 0$. The dependence on the fermion representation and the pattern of chiral-symmetry breaking is studied for various gauge groups in [144]. A (finite-temperature) study with $n_f = 2$ dynamical fermions is given in [145].

A recent study [146] using related finite-size techniques with Neuberger's Dirac operator (8.93) in quenched $SU(3)$ at $\beta = 5.85$ gave the result $a^3\Sigma = 0.0032(4)$. A further non-perturbative computation [147] of the appropriate multiplicative renormalization factor then allows conversion value $\Sigma_{\overline{\rm MS}}(\mu = 2\,{\rm GeV}) \approx (270\,{\rm MeV})^3$ in the MS-bar scheme.

8.6 Chiral gauge theory

In QED and QCD the representation of the gauge group carried by all left- and right-handed fields is *real* up to equivalence. For example, in QCD, let Ω be the fundamental representation of $SU(3)$. The left-handed fields are $\psi_L = P_L\psi$ and $(\bar{\psi}_R C)^{\rm T} = P_L(\bar{\psi}C)^{\rm T}$, with C the charge-conjugation matrix (cf. appendix D), while the right-handed fields are $\psi_R = P_R\psi$ and $(\bar{\psi}_L C)^{\rm T} = P_R(\bar{\psi}C)^{\rm T}$. The fields transform as

$$\psi_L \to \Omega\psi_L, \quad (\bar{\psi}_R C)^{\rm T} \to \Omega^*(\bar{\psi}_R C)^{\rm T}, \quad {\rm left}; \tag{8.105}$$

$$\psi_R \to \Omega\psi_R, \quad (\bar{\psi}_L C)^{\rm T} \to \Omega^*(\bar{\psi}_L C)^{\rm T}, \quad {\rm right}. \tag{8.106}$$

Taking ψ_L and $\bar{\psi}_R$ in pairs, the representation of the gauge group has the form of a direct sum $\Omega \oplus \Omega^*$, which is real up to the equivalence transformation $\Omega \oplus \Omega^* \to \Omega^* \oplus \Omega$.

The fundamental representation of $U(1)$, a phase factor, is evidently complex, but the fundamental representation of $SU(2)$ is real up to equivalence: $\Omega^* = \exp(i\omega_k \frac{1}{2}\sigma_k)^* = \sigma_2 \Omega \sigma_2$. It is not difficult to see, e.g. by looking at the element $\exp(i\omega_8 \frac{1}{2}\lambda_8)$, that the fundamental representation of $SU(3)$ is complex. The adjoint representation of $SU(n)$ is real for all n.

Chiral gauge theories are models in which the representation of the gauge group is truly *complex* (no reality up to equivalence). The Standard Model, which has the gauge group $U(1) \times SU(2) \times SU(3)$, is a chiral gauge theory, as can be seen by looking at the $U(1)$ charges of the left- and right-handed fields. Since this model is able to describe all known interactions up till now, it is evidently desirable to give it a non-perturbative lattice formulation.† This turns out to be very difficult.

To get a glimpse of the problem, consider a $U(1)$ model with continuum action

$$S_{\rm F} = - \int d^4x \, \bar{\psi}_{\rm L} \gamma^\mu (\partial_\mu - i g q_{\rm L} A_\mu)\psi_{\rm L} + {\rm L} \to {\rm R}, \qquad (8.107)$$

assuming for the moment no further quantum numbers (no 'flavors'). The fields transform as

$$\psi_{\rm L} \to e^{i\omega q_{\rm L}}\psi_{\rm L}, \quad \bar{\psi}_{\rm R} \to e^{-i\omega q_{\rm R}}\bar{\psi}_{\rm R}, \text{ left;} \qquad (8.108)$$

$$\psi_{\rm R} \to e^{i\omega q_{\rm R}}\psi_{\rm R}, \quad \bar{\psi}_{\rm L} \to e^{-i\omega q_{\rm L}}\bar{\psi}_{\rm L}, \text{ right,} \qquad (8.109)$$

and we see, e.g. from the pair $\psi_{\rm L}$ and $\bar{\psi}_{\rm R}$, that the model is chiral if the charges $q_{\rm R}$ and $q_{\rm L}$ are not equal. Assuming this to be the case, it follows that $\bar{\psi}\psi = \bar{\psi}_{\rm R}\psi_{\rm L} + \bar{\psi}_{\rm L}\psi_{\rm R}$ is not gauge invariant. Consequently there can be no mass term for the fermions. We also cannot use $P_{\rm L} + P_{\rm R} = 1$ and eliminate γ_5 from the action. So the gauge-field couples also to an axial-vector current (there is a term $\bar{\psi}i\gamma^\mu\gamma_5\psi A_\mu$ in the action), instead of only to vector currents as in QED and QCD. These features are generic for chiral gauge theories: no mass terms and axial-vector currents that are dynamical (rather than being just symmetry currents of global chiral symmetry). With γ_5 prominent in the vertex functions we may expect chiral anomalies to play a role. This has been analyzed in perturbation theory in the continuum, with the conclusion that the above model is unsatisfactory because gauge invariance is spoilt by anomalies due to contributions involving triangle diagrams (cf. figure 8.1). These problems can be avoided by extending the model to contain more than one 'flavor', with charges $q_{{\rm L}f}$ and $q_{{\rm R}f}$, such that the anomalies cancel out between the different flavors, which requires $\sum_f (q_{{\rm L}f}^3 - q_{{\rm R}f}^3) = 0$. The model

---

† We consider $U(1)$-neutral right-handed neutrino fields $\psi_{\rm R}$ (and $\bar{\psi}_{\rm R}$) as part of the Standard Model.

and its representation of the gauge group are then called 'anomaly-free'. Such considerations played an important role in the construction of the Standard Model. It was noticed that the anomalies in the lepton sector could cancel out against those in the quark sector [148, 151]. This is how the Standard Model is anomaly-free.

We continue with the choice of integer $q_L = q$, $q_R = 0$. With just one flavor the continuum model is then anomalous. Let us see what happens if we put the above model naively on the lattice.

A Euclidean naive lattice action is easy to write down:

$$S_F = -\sum_{x\mu} \tfrac{1}{2}\big[\bar\psi_x\gamma_\mu(U_{\mu x}P_L + P_R)\psi_{x+\hat\mu} - \bar\psi_{x+\mu}\gamma_\mu(U_{\mu x}P_L + P_R)\psi_x\big],$$

(8.110)

with a path integral

$$Z = \int D\bar\psi D\psi\, DU\, e^S,$$

(8.111)

in which $S = S_F + S_U$ with S_U the usual plaquette action. The lattice action and measure are gauge invariant (for the fermion measure this follows from (8.82) with $V_{Lx} = \exp(i\omega_x)$, $V_{Rx} = 1$). In this model the right-handed ψ_R and the left-handed $(\bar\psi_R C)^T$ are just free fields, they are not coupled to the gauge fields. However, the species-doubling phenomenon induces 16 fermion flavors in the scaling region. What are the charges of these fermions?

To answer this question consider a fermion line in a diagram with a gauge-field line attached to it. The corresponding mathematical expression is

$$\cdots S(p)V_\mu(p,q;k)S(q)\cdots,$$

(8.112)

where $S(p)$ is the massless naive fermion propagator and $V_\mu(p,q;k)$ the bare vertex function for the model ($p = q+k$). Such vertex functions have been determined in problem (i) in section 6.6 for the case of QED, and to get these for the present case we only have to make the substitution $g\gamma_\mu \to g\gamma_\mu P_L$ in (6.99), giving

$$V_\mu(p,q;k) = ig\gamma_\mu P_L \tfrac{1}{2}\big(e^{iaq_\mu} + e^{iap_\mu}\big).$$

(8.113)

To interpret this expression in the scaling region for fermion species A we use (6.26) and (6.31) and substitute $p \to k_A + p$ and $q \to k_A + q$ into

(8.112) $(k_A = \pi_A/a)$,

$$\cdots S(k_A + p)V_\mu(k_A + p, k_A + q; k)S(k_A + q) \cdots \tag{8.114}$$

$$= \cdots S_A^\dagger \left[\frac{-i\gamma_\kappa p_\kappa}{p^2} \, gi\gamma_\mu \tfrac{1}{2}(1 - \epsilon_A\gamma_5) \frac{-i\gamma_\lambda q_\lambda}{q^2} + O(a) \right] S_A \cdots,$$

where we used (6.30) and (6.28) for the terms not involving γ_5 and

$$\epsilon_A\gamma_\mu\gamma_5 = S_A\gamma_\mu\gamma_5 S_A^\dagger \, \cos(\pi_{A\mu}). \tag{8.115}$$

Using (6.29) we find $\epsilon_A = +1$ for $\pi_A = \pi_0$, $\epsilon_A = -1$ for $\pi_A = \pi_1, \ldots,$
π_4, $\epsilon_A = +1$ for $\pi_A = \pi_{12}, \ldots, \pi_{34}$, $\epsilon_A = -1$ for $\pi_A = \pi_{123}, \ldots, \pi_{234}$ and
$\epsilon_A = +1$ for $\pi_A = \pi_{1234}$, such that

$$\sum_A \epsilon_A = 1 - 4 + 6 - 4 + 1 = 0. \tag{8.116}$$

From (8.114) we conclude that in the scaling region we have eight continuum fields with $q_L^{\text{cont}} = 1$ ($\epsilon_A = 1$), and eight with $q_R^{\text{cont}} = 1$ ($\epsilon_A = -1$), in addition to the uncharged fields: the lattice has produced flavors (the species doublers) such that the anomalies cancel out.[5] However, since all the q_L^{cont} and q_R^{cont} are equal, the model is not a chiral gauge theory! It is just QED with eight equal-mass Dirac fermions (plus eight neutral Dirac fermions).

A natural suggestion for a lattice formulation of the Standard Model is to give the doubler fermions masses of order of the lattice cutoff through Wilson-type Yukawa couplings with the Higgs field [119, 149, 150]. Because the Standard Model is anomaly-free the set of doublers in such a formulation is anomaly-free too: the set of 15 doublers of some fermion contributes to anomalies with the same strength as this fermion (opposite in sign, $\sum_{A=2}^{16} \epsilon_A = -1$). Insofar as anomalies are concerned there is no objection to the decoupling of the doublers. Other objections [151, 152], namely that masses of the order of the cutoff might not be possible because renormalized couplings cannot be arbitrarily strong (triviality is expected to play a role here), do not apply if new phases come into play. This is indeed the case. On turning on the Wilson–Yukawa couplings one runs into a new phase, called the paramagnetic strong-coupling (PMS) phase [153]. Unfortunately, in this phase the doublers bind with the Higgs field to give right-handed fields transforming in the same representation as the left-handed fields, or vice-versa, and the result is a non-chiral (vector) gauge theory in the scaling region [154, 155]. Other models [156] (see also [157]) which can be put into this Wilson–Yukawa framework have been argued to fare the same fate [158]. Another approach is to keep the doublers as heavy physical particles in mirror fermion models [159].

How to formulate a lattice chiral gauge model? This problem is difficult because of the peculiar symmetry breaking of chiral anomalies. We want them to be there without interfering with gauge invariance. Nielson and Ninomiya [160] formulated a no-go theorem that has to be overcome first. They used a Hamiltonian description (continuous time and spatial lattice), and, loosely speaking, the theorem states that, under cherished conditions such as translation invariance, locality and Hermiticity, a free-fermion lattice model with a $U(1)$ invariance has always an equal number of left- and right-handed fermions of a given $U(1)$ charge. The $U(1)$ is supposed to be contained in the gauge group and the implication is that the model can be extended only into an interacting gauge theory that is 'vector' and not chiral. A simpler Euclidean formulation is given in [161]. An extension to an effective action formulation is given in [129].

The Euclidean reasoning runs as follows. Suppose that we replace $\sin k_\mu \to F_\mu(k)$ in the naive fermion propagator. This corresponds to the translation-invariant action of the form (ignoring possible neutral fields)

$$S_{\rm F} = -\sum_{xy\mu} \bar{\psi}_x \gamma_\mu P_{\rm L} \tilde{F}_\mu(x-y)\psi_y, \quad iF_\mu(k) = \sum_x \exp(-ikx)\tilde{F}_\mu(x),$$

$$(8.117)$$

which has a $U(1)$ invariance $\psi \to \exp(i\omega q)\,\psi$, $\bar{\psi} \to \exp(-i\omega q)\,\bar{\psi}$. Hermiticity is easy to state in the Hamiltonian formulation: $\hat{H}^\dagger = \hat{H}$. In the Euclidean formulation we require the spatial part of the action ($\mu = 1, 2, 3$) to be Hermitian and extend this to $\mu = 4$ by covariance. Then Hermiticity means that $F_\mu(x)^* = -F_\mu(-x)$, so $F_\mu(k)$ is real. Locality means that $\tilde{F}_\mu(x)$ approaches zero sufficiently fast as $|x| \to \infty$. This implies that its Fourier transform is not singular and we shall assume $F_\mu(k)$ to be smooth, i.e. it and all its derivatives are continuous. If $F_\mu(k)$ has isolated zeros of first order then the model has a particle interpretation. Near a zero at $k = \bar{k}$,

$$F_\mu(k) = Z_{\mu\nu}(k_\nu - \bar{k}_\nu) + O((k - \bar{k})^2), \quad (8.118)$$

with coefficients $Z_{\mu\nu}$ forming a matrix Z with $\det Z \neq 0$. We write

$$Z = RP, \quad (8.119)$$

with R an orthogonal matrix and P a symmetric positive matrix. The matrix R can be absorbed in a unitary transformation,

$$\gamma_\mu(1 - \gamma_5)R_{\mu\nu} = \Lambda^\dagger \gamma_\nu(1 - \epsilon\gamma_5)\Lambda, \quad \epsilon = \det R = \pm 1. \tag{8.120}$$

For $\epsilon = 1$, Λ is a rotation $\exp(\frac{1}{2}\varphi_{\mu\nu}[\gamma_\mu, \gamma_\nu])$, for $\epsilon = -1$, Λ is e.g. γ_4 times a rotation (cf. appendix D). So for k near $\bar{k}$ the fermion propagator is equivalent to the continuum expression

$$S(k) \approx \frac{-i\gamma_\mu p_\mu}{p^2}\frac{1 - \epsilon\gamma_5}{2}, \quad p_\mu \equiv P_{\mu\nu}(k - \bar{k})_\nu, \tag{8.121}$$

which corresponds to a left- ($\epsilon = +1$) or right-handed ($\epsilon = -1$) fermion field.

Now comes input from topology: ϵ is the index of the vector field $F_\mu(k)$ of its zero at $k = \bar{k}$, i.e. the degree of the mapping $F_\mu/|F| = R_{\mu\nu}p_\nu/|p|$ onto S^4. The Poincaré–Hopf theorem states that the global sum of the indices equals the Euler characteristic χ_E of the manifold on which the vector field is defined: $\sum \epsilon = \chi_E$. In our case this manifold is the momentum-space torus T^4, for which $\chi_E = 0$. Hence, there must be an even number of zeros and in the continuum limit we have an equal number of left- and right-handed fermion fields with the same charge. The naive $U(1)$ model above is a typical illustration of the theorem.

To avoid these theorems we have to avoid some of their assumptions (including hidden assumptions). Giving up translation invariance (e.g. using a random lattice), Hermiticity (e.g. $S_F = -\sum_{x\mu}\bar{\psi}_x\gamma_\mu\partial_\mu\psi_x$, which gives the complex $F_\mu(k) = (e^{ik_\mu} - 1)/i$), or locality (e.g. the discontinuous $F_\mu(k) = 2\sin(k_\mu/2)$ (mod 2π) has only a zero at the origin but corresponds to $F_\mu(x)$ falling only like $|x|^{-1}$) tends to lead to other trouble (for a review, see [162]). The basic reason is that, with an exactly gauge-invariant action and fermion measure, there can be no anomaly, which means that it cancels out in one way or another, generically without the desired particle interpretation.

One line of approach is to give up gauge invariance at finite lattice spacing by working in a fixed gauge and adding counterterms such that gauge invariance is restored in the continuum limit [163, 164]. However, non-perturbative gauge fixing has its own complications, not least the existence of Gribov copies, i.e. configurations differing by a gauge transformation satisfying the same gauge condition. A gauge-fixed $U(1)$ model appears to have passed basic tests [166]. For a review see [168]. One may try to keep the fermions in the continuum, or on a finer lattice than the gauge-field lattice, and invoking restoration of gauge symmetry by the mechanism of [167]. See [168] for a review. Gauge-symmetry restoration was also invoked in models gauging non-invariant models, using Wilson fermions or gauging the staggered flavors [162, 169] but it failed in its simplest realization [170]. Further information can be found in the reviews presented at the Lattice meetings [171].

New developments that constitute a major advancement can be classified under the heading 'overlap' and 'Ginsparg–Wilson' fermions. Theorems of the Nielson–Ninomiya type are avoided by having an infinite number of fermion field components ('overlap'), and changing the definition of γ_5, such that it is as usual for $\bar{\psi}$ but for ψ it involves replacing γ_5 by

$$\hat{\gamma}_5 = \gamma_5(1 - aD), \qquad (8.122)$$

where D is a Ginsparg–Wilson Dirac operator, together with an elaborate definition of the fermion measure in the path integral (apparently giving up Hermiticity on the lattice) [172]. The subject is beautiful and erudite and the reader is best introduced by the reviews [173, 174] ('overlap') and [175] ('Ginsparg–Wilson'). One may feel uncomfortable, though, about using formulations with an infinite number of field components; it runs contrary to the basic idea of being able to approach infinity from the finite.

8.7 Outlook

There is of course a lot more to lattice field theory than has been presented here. An introduction to finite temperature can be found in [9]. Simulation algorithms are introduced in [4, 10]; improved actions and electroweak matrix elements are discussed in [14, 15]. See also [16] for advanced material. For an introduction to simplicial gravity[6] see [17]. Non-perturbative lattice formulations of quantum fields out of equilibrium are still in their infancy.[7] For the current status of all this, see the proceedings of the 'Lattice' meetings.

8.8 Problems

(i) *The pion–nucleon σ model*

Consider an effective nucleon field N that is a doublet in terms of Dirac proton (p) and neutron (n) fields

$$N(x) = \begin{pmatrix} p(x) \\ n(x) \end{pmatrix}. \qquad (8.123)$$

The effective action of the pion–nucleon sigma model is given by

$$S_{\text{eff}} = -\int d^4x \left[\bar{N}\gamma^\mu \partial_\mu N + G\bar{N}(\phi P_R + \phi^\dagger P_L)N\right] + S_{O(4)}, \quad (8.124)$$

where $S_{O(4)}$ is the scalar field action of the $O(4)$ model (equations (3.1) and (3.4)) and ϕ is a matrix field constructed out of the scalar fields,

$$\phi = \varphi^0 \mathbb{1} + i\sum_{k=1}^{3} \varphi^k \tau_k. \quad (8.125)$$

The τ_k are the three Pauli matrices, which act on the p and n components of N and G is the pion–nucleon coupling constant.

Show that the action is invariant under $SU(2) \times SU(2)$ transformations

$$N \to VN, \quad \bar{N} \to \bar{N}\bar{V}, \quad \phi \to V_L \phi V_R^\dagger, \quad V_{L,R} \in SU(2). \quad (8.126)$$

Verify that the transformation on the matrix scalar field ϕ is equivalent to an $SO(4)$ rotation on the φ^α. Hint: check that $\phi^\dagger \phi = \varphi^2 \mathbb{1}$, $\det \phi = \varphi^2$; and hence that ϕ may be written as $\phi = \sqrt{\varphi^2}\, U$, $U \in SU(2)$.

This chiral invariance of the sigma-model action is a nice expression of the symmetry properties of the underlying quark–gluon theory. When the symmetry is spontaneously broken, such that the ground-state value of the scalar field is $\phi_g = f\mathbb{1}$, $f = \varphi_g^0$, the action acquires a mass term $Gf\bar{N}N$: the nucleon gets its mass from spontaneous breaking of chiral symmetry, $m_N = Gf$. This relation is in fair agreement with experiment. On introducing the weak interactions into the model one finds that f equals the pion decay constant, $f = f_\pi \approx 93$ MeV, while $G \approx 13$ from pion–nucleon-scattering experiments, so with $m_N = 940$ MeV we have to compare $m_N/f \approx 10$ with 13.

The field φ^0 is often denoted by σ, and φ^k by π^k, the sigma and pion fields. The pions are stable within the strong interactions but the σ is a very unstable particle with mass m_σ in the range 600–1200 MeV. Given $m_\pi = 140$ MeV and $m_\sigma = 900$ MeV, determine the other parameters in the action.

Reanalyze the model in 'polar coordinates' $\phi = \rho U$, $U \in SU(2)$ with ρ a single-component scalar field. Note that ρ plays the role of the matrix field H introduced in (8.19). What is its mass?

In sections 8.1 the effective field ϕ is a general complex 4×4 matrix, which has eight independent real parameters, whereas the above ϕ has only the four real φ^α, which cannot incorporate chiral $U(1)$ transformations. Verify this, and work out a generalization in which ϕ has the general form. (Include in the action terms that break chiral $U(1)$.)

(ii) *Free fermion* $\langle \bar{\psi}\psi \rangle$

Consider free 'naive' fermions on the lattice (one flavor). Show that

$$\Sigma \equiv -\langle \bar{\psi}\psi \rangle = a^{-3} \int_{-\pi}^{\pi} \frac{d^4k}{(2\pi)^4} \frac{4am}{a^2m^2 + \sum_\mu \sin^2 k_\mu}, \qquad (8.127)$$

and that it has the expansion

$$\Sigma = c_1 m a^{-2} + m^3 [c_3 \ln(am) + c_3'] + \cdots, \qquad (8.128)$$

where the $\cdots$ vanish as $a \to 0$. Hint: use (3.66).

Now consider free Wilson fermions. Show that for this case the expansion takes the form

$$\Sigma = c_0 a^{-3} + c_1 m a^{-2} + c_2 m^2 a^{-1} + m^3 [c_3 \ln(am) + c_3'] + \cdots, \qquad (8.129)$$

where $m = M - 4r/a$. Find expressions for the coefficients c_k.

(iii) *Research project: the index theorem*

Go through the following formal arguments.

In the continuum, let $D = \gamma_\mu [\partial_\mu - iG_\mu(x)]$ be the Dirac operator in an external gauge field G_μ in a finite volume with periodic (up to gauge transformations) boundary conditions. Consider the divergence equation for the flavor-singlet axial current

$$\partial_\mu \bar{\psi} i \gamma_\mu \gamma_5 \psi = 2m \bar{\psi} i \gamma_5 \psi + 2iq, \qquad (8.130)$$

where we assumed that there is only one flavor. Taking the fermionic average and integrating over (Euclidean) space–time gives

$$0 = -2m \operatorname{Tr} [\gamma_5 (m + D)^{-1}] + 2i\nu, \qquad (8.131)$$

where the trace is over space–time and Dirac indices and $\nu = Q_{\text{top}}$ is the topological charge.

Verify that iD is a Hermitian operator, $(iD)^\dagger = iD$.

Let f_s be the eigenvectors of D with (purely imaginary) eigenvalues $i\lambda_s$,

$$Df_s = i\lambda_s f_s, \quad D\gamma_5 f_s = -i\lambda_s \gamma_5 f_s, \qquad (8.132)$$

and assume the eigenvectors to be orthogonal and complete,

$$f_s^\dagger f_t = \delta_{st}, \quad \sum_s f_s f_s^\dagger = 1. \qquad (8.133)$$

Because f_s and $\gamma_5 f_s$ correspond generically to different eigenvalues,

$$f_s^\dagger \gamma_5 f_s = 0, \quad \lambda_s \neq 0. \qquad (8.134)$$

For $\lambda_s = 0$, $[D, \gamma_5] f_s = 0$, so in this subspace we can look for simultaneous eigenvectors of D and γ_5. The eigenvalues of γ_5 are ± 1,

$$\gamma_5 f_s = \pm f_s, \quad \lambda_s = 0. \qquad (8.135)$$

It follows that

$$\nu = m \sum_s \frac{\text{Tr}\,(\gamma_5 f_s f_s^\dagger)}{m + i\lambda_s} = \sum_{s,\,\lambda_s = 0} f_s^\dagger \gamma_5 f_s = n_+ - n_-, \qquad (8.136)$$

with $n_\pm$ the number of zero modes with chirality $\gamma_5 = \pm 1$.

Periodicity modulo gauge transformations is needed in order to allow non-zero topological charge. For the proper mathematical setting in the continuum, see e.g. [12]. Lattice studies using Wilson and staggered fermions are in [107, 108, 114, 115, 90], while [135] gives an introduction to Ginsparg–Wilson fermions. Choose one of these studies and reproduce (and possibly extend) its results.

(iv) *Research project: the theta parameter of QCD*
Consider the QCD action with generalized mass term

$$\int d^4x \, \bar{\psi}' m \psi', \quad m = m_{\rm L} P_{\rm L} + m_{\rm L}^\dagger P_{\rm R}, \qquad (8.137)$$

in which $m_{\rm L}$ is a fairly arbitrary complex matrix. Assume that it can be transformed into a diagonal matrix by the transformation

$$V_{\rm L}^\dagger m_{\rm L} V_{\rm R} = m_{\rm diag} = \text{diagonal with entries} \geq 0. \qquad (8.138)$$

Suppose this transformation is the result of a chiral transformation on the fermion fields (cf. (8.2)),

$$\psi' = V\psi, \quad \bar{\psi}' = \bar{\psi}\bar{V}. \qquad (8.139)$$

In continuum treatments the fermion measure is not invariant under such a transformation, but is produces the chiral anomaly in the form [132]

$$D\bar{\psi}'D\psi' = D\bar{\psi}D\psi\, e^{i\theta \int d^4x\, q(x)}, \qquad \theta = \arg(\det m_{\rm L}). \qquad (8.140)$$

So in terms of the un-primed fermion fields we have an additional term in the (Euclidean) action proportional to the topological charge,

$$S = -\int d^4x \left[\frac{1}{2g^2}\,\mathrm{Tr}\,(G_{\mu\nu}G_{\mu\nu}) + \bar{\psi}\gamma_\mu D_\mu\psi + \bar{\psi}m_{\rm diag}\psi - i\theta q \right]. \qquad (8.141)$$

The original mass m may be the result of electroweak symmetry breaking. Experiments constrain the value of θ, which violates CP invariance, to be less than 10^{-9} in magnitude.

Our problem is to give a rigorous version of the above reasoning using the lattice regularization. With Wilson's fermion method the following steps get us going.

Consider the fermion determinant $\exp[\mathrm{Tr}\,\ln(\slashed{D} - W + M)]$, where M is arbitrary. In the scaling region M is close to the critical value $M_{\rm c}$; if not, then there is no continuum physics. So assume that $M = M_{\rm c} + m$, with $M_{\rm c} \propto r\mathbb{1}$ and m arbitrary as in the above continuum outline. With Wilson's fermion method the fermion measure is invariant under chiral transformations and the anomaly comes from the non-invariant term $\bar{\psi}(W - M_{\rm c})\psi$ in the action. So we have

$$\mathrm{Tr}\,[\ln(\slashed{D} - W + M_{\rm c} + m)] = \mathrm{Tr}\,\{\ln[\slashed{D} + \bar{V}(M_{\rm c} - W)V + m_{\rm diag}]\}. \qquad (8.142)$$

To evaluate this consider a change δV of V. Then the above expression changes by

$$\mathrm{Tr}\,\{[\delta\bar{V}(M_{\rm c} - W)V + \bar{V}(M_{\rm c} - W)\,\delta V] \\ \times [\slashed{D} + \bar{V}(M_{\rm c} - W)V + m_{\rm diag}]^{-1}\}. \qquad (8.143)$$

Expanding this expression in terms of the gauge field leads to an infinite number of diagrams with external gauge-field lines impinging upon a closed fermion loop. The crucial point is now that the factor $M_{\rm c} - W$ in the numerator above suppresses the region of loop-momentum integration where $m_{\rm diag}$ has any influence. For example in momentum space at lowest order, $M_{\rm c} - W \to$

$ra^{-1}\sum_\mu[1 - \cos(ak_\mu)]$, and we therefore need a loop momentum k of order $a^{-1} \gg m_{\text{diag}}$ to give a non-vanishing contribution. See [70] for an explicit computation of the triangle-diagram-like contributions. So we may as well set $m_{\text{diag}} = 0$ in (8.143). Then (8.143) can be rewritten in the form

$$\text{Tr}\,[\bar{V}^{-1}\,\delta\bar{V}(M_c - W)(\slashed{D} + M_c - W)^{-1}$$
$$+ \delta V\,V^{-1}(\slashed{D} + M_c - W)^{-1}(M_c - W)]$$
$$= \text{Tr}\,[(V_L\,\delta V_L^{-1} - V_R\,\delta V_R^{-1})\,\gamma_5\,(M_c - W)(\slashed{D} + M_c - W)^{-1}],$$
$$(8.144)$$

where we used the fact that $\slashed{D}$, M_c and W are all flavor diagonal, and the cyclic property of the trace. Denoting the trace over space–time plus Dirac indices (excluding the flavor indices) by Tr_{st} we have the result [70, 133, 134]

$$\text{Tr}_{\text{st}}\,[\gamma_5\,(M_c - W)(\slashed{D} + M_c - W)^{-1}] = Q_{\text{top}}, \quad a \to 0. \quad (8.145)$$

Note that this result is *independent* of the r parameter [70], as long as it is non-zero. The coefficient of Q_{top} is given by

$$\text{Tr}_{\text{flavor}}\,(V_L\,\delta V_L^{-1} - V_R\,\delta V_R^{-1}) = \delta\ln[\det(V_R V_L^{-1})] = i\,\delta\arg(\det m_L). \quad (8.146)$$

So one concludes that, in the continuum limit,

$$\exp\{\text{Tr}\,[\ln(\slashed{D} + M_c - W + m)]\}$$
$$= e^{i\theta Q_{\text{top}}}\,\exp\{\text{Tr}\,[\ln(\slashed{D} + M_c - W + m_{\text{diag}})]\}, \quad (8.147)$$
$$\theta = \arg(\det m_L), \quad (8.148)$$

which is equivalent to the continuum result.

By taking the continuum *limit* we have happily been able to ignore finite renormalization factors κ ($\kappa = 1 + O(g^2) \to 1$, g^2 is the bare gauge coupling).

The problem with the above reasoning, taken from [149], is how to improve it such that it applies in a practical scaling region with g^2 not much less than 1.

Appendix A
$SU(n)$

A.1 Fundamental representation of $SU(n)$

In the following appendices we record some properties of the representations of the group $SU(n)$. First we review the construction of a complete basis set of Hermitian traceless $n \times n$ matrices, similar to the $n = 2, 3$ examples. We shall denote these matrices by λ_k, $k = 1, 2, \ldots, n^2 - 1$. The symmetric off-diagonal matrices have the form

$$(\lambda_k)_{ab} = \delta_{am}\delta_{an} + \delta_{bm}\delta_{an} \quad k \leftrightarrow \{m, n\} \tag{A.1}$$

and the antisymmetric matrices are given by

$$(\lambda_k)_{ab} = i(\delta_{am}\delta_{an} - \delta_{bm}\delta_{an}), \tag{A.2}$$

where $a, b, m, n = 1, 2, \ldots, n$, $m > n$. The non-zero elements of the diagonal matrices may be taken as

$$(\lambda_k)_{aa} = \sqrt{\frac{2}{m + m^2}} \quad a = 1, \ldots, m, \tag{A.3}$$

$$= -m\sqrt{\frac{2}{m + m^2}} \quad a = m + 1, \tag{A.4}$$

where $m = 1, 2, \ldots, n - 1$. We add the multiple of the unit matrix

$$\lambda_0 = \sqrt{\frac{2}{n}}\, \mathbb{1}, \tag{A.5}$$

such that the $k = 0, 1, \ldots, n^2 - 1$ matrices form a complete set of $n \times n$ matrices. They satisfy

$$\lambda_k = \lambda_k^\dagger, \tag{A.6}$$

$$\mathrm{Tr}\,(\lambda_k \lambda_l) = 2\delta_{kl}, \tag{A.7}$$

229

and either $\lambda_k = \lambda_k^T = \lambda_k^*$ or $\lambda_k = -\lambda_k^T = -\lambda_k^*$. An arbitrary matrix X can be written as a superposition of the λ's,

$$X = X_k \lambda_k, \tag{A.8}$$

$$X_k = \tfrac{1}{2} \operatorname{Tr} (X\lambda_k). \tag{A.9}$$

For instance

$$\lambda_k \lambda_l = \Lambda_{klm} \lambda_m, \tag{A.10}$$

$$\Lambda_{klm} = \tfrac{1}{2} \operatorname{Tr} (\lambda_k \lambda_l \lambda_m). \tag{A.11}$$

Let

$$\Lambda_{klm} = d_{klm} + i f_{klm}, \tag{A.12}$$

where d_{klm} and f_{klm} are real. Then

$$
\begin{aligned}
d_{klm} &= \tfrac{1}{4} \operatorname{Tr} (\lambda_k \lambda_l \lambda_m + \lambda_k^* \lambda_l^* \lambda_m^*) = \tfrac{1}{4} \operatorname{Tr} (\lambda_k \lambda_l \lambda_m + \lambda_k^T \lambda_l^T \lambda_m^T) \\
&= \tfrac{1}{4} \operatorname{Tr} (\lambda_k \lambda_l \lambda_m + \lambda_m \lambda_l \lambda_k) = \tfrac{1}{4} \operatorname{Tr} (\lambda_k \lambda_l \lambda_m + \lambda_l \lambda_k \lambda_m) \\
&= \tfrac{1}{4} \operatorname{Tr} (\{\lambda_k, \lambda_l\} \lambda_m),
\end{aligned}
\tag{A.13}
$$

and similarly,

$$i f_{klm} = \tfrac{1}{4} \operatorname{Tr} ([\lambda_k, \lambda_l] \lambda_m). \tag{A.14}$$

These representations of the d's and f's and the cyclic properties of the trace imply that d_{klm} is totally symmetric under interchange of any of its labels. Likewise f_{klm} is totally antisymmetric. Hence, (A.10) and (A.12) imply

$$[\lambda_k, \lambda_l] = 2i f_{klm} \lambda_m, \tag{A.15}$$

$$\{\lambda_k, \lambda_l\} = 2 d_{klm} \lambda_m. \tag{A.16}$$

We note in passing that

$$\lambda_0 \lambda_l = \sqrt{\frac{2}{n}} \lambda_l \rightarrow d_{0lm} = \sqrt{\frac{2}{n}} \delta_{lm}, \quad f_{0lm} = 0. \tag{A.17}$$

A standard choice for the generators t_k of the group $SU(n)$ in the fundamental (defining) representation is given by

$$t_k = \tfrac{1}{2} \lambda_k, \quad k = 1, 2, \ldots, n^2 - 1. \tag{A.18}$$

In the exponential parameterization an arbitrary group element can be written as

$$U = \exp(i\alpha^k t_k), \tag{A.19}$$

where the α^k are $n^2 - 1$ real parameters. From their occurence in the commutation relations

$$[t_k, t_l] = i f_{klm} t_m, \tag{A.20}$$

the f_{klm} are called the structure constants of the group.

Next we calculate the value C_2 of the quadratic Casimir operator $t_k t_k$ in the defining representation. For this we need a useful formula that follows from expanding the matrix $X_{ab}^{(cd)} \equiv 2\delta_{ad}\delta_{bc}$ in terms of $(\lambda_k)_{ab}$. According to (A.8) and (A.9) we have the expansion coefficients $X_k^{(cd)} = \mathrm{Tr}\,(X^{(cd)}\lambda_k)/2 = \delta_{ad}\delta_{bc}(\lambda_k)_{ba} = (\lambda_k)_{cd}$, hence,

$$(\lambda_k)_{ab}(\lambda_k)_{cd} = 2\delta_{ad}\delta_{bc}, \tag{A.21}$$

where the summation is over $k = 0, 1, \ldots, n^2 - 1$ on the left-hand side. It follows that

$$\begin{aligned}
(t_k)_{ab}(t_k)_{cd} &= \frac{1}{4}(\lambda_k)_{ab}(\lambda_k)_{cd} - \frac{1}{4}(\lambda_0)_{ab}(\lambda_0)_{cd} \\
&= \frac{1}{2}\delta_{ad}\delta_{bc} - \frac{1}{2n}\delta_{ab}\delta_{cd}
\end{aligned} \tag{A.22}$$

(note that $k = 0$ is lacking for the t_k). Contraction with δ_{bc} gives

$$(t_k t_k)_{ad} = \frac{1}{2}\left(n - \frac{1}{n}\right)\delta_{ad} \equiv C_2\,\delta_{ad}, \tag{A.23}$$

or

$$C_2^{\mathrm{fund}} = \frac{1}{2}\left(n - \frac{1}{n}\right). \tag{A.24}$$

For $n = 2$, $C_2^{\mathrm{fund}} = \frac{3}{4}$ which is just the usual value $j(j+1)$ for the $j = \frac{1}{2}$ representation of $SU(2)$.

A.2 Adjoint representation of $SU(n)$

The adjoint (regular) representation R is the representation carried by the generators,

$$U^\dagger t_k U = R_{kl} t_l. \quad U \in SU(n). \tag{A.25}$$

Note that $\mathrm{Tr}\,(U^\dagger t_k U) = \mathrm{Tr}\,t_k = 0$, so that $U^\dagger t_k U$ can indeed be written as a linear superposition of the t_k. By eq. (A.9) we have the explicit representation in terms of the group elements

$$R_{kl} = 2\,\mathrm{Tr}\,(U^\dagger t_k U t_l). \tag{A.26}$$

We shall now calculate R in terms of the parameters α^k of the exponential parameterization of U. Let

$$U(y) = \exp(iy\alpha^p t_p), \quad R_{kl}(y) = 2\operatorname{Tr}(U^\dagger(y)t_k U(y)t_l). \tag{A.27}$$

Then

$$\begin{aligned}
\frac{\partial}{\partial y} R_{kl}(y) &= -i\alpha^p\, 2\operatorname{Tr}(U^\dagger(y)[t_p, t_k]U(y)t_l)\\
&= \alpha^p f_{pkn}\, 2\operatorname{Tr}(U^\dagger(y)t_n U(y)t_l)\\
&= i\alpha^p (F_p)_{kn} R_{nl},
\end{aligned} \tag{A.28}$$

where

$$(F_p)_{mn} = -if_{pmn}. \tag{A.29}$$

In matrix notation (A.28) reads

$$\frac{\partial}{\partial y} R(y) = i\alpha^p F_p R(y), \tag{A.30}$$

which differential equation is solved by

$$R(y) = \exp(iy\alpha^p F_p), \tag{A.31}$$

using the boundary condition $R(0) = 1$. Hence,

$$R = \exp(i\alpha^p F_p), \tag{A.32}$$

and we see that the F_p are the generators in the adjoint representation. By the antisymmetry of the structure constants we have

$$F_p = -F_p^* = -F_p^{\mathrm{T}}, \tag{A.33}$$

and it follows that the matrices R are real and orthogonal,

$$R = R^*, \quad R^{\mathrm{T}} = R^{-1}. \tag{A.34}$$

Notice that the derivation of (A.28) uses only the commutation relations of the generators, so that we have for an arbitrary representation $D(U)$

$$D(U)^{-1}T_k D(U) = R_{kl}T_l, \tag{A.35}$$

where the T_k are the generators in this representation D.

Next we calculate the value of the Casimir operator in the adjoint representation, $F_p F_p$, using the results of the previous appendix:

$$\begin{aligned}
(F_p F_p)_{km} &= if_{kpl}\, if_{lpm}\\
&= 4\operatorname{Tr}(t_k t_p t_l)\, if_{lpm} = 8\operatorname{Tr}(t_p t_l t_k)\operatorname{Tr}([t_m, t_l]t_p)\\
&= 8(t_p)_{ab}(t_l t_k)_{ba}\,[t_m, t_l]_{dc}(t_p)_{cd}.
\end{aligned} \tag{A.36}$$

With (A.22) for $(t_p)_{ab}(t_p)_{cd}$, this gives

$$(F_p F_p)_{km} = 4\,\mathrm{Tr}\,(t_l t_k [t_m, t_l]), \tag{A.37}$$

and using (A.22) again and $t_l t_l = [(n^2 - 1)/2n]\mathbb{1}$ gives finally

$$F_p F_p = n\mathbb{1}, \quad C_2^{\mathrm{adj}} = n. \tag{A.38}$$

The matrix $S_k(\alpha)$ introduced in (4.41) can be calculated as follows. We write $D(U(\alpha)) = D(\alpha)$ and consider (4.42),

$$M(y) = D(y\alpha)D(y\alpha + y\epsilon)^{-1} = 1 - i\epsilon^k S_k(\alpha) + O(\epsilon^2) \tag{A.39}$$

$$= e^{iy\alpha^k T_k}\, e^{-iy(\alpha^k + \epsilon^k)T_k}. \tag{A.40}$$

Then

$$\frac{\partial}{\partial y} M(y) = D(y\alpha)[i\alpha^k T_k - i(\alpha^k + \epsilon^k)T_k]D(y\alpha + y\epsilon)^{-1}$$

$$= -i\epsilon^k D(y\alpha)T_k D(y\alpha)^{-1} + O(\epsilon^2)$$

$$= -i\epsilon^k R_{kl}^{-1}(y\alpha)T_l + O(\epsilon^2). \tag{A.41}$$

This differential equation can be integrated with the boundary condition $M(0) = 1$, using $R^{-1}(y\alpha) = \exp(-iy\alpha)$, $\alpha \equiv \alpha^p F_p$,

$$M(y) = 1 - i\epsilon^k \left(\frac{1 - e^{-iy\alpha}}{i\alpha}\right)_{kl} T_l + O(\epsilon^2). \tag{A.42}$$

Setting $y = 1$ we find $S_k(\alpha) = S_{kl}(\alpha)T_l$ with

$$S_{kl}(\alpha) = \left(\frac{1 - e^{-i\alpha}}{i\alpha}\right)_{kl}, \quad \alpha = \alpha^p F_p. \tag{A.43}$$

We end this appendix with an expression for $\mathrm{Tr}\, T_k T_l$ in an arbitrary representation D. The matrix

$$I_{kl} = \mathrm{Tr}\,(T_k T_l) \tag{A.44}$$

is invariant under transformations in the adjoint representation,

$$R_{kk'} R_{ll'} I_{k'l'} = \mathrm{Tr}\,(D^{-1}T_k D\, D^{-1}T_l D) = I_{kl}. \tag{A.45}$$

By Schur's lemma, I_{kl} must be a multiple of the identity matrix,

$$I_{kl} = \rho\,\delta_{kl}. \tag{A.46}$$

Putting $k = l$ and summing over k gives the relation

$$(n^2 - 1)\rho(D) = C_2(D)\,\mathrm{dimension}(D). \tag{A.47}$$

For the fundamental and adjoint representations we have

$$\rho_{\text{fund}} = \tfrac{1}{2}, \tag{A.48}$$

$$\rho_{\text{adj}} = n. \tag{A.49}$$

A.3 Left and right translations in $SU(n)$

Let Ω and U be elements of $SU(n)$. We define left and right transformations by

$$U'(L) = \Omega U, \quad U'(R) = U\Omega, \tag{A.50}$$

respectively, which may be interpreted as translations in group space, $U \to U'$. In a parameterization $U = U(\alpha)$, $\Omega = \Omega(\varphi)$, this implies transformations of the α's,

$$\alpha'^k(L) = f^k(\alpha, \varphi, L), \tag{A.51}$$

and similarly for R. We shall first concentrate on the L case. For Ω near the identity we can write,

$$\Omega = 1 + i\varphi^m t_m + \cdots, \tag{A.52}$$

$$\alpha'^k(L) = \alpha^k + \varphi^m S^k{}_m(\alpha, L) + \cdots, \tag{A.53}$$

$$S^k{}_m(\alpha, L) = \frac{\partial}{\partial \varphi^m} f^k(\alpha, \varphi, L)_{|\varphi=0}. \tag{A.54}$$

The $S^k{}_m(\alpha, L)$ (which are analogous to the tetrad or 'Vierbein' in General Relativity) can found in terms of the $S_{km}(\alpha)$ as follows,

$$U'(L) = (1 + i\varphi^m t_m + \cdots)U, \tag{A.55}$$

$$t_m U = -i \frac{\partial}{\partial \varphi^m} U_{|\varphi=0} = -i \frac{\partial U}{\partial \alpha^k} \frac{\partial \alpha^k}{\partial \varphi^m}_{|\varphi=0}$$

$$= -i \frac{\partial U}{\partial \alpha^k} S^k{}_m(\alpha, L). \tag{A.56}$$

Differentiating $UU^\dagger = 1$ gives

$$\frac{\partial U}{\partial \alpha^k} = -U \frac{\partial U^\dagger}{\partial \alpha^k} U, \tag{A.57}$$

and using this in (A.56) we get

$$t_m U = iU \frac{\partial U^\dagger}{\partial \alpha^k} U S^k{}_m(\alpha, L), = S_k(\alpha, L) U S^k{}_m(\alpha, L), \tag{A.58}$$

where

$$S_k(\alpha, L) \equiv iU \frac{\partial U^\dagger}{\partial \alpha^k} \tag{A.59}$$

is the S_k introduced earlier in (4.41). The factor U can be canceled out from the above equation,

$$t_m = S_k(\alpha, L) S^k{}_m(\alpha, L). \tag{A.60}$$

We have already shown in (4.43) that S_k is a linear superposition of the generators, $S_k(\alpha, L) = S_{kn}(\alpha, L)t_n$, so we get

$$t_m = t_n S_{kn}(\alpha, L) S^k{}_m(\alpha, L) \tag{A.61}$$

or

$$\delta_{mn} = S_{kn}(\alpha, L) S^k{}_m(\alpha, L). \tag{A.62}$$

Thus $S^k{}_m(\alpha, L)$ is the inverse (in the sense of matrices) of $S_{km}(\alpha, L)$. Introducing the differential operators

$$X_m(L) = S^k{}_m(\alpha, L) \frac{\partial}{i \, \partial \alpha^k} \tag{A.63}$$

we can rewrite (A.56) in the form

$$X_m(L)U = t_m U. \tag{A.64}$$

It follows from this equation that the $X_m(L)$ have the commutation relations

$$[X_m(L), X_n(L)] = -i f_{mnp} X_p(L). \tag{A.65}$$

These differential operators may be called the generators of left translations.

For the right translations we get in similar fashion

$$U t_m = -i \frac{\partial U}{\partial \alpha^k} S^k{}_m(\alpha, R) = U S_k(\alpha, R) S^k{}_m(\alpha, R), \tag{A.66}$$

$$S_k(\alpha, R) \equiv -i U^\dagger \frac{\partial U}{\partial \alpha^k}$$
$$= U^\dagger S_k(\alpha, L) U = S_{kn}(\alpha, L) U^\dagger t_n U$$
$$= S_{kp}(\alpha, L) R_{pn} t_n, \tag{A.67}$$

$$S_k(\alpha, R) = S_{kn}(\alpha, R) t_n, \tag{A.68}$$

$$S_{kn}(\alpha, R) = S_{kp}(\alpha, L) R_{pn}, \tag{A.69}$$

$$\delta_{mn} = S_{kn}(\alpha, R) S^k{}_m(\alpha, R), \tag{A.70}$$

$$X_m(R) = S^k{}_m(\alpha, R) \frac{\partial}{i \, \partial \alpha^k}, \tag{A.71}$$

$$X_m(R)U = U t_m, \tag{A.72}$$

$$[X_m(R), X_n(R)] = +i f_{mnp} X_p(R) \tag{A.73}$$

The left and right generators commute,

$$[X_m(L), X_n(R)] = 0, \tag{A.74}$$

which follows directly from (A.64) and (A.72), and their quadratic Casimir operators are equal,

$$X^2(L) = X_m(L)X_m(L), \quad X^2(R) = X_m(R)X_m(R), \tag{A.75}$$

$$X^2(R)U = U t_m t_m = C_2 U = t_m t_m U = X^2(L)U. \tag{A.76}$$

The differential operator $X^2 = X^2(L) \;'= X^2(R)$ is invariant under coordinate transformations on group space and is also known as a Laplace–Beltrami operator.

Finally, the metric introduced in (4.91) can be expressed in terms of the analogs of the tetrads,

$$g_{kl}(\alpha) = S_{kp}(\alpha, L)S_{lp}(\alpha, L) = S_{kp}(\alpha, R)S_{lp}(\alpha, R), \tag{A.77}$$

$$S_{kp}(\alpha, L) = g_{kl}(\alpha)S^l{}_p(\alpha, L), \quad S_{kp}(\alpha, R) = g_{kl}(\alpha)S^l{}_p(\alpha, R). \tag{A.78}$$

For a parameterization that is regular near $U = 1$ (such as $\exp(i\alpha^k t_k)$),

$$U = 1 + i\alpha^k t_k + O(\alpha^2), \tag{A.79}$$

it is straightforward to derive that

$$S^k{}_p(\alpha, L) = \delta_{kp} - \tfrac{1}{2} f_{kpl}\alpha^l + O(\alpha^2), \tag{A.80}$$

$$S^k{}_p(\alpha, R) = \delta_{kp} + \tfrac{1}{2} f_{kpl}\alpha^l + O(\alpha^2), \tag{A.81}$$

$$g_{kl}(\alpha) = \delta_{kl} + O(\alpha^2). \tag{A.82}$$

A.4 Tensor method for $SU(n)$

It is sometimes useful to view the matrices U representing the fundamental representation of $SU(n)$ as tensors. Products of U's then transform as tensor products and integrals over the group reduce to invariant tensors. It will be useful to write the matrix elements with upper and lower indices, $U_{ab} \rightarrow U^a_b$. We start with the simple integral

$$\int dU \, U^a_b U^{\dagger p}_q = I^{ap}_{bq}. \tag{A.83}$$

By making the transformation of variables $U \rightarrow VUW^\dagger$, it follows that the right-hand side above is an invariant tensor in the following sense:

$$I^{ap}_{bq} = V^a_{a'} W^p_{p'} V^{\dagger q'}_q W^{\dagger b'}_b I^{a'p'}_{b'q'}. \tag{A.84}$$

Here V and W are arbitrary elements of $SU(n)$ and similarly for their matrix elements in the fundamental representation and their complex conjugates $V^\dagger$ and $W^\dagger$. We are using a notation in which matrix indices of U are taken from the set a, b, c, d, ..., while those of $U^\dagger$ are taken from p, q, r, s, ... Upper indices in the first set transform with V, upper indices in the second set transform with W; lower indices in the first set transform with $W^\dagger$, lower indices in the second set transform with $V^\dagger$, as in

$$U_b^a \rightarrow V_{a'}^a W_b^{\dagger b'} U_{b'}^{a'}, \qquad U_q^{\dagger p} \rightarrow W_{p'}^p V_q^{\dagger q'} U_{q'}^{\dagger p'}. \tag{A.85}$$

This notation suffices for not-too-complicated expressions.

Returning to the above group integral, there is only one such invariant tensor: $I_{bq}^{ap} = c\delta_q^a \delta_b^p$, which is a simple product of Kronecker deltas. The constant c can be found by contracting the left- and right-hand sides with δ_b^p, with the result

$$\int dU \, U_b^a U_q^{\dagger p} = \frac{1}{n} \delta_q^a \delta_b^p. \tag{A.86}$$

Invariant tensors have to be linear combinations of products of Kronecker tensors and the Levi-Civita tensors

$$\epsilon^{a_1 \cdots a_n} = +1, \quad \text{even permutation of } 1, \ldots, n$$
$$= -1, \quad \text{odd permutation of } 1, \ldots, n, \tag{A.87}$$

and similarly for $\epsilon_{a_1 \cdots a_n}$, etc. They are invariant because

$$V_{a_1'}^{a_1} \cdots V_{a_1'}^{a_1} \epsilon^{a_1' \cdots a_n'} = \det V \, \epsilon^{a_1 \cdots a_n}. \tag{A.88}$$

These tensors appear in

$$\int dU \, U_{b_1}^{a_1} \cdots U_{b_n}^{a_n} = \frac{1}{n!} \epsilon^{a_1 \cdots a_n} \epsilon_{b_1 \cdots b_n} \tag{A.89}$$

$$= \frac{1}{n!} \sum_{\text{perm} \pi} (-1)^\pi \delta_{b_{\pi 1}}^{a_1} \cdots \delta_{b_{\pi n}}^{a_n}. \tag{A.90}$$

The coefficient can be checked by contraction with $\epsilon_{a_1 \cdots a_n}$.

In writing down possible invariant tensors for group integrals we have to keep in mind that, according to (A.85), there can be only Kronecker deltas with one upper and one lower index, and furthermore one index should correspond to a U and the other index to a $U^\dagger$, i.e. they should be of the type δ_p^a or δ_a^p. It is now straightforward to derive identities for integrals of the next level of complication:

$$\int dU\, U_b^a U_d^c U_f^e = 0, \quad n > 3, \tag{A.91}$$

$$\int dU\, U_b^a U_d^c U_q^{\dagger p} U_s^{\dagger r} = \frac{1}{n^2 - 1} \left(\delta_q^a \delta_s^c \delta_b^p \delta_d^r + \delta_s^a \delta_q^c \delta_b^r \delta_d^p \right)$$
$$- \frac{1}{n(n^2 - 1)} \left(\delta_s^a \delta_q^c \delta_b^p \delta_d^r + \delta_q^a \delta_s^c \delta_b^r \delta_d^p \right), \quad n > 2. \tag{A.92}$$

Note the symmetry under $(a, b) \leftrightarrow (c, d)$ and $(p, q) \leftrightarrow (r, s)$ in (A.92). The coefficients follow, e.g. by contraction with δ_d^p. By contracting (A.92) with the generators $(t_k)_c^s (t_l)_r^d$ we get an identity needed in the main text:

$$\int dU\, U_b^a U_q^{\dagger p} R_{kl}(U) = \frac{2}{n^2 - 1} (t_k)_q^a (t_l)_b^p, \quad n > 2. \tag{A.93}$$

where $R_{kl}(U)$ is the adjoint representation of U (cf. (A.26)).

Appendix B
Quantization in the temporal gauge

Gauge-field quantization in the temporal gauge in the continuum is often lacking in text books. Here follows a brief outline. Consider the action of $SU(n)$ gauge theory,

$$S = -\int d^4x \, \frac{1}{4g^2} G^p_{\mu\nu} G^{\mu\nu p}. \tag{B.1}$$

The stationary action principle leads to the equations of motion

$$D_\mu G^{\mu\nu p} = \partial_\mu G^{\mu\nu p} + f_{pqr} G^q_\mu G^{\mu\nu r} = 0. \tag{B.2}$$

where D_μ is the covariant derivative in the adjoint representation. Note that we are using a Minkowski-space metric with signature $(-1,1,1,1)$, e.g. $G^{0np} = -G_0{}^{np} = -G^p_{0n}$. The Lagrangian is given by

$$L(G,\dot{G}) = \int d^3x \left(\frac{1}{2g^2} G^p_{0n} G^p_{0n} - \frac{1}{4g^2} G^p_{mn} G^p_{mn} \right), \tag{B.3}$$

where

$$G^p_{0n} = \dot{G}^p_n - \partial_n G^p_0 + f_{pqr} G^q_0 G^r_n, \tag{B.4}$$

and the canonical momenta are given by

$$\Pi^p_0 \equiv \frac{\delta L}{\delta \dot{G}^p_0} = 0, \tag{B.5}$$

$$\Pi^p_n \equiv \frac{\delta L}{\delta \dot{G}^p_n} = \frac{1}{g^2} G^p_{0n}. \tag{B.6}$$

The fact that L is independent of $\dot{G}^p_0$ and consequently the canonical momentum of G^p_0 vanishes is incompatible with the presumed canonical Poisson brackets $(G^p_0, \Pi^q_0) \overset{?}{=} \delta_{pq}\delta(\mathbf{x} - \mathbf{y})$, unless we eliminate G^p_0 as variable by a choice of gauge. This is the 'temporal gauge'

$$G_0^p = 0. \tag{B.7}$$

The Hamiltonian in the temporal gauge is given by

$$
\begin{aligned}
H(G,\Pi) &= \int d^3x \, \Pi_m^p \dot{G}_m^p - L \\
&= \int d^3x \left(\frac{g^2}{2} \Pi_m^p \Pi_m^p + \frac{1}{4g^2} G_{mn}^p G_{mn}^p \right).
\end{aligned} \tag{B.8}
$$

However, one does not want to lose the time component ($\nu = 0$) of the equations of motion (B.2). In canonical variables this equation reads

$$T^p \equiv \partial_m \Pi_m^p + f_{pqr} G_m^q \Pi_m^r = 0, \tag{B.9}$$

and we see that it does not contain a time derivative. It is a constraint equation for every space–time point. Imposing it at one time, the question of whether it is compatible with Hamilton's equations arises.

Let us address this question directly in the quantized case, assuming the canonical commutation relations

$$[\hat{G}_m^p(\mathbf{x}), \hat{\Pi}_n^q(\mathbf{y})] = \delta_{pq}\delta(\mathbf{x}-\mathbf{y}), \quad [\hat{G}_m^p(\mathbf{x}), \hat{G}_n^q(\mathbf{y})] = 0 = [\hat{\Pi}_m^p(\mathbf{x}), \hat{\Pi}_n^q(\mathbf{y})]. \tag{B.10}$$

Now it is straightforward to check that the $\hat{T}^p$ defined in (B.9) generate time-independent gauge transformations, e.g. $\hat{\Omega}^\dagger \hat{G}_m^p \hat{\Omega} =$ infinitesimally gauge-transformed $\hat{G}_m^p$, where $\hat{\Omega} = 1 + i \int d^3x \, \omega^p(\mathbf{x}) \hat{T}^p(\mathbf{x}) + O(\omega^2)$. The Hamiltonian is gauge invariant,

$$[\hat{T}^p, \hat{H}] = 0, \tag{B.11}$$

and the constraints are compatible with the Heisenberg equations of motion. A formal Hilbert-space realization of the canonical commutation relations (B.10) is given by the coordinate representation

$$\langle G|\hat{G}_m^p(\mathbf{x})|\Psi\rangle = G_m^p(\mathbf{x})\langle G|\Psi\rangle, \tag{B.12}$$

$$\langle G|\hat{\Pi}_m^p(\mathbf{x})|\Psi\rangle = \frac{\delta}{i\delta G_m^p(\mathbf{x})} \langle G|\Psi\rangle, \tag{B.13}$$

with wave functionals $\Psi(G) = \langle G|\Psi\rangle$. Unlike quantization in other gauges, there are no negative norm states here, but physical states have to be gauge invariant,

$$\hat{T}^p(\mathbf{x})\,|\Psi\rangle_{\text{phys}} = 0. \tag{B.14}$$

Such states can be formally written as a superposition of Wilson loops and this is useful for analytic calculations at strong coupling (on the lattice, of course, to make it well defined), but not at weak coupling.

Finally, the analogy with QED may be stressed in the notation by writing

$$E_k^p = \frac{1}{g} G^{0kp} = -g\Pi_m^p, \quad B_k^p = \frac{1}{2g} \epsilon_{klm} G_{lm}^p, \tag{B.15}$$

in terms of which

$$H = \int d^3x \left(\tfrac{1}{2} E^2 + \tfrac{1}{2} B^2 \right). \tag{B.16}$$

In case other fields are present, there are additional contributions to T^p that act as generators for these fields, e.g. for QCD, $\rho_p = \psi^+ \lambda_p \psi / 2$, and (B.9) becomes the non-Abelian version of Gauss's law:

$$D_k E_k^p = g\rho_p. \tag{B.17}$$

Appendix C
Fermionic coherent states

In this appendix we derive the field representation for fermion operators. In the Bose case the field representation was just the coordinate representation which is also much used in quantum mechanics. For Fermi operators the analog leads to the so-called Grassmann variables. This means that the Fermi operator fields $\hat{\psi}(x)$ will be represented by 'numbers' $\psi(x)$, which have to be *anticommuting*. As this might not be so familiar, we shall first describe how this works.

Consider the quantum Fermi operators satisfying the commutation relations

$$\{\hat{a}_k, \hat{a}_l\} = 0, \quad \{\hat{a}_k^\dagger, \hat{a}_l^\dagger\} = 0, \quad \{\hat{a}_k, \hat{a}_l^\dagger\} = \delta_{kl}, \qquad (C.1)$$

where $\{A, B\} = AB + BA$. In the following we shall consider a finite number n of such operators, $k = 1, 2, \ldots, n$. (In the continuum limit of a fermionic lattice field theory $n \to \infty$.) It is sometimes convenient to use the $2n$ equivalent Hermitian operators

$$\hat{a}_k^1 = (\hat{a}_k + \hat{a}_k^\dagger)/\sqrt{2}, \quad \hat{a}_k^2 = (\hat{a}_k - \hat{a}_k^\dagger)/i\sqrt{2}, \qquad (C.2)$$

with the commutation relations

$$\{\hat{a}_k^p, \hat{a}_l^q\} = \delta_{pq}\delta_{kl}, \quad p, q = 1, 2. \qquad (C.3)$$

The non-Hermitian operators are used more often.

It is clarifying to look at a representation in Hilbert space. For $n = 1$ we have the 'no-quantum state' $|0\rangle$ which is by definition the eigenstate of $\hat{a}$ with eigenvalue 0, $\hat{a}|0\rangle = 0$, and the one-quantum state $|1\rangle$ obtained from $|0\rangle$ by the application of $\hat{a}^\dagger$, $|1\rangle = \hat{a}^\dagger|0\rangle$. Further application of $\hat{a}^\dagger$ on $|0\rangle$ gives zero, since $(\hat{a}^\dagger)^2 = 0$ because of (C.1) (note that $|1\rangle$ is the 'no-quantum state' for $\hat{a}^\dagger$). So a pair of Fermi operators $(\hat{a}, \hat{a}^\dagger)$ can be represented in a simple two-dimensional Hilbert space,

$$|0\rangle \rightarrow \begin{pmatrix} 0 \\ 1 \end{pmatrix}, \quad |1\rangle \rightarrow \begin{pmatrix} 1 \\ 0 \end{pmatrix}, \quad \hat{a} \rightarrow \begin{pmatrix} 0 & 0 \\ 1 & 0 \end{pmatrix}, \quad \hat{a}^\dagger \rightarrow \begin{pmatrix} 0 & 1 \\ 0 & 0 \end{pmatrix}. \tag{C.4}$$

For $n > 1$ we can take a tensor product of these representations. A basis in Hilbert space is provided by

$$|k_1 \cdots k_p\rangle = \hat{a}_{k_1}^\dagger \cdots \hat{a}_{k_p}^\dagger |0\rangle, \quad p = 1, \ldots, n, \tag{C.5}$$

with the properties

$$\sum_{p=0}^{n} \frac{1}{p!} \sum_{k_1 \cdots k_p} |k_1 \cdots k_p\rangle\langle k_1 \cdots k_p| = 1, \tag{C.6}$$

$$\langle k_1 \cdots k_p | l_1 \cdots l_q \rangle = \delta_{pq} \delta_{l_1 \cdots l_q}^{k_1 \cdots k_p}, \tag{C.7}$$

where

$$\delta_{l_1 \cdots l_q}^{k_1 \cdots k_p} = \sum_{\text{perm } \pi} (-1)^\pi \delta_{\pi l_1}^{k_1} \cdots \delta_{\pi l_p}^{k_p}. \tag{C.8}$$

An arbitrary state $|\psi\rangle$ can be written as†

$$|\psi\rangle = \psi(\hat{a}^\dagger)|0\rangle, \tag{C.9}$$

$$\psi(\hat{a}^\dagger) = \sum_{p=0}^{n} \frac{1}{p!} \psi_{k_1 \cdots k_p} \hat{a}_{k_1}^\dagger \cdots \hat{a}_{k_p}^\dagger, \tag{C.10}$$

where $\psi_{k_1 \cdots k_p}$ is totally antisymmetric in $k_1 \cdots k_p$, and we sum over repeated indices unless indicated otherwise. An arbitrary operator $\hat{A}$ can be written as

$$\hat{A} = \sum_{pq} \frac{1}{p!q!} A_{k_1 \cdots k_p, l_1 \cdots l_q} \hat{a}_{k_1}^\dagger \cdots \hat{a}_{k_p}^\dagger \hat{a}_{l_q} \cdots \hat{a}_{l_1}, \tag{C.11}$$

where all creation operators are ordered to the left of all annihilation operators. This is called the *normal ordered* form of $\hat{A}$. A familiar example is the number operator

$$\hat{N} = \hat{a}_k^\dagger \hat{a}_k, \tag{C.12}$$

which has eigenvectors $|k_1 \cdots k_p\rangle$ with eigenvalue p. Note that $A_{k_1 \cdots k_p, l_1 \cdots l_q}$ is in general not equal to $\langle k_1 \cdots k_p | \hat{A} | l_1 \cdots l_q \rangle$. Note also that the coefficients $A_{k_1 \cdots k_p, l_1 \cdots l_q}$ may themselves be elements of a

---

† Recall that repeated indices are summed, i.e. $\psi_{k_1 \cdots k_p} \hat{a}_{k_1}^\dagger \cdots \hat{a}_{k_p}^\dagger = \sum_{k_1=1}^{n} \cdots \sum_{k_p=1}^{n} \psi_{k_1 \cdots k_p} \hat{a}_{k_1}^\dagger \cdots \hat{a}_{k_p}^\dagger.$

Grassmann algebra, e.g. $\hat{A} = c_k^+ \hat{a}_k + \hat{a}_k^\dagger c_k$, with anticommuting c and c^+.

Suppose now that there are eigenstates $|a\rangle$ of the $\hat{a}_k$ with eigenvalue a_k. Then it follows that the a_k have to be anticommuting:

$$a_k a_l = -a_l a_k. \tag{C.13}$$

To see this, assume

$$\hat{a}_k a_l = \epsilon a_l \hat{a}_k, \tag{C.14}$$

with ϵ some number $\neq 0$. Then

$$\begin{aligned}
\hat{a}_k \hat{a}_l |a\rangle = \hat{a}_k a_l |a\rangle = \epsilon a_l \hat{a}_k |a\rangle = \epsilon a_l a_k |a\rangle \\
= -\hat{a}_l \hat{a}_k |a\rangle = -\epsilon a_k a_l |a\rangle. \tag{C.15}
\end{aligned}$$

Hence (C.13) has to hold. The a_k cannot be ordinary numbers. Assuming $a_k |a\rangle = +|a\rangle a_k$ leads to

$$\begin{aligned}
\hat{a}_k a_l |a\rangle = \hat{a}_k |a\rangle a_l = a_k |a\rangle a_l = a_k a_l |a\rangle \\
= \epsilon a_l \hat{a}_k |a\rangle = \epsilon a_l a_k |a\rangle, \tag{C.16}
\end{aligned}$$

and it follows that

$$\epsilon = -1. \tag{C.17}$$

So the 'numbers' a_k have to anticommute with the fermionic operators as well.

We also introduce independent conjugate anticommuting a_k^+, assume these to anticommute with the a_k and the Fermi operators, and impose the usual rules of Hermitian conjugation,

$$\hat{a}_k \xrightarrow{\dagger} \hat{a}_k^\dagger, \quad a_k \xrightarrow{\dagger} a_k^+, \quad |a\rangle \xrightarrow{\dagger} \langle a|, \quad \langle a|\hat{a}_k^\dagger = \langle a|a_k^+, \tag{C.18}$$

$$a_k a_l \xrightarrow{\dagger} a_l^+ a_k^+, \quad \{a_k^+, a_l^+\} = 0. \tag{C.19}$$

The anticommuting a_k^+ are on the same footing as the a_k.

The a_k and a_k^+ together with the unit element 1 generate a *Grassmann algebra*. An arbitrary element f of this algebra has the form

$$\begin{aligned}
f(a^+, a) = f_{0,0} + f_{k,0} a_k^+ + f_{0,l} a_l + \frac{1}{2!} f_{k_1 k_2, 0} a_{k_1}^+ a_{k_2}^+ \\
+ f_{k,l} a_k^+ a_l + \cdots + f_{1 \cdots n, 1 \cdots n} a_1^+ \cdots a_n^+ a_n \cdots a_1,
\end{aligned}$$

$$\tag{C.20}$$

where the f's are complex numbers.

We have extended Hilbert space into a vector space over the elements of a Grassmann algebra. The a_k and a_k^+ are called Grassmann variables and $f(a^+, a)$ is called a function of the Grassmann variables. This nomenclature could be somewhat misleading – the generators a_k and a_k^+ are fixed objects and it is only the indices 'k' and '$+$' that vary. However, we will also be using other generators b_k, b_k^+, c_k, ..., and so effectively we draw elements from a Grassmann algebra with an infinite number of generators. It is straightforward to construct a matrix representation of these generators, but this does not seem to be useful because the rules above are sufficient for our derivations.

We now express the $|a\rangle$ in terms of the basis vectors (C.5). The state $|a\rangle$ is given by

$$|a\rangle = e^{-a_k \hat{a}_k^\dagger}|0\rangle. \tag{C.21}$$

Indeed, since $(a_k)^2 = 0$,

$$e^{-a_k \hat{a}_k^\dagger} = \prod_k e^{-a_k \hat{a}_k^\dagger} = \prod_k (1 - a_k \hat{a}_k^\dagger), \tag{C.22}$$

and using $\hat{a}_k(1 - a_k \hat{a}_k^\dagger)|0\rangle = a_k \hat{a}_k \hat{a}_k^\dagger |0\rangle = a_k|0\rangle$ (no summation over k) gives

$$\hat{a}_k|a\rangle = \left[\prod_{l \neq k}(1 - a_l \hat{a}_l^\dagger)\right] \hat{a}_k(1 - a_k \hat{a}_k^\dagger)|0\rangle = \left[\prod_{l \neq k}(1 - a_l \hat{a}_l^\dagger)\right] a_k|0\rangle$$

$$= a_k|a\rangle. \tag{C.23}$$

Note that a_k *commutes* with pairs of fermion objects, e.g. $[a_k, a_l \hat{a}_m^\dagger] = 0$. Two states $|a\rangle$ and $|b\rangle$ have the inner product

$$\langle a|b\rangle = \langle 0|(1 - \hat{a}_1 a_1^+) \cdots (1 - \hat{a}_n a_n^+)(1 - b_n \hat{a}_n^\dagger) \cdots (1 - b_1 \hat{a}_1^\dagger)|0\rangle$$

$$= \prod_k (1 + a_k^+ b_k)$$

$$= e^{a^+ b}, \tag{C.24}$$

where

$$a^+ b \equiv a_k^+ b_k. \tag{C.25}$$

We would like a completeness relation of the form

$$\hat{1} = \int da^+ \, da \, \frac{|a\rangle\langle a|}{\langle a|a\rangle}. \tag{C.26}$$

For $n = 1$ this relation reads

$$
\begin{aligned}
\hat{1} &= |0\rangle\langle 0| + \hat{a}^\dagger |0\rangle\langle 0|\hat{a} \\
&= \int da^+ \, da \, (1 - a^+ a)(1 - a\hat{a}^\dagger)|0\rangle\langle 0|(1 - \hat{a}a^+) \\
&= \int da^+ \, da \, [\, (1 - a^+ a)|0\rangle\langle 0| - a\hat{a}^\dagger |0\rangle\langle 0| \\
&\qquad\qquad + a^+ |0\rangle\langle 0|\hat{a} + aa^+ \hat{a}^\dagger |0\rangle\langle 0|\hat{a} \,],
\end{aligned}
\tag{C.27}
$$

which is satisfied if we define the Berezin 'integral':

$$
\int da = 0, \quad \int da^+ = 0, \quad \int da \, a = 1, \quad \int da^+ a^+ = 1,
\tag{C.28}
$$

where da and da^+ are taken anticommuting. For general n we define

$$
da = da_1 \cdots da_n, \quad da^+ = da_n^+ \cdots da_1^+,
\tag{C.29}
$$

$$
\int da_k = 0, \quad \int da_k \, a_k = 1, \quad \int da_k^+ = 0, \quad \int da_k^+ a_k^+ = 1
\tag{C.30}
$$

(no summation over k; anticommuting da's and da^+'s). The integral sign symbolizes Grassmannian integration, which has some similarities to ordinary integration (and differentiation, see (C.42)). Cumbersome checking of minus signs can be avoided by combining every da_k with da_k^+ into commuting pairs, as in the notation

$$
da^+ \, da \equiv \prod_{k=1}^{n} da_k^+ \, da_k,
\tag{C.31}
$$

which we shall use in the following. Similar pairing will be done repeatedly in the following.

We check the completeness relation (C.26) for general n by verifying that it gives the right answer for an arbitrary inner product $\langle \psi | \phi \rangle$. Multiplying (C.9) by (C.26), we get

$$
|\psi\rangle = \int da^+ \, da \, e^{-a^+ a} \, \psi(a^+)|a\rangle,
\tag{C.32}
$$

$$
\psi(a^+) = \langle a | \psi \rangle = \sum_p \frac{1}{p!} \psi_{k_1 \cdots k_p} a_{k_1}^+ \cdots a_{k_p}^+.
\tag{C.33}
$$

The inner product takes the form

$$\langle \psi | \phi \rangle = \int da^+ \, da \, e^{-a^+ a} \, \psi(a^+)^\dagger \phi(a^+)$$

$$= \sum_{pq} \frac{1}{p!q!} \psi^*_{k_1 \cdots k_p} \phi_{l_1 \cdots l_q}$$

$$\times \int da^+ \, da \, e^{-a^+ a} \, a_{k_p} \cdots a_{k_1} a^+_{l_1} \cdots a^+_{l_q}. \qquad (C.34)$$

By (C.28) the integral is non-zero only if $p = q$ and $(k_1, \ldots, k_p) = (l_1, \ldots, l_p)$ up to a permutation,

$$\int da^+ \, da \, e^{-a^+ a} a_{k_p} \cdots a_{k_1} a^+_{k_1} \cdots a^+_{k_p} = \prod_{l \neq k_1, \cdots, k_p} \int da^+_l \, da_l \, e^{-a^+_l a_l}$$

$$\times \prod_{m = k_1, \cdots, k_p} \int da^+_m \, da_m \, a_m a^+_m$$

$$= 1, \qquad (C.35)$$

and

$$\int da^+ \, da \, e^{-a^+ a} a_{k_p} \cdots a_{l_1} a^+_{k_1} \cdots a^+_{l_q} = \delta_{pq} \delta^{k_1 \cdots k_p}_{l_1 \cdots l_q}. \qquad (C.36)$$

Hence, (C.34) gives

$$\langle \psi | \phi \rangle = \sum_p \frac{1}{p!} \psi^*_{k_1 \cdots k_p} \phi_{k_1 \cdots k_p}, \qquad (C.37)$$

which is the right answer. Therefore (C.26) is correct for general n.

The connection between Grassmannian integration and differentiation can be seen as follows. Left and right differentiation can be defined by looking at terms linear in a translation over fermion b_k,

$$f(a + b) = f(a) + b_k f^{\rm L}_k(a) + \tfrac{1}{2} b_k b_l f^{\rm L}_{kl}(a) + \cdots \qquad (C.38)$$

$$= f(a) + f^{\rm R}_k(a) b_k + \tfrac{1}{2} f^{\rm R}_{kl}(a) b_k b_l + \cdots, \qquad (C.39)$$

which suggest

$$\frac{\partial}{\partial a_k} f(a) := f^{\rm L}_k(a), \qquad (C.40)$$

$$f(a) \frac{\overleftarrow{\partial}}{\partial a_k} := f^{\rm R}_k(a) \qquad (C.41)$$

(the extension to functions of both a and a^+ is obvious). It follows that 'integration' is left differentiation:

$$\int da_k \, f(a) = \frac{\partial}{\partial a_k} f(a). \tag{C.42}$$

We shall now derive some further important properties of Grassmannian integration. Let $f(a^+, a)$ be an arbitrary element of the Grassmann algebra of the form by (C.20). Then

$$\int da^+ \, da \, f(a^+, a) = f_{1\cdots n, 1\cdots n}. \tag{C.43}$$

It follows that the integration is translation invariant,

$$\int da^+ \, da \, f(a^+ + b^+, a + b) = \int da^+ \, da \, f(a^+, a). \tag{C.44}$$

Furthermore, for an arbitrary matrix M,

$$\int da^+ \, da \, e^{-a^+ M a} = \int da^+ \, da \, \frac{(-1)^n}{n!} (a^+ M a)^n$$

$$= \int da^+ \, da \, \frac{1}{n!} M_{k_1 l_1} \cdots M_{k_n l_n} a_{l_1} a_{k_1}^+ \cdots a_{l_n} a_{k_n}^+$$

$$= \frac{1}{n!} M_{k_1 l_1} \cdots M_{k_n l_n} \delta_{l_1 \cdots l_n}^{k_1 \cdots k_n}. \tag{C.45}$$

Using the identity

$$\epsilon_{k_1 \cdots k_n} \epsilon_{l_1 \cdots l_n} = \delta_{l_1 \cdots l_n}^{k_1 \cdots k_n}, \tag{C.46}$$

where $\epsilon_{k_1 \cdots k_n}$ is the n-dimensional ϵ tensor (with $\epsilon_{1\cdots n} = +1$) we obtain the formula

$$\int da^+ \, da \, e^{-a^+ M a} = \det M, \tag{C.47}$$

since

$$\det M = M_{1 l_1} \cdots M_{n l_n} \epsilon_{l_1 \cdots l_n}. \tag{C.48}$$

The more general formula

$$\int da^+ \, da \, e^{-a^+ M a + a^+ b + b^+ a} = \det M \, e^{b^+ M^{-1} b}. \tag{C.49}$$

follows from the translation invariance (C.44) by making the translation $a^+ \to a^+ + b^+ M^{-1}$, $a \to a + M^{-1} a$. Note that (C.49) remains well defined if $\det M \to 0$.

We can interpret (C.44) as a translation invariance of the fermionic 'measure',

$$da^+ = d(a^+ + b^+), \quad da = d(a + b). \tag{C.50}$$

A linear multiplicative transformation of variables

$$a_k \to T_{kl} a_l, \quad a_k^+ \to a_l^+ S_{lk}, \tag{C.51}$$

has the effect

$$d(a^+ S) = (\det S)^{-1} da^+, \quad d(Ta) = (\det T)^{-1} da, \tag{C.52}$$

i.e.

$$\int da^+ \, da \, f(a^+ S, Ta) = \det(ST) \int da^+ \, da \, f(a^+, a). \tag{C.53}$$

This follows easily from (C.43) and (C.48). According to (C.52), the fermionic measure transforms inversely to the bosonic measure dx: $d(Tx) = \det T \, dx$.

We note in passing the formula

$$\int da \, e^{-\frac{1}{2} a^T M a} = \pm \sqrt{\det M}, \tag{C.54}$$

where T denotes transposition and M is an antisymmetric matrix (in this case only the antisymmetric part of M contributes anyway). This formula follows from (C.47), by making the transformation of variables

$$\begin{pmatrix} a_k \\ a_k^+ \end{pmatrix} = \frac{1}{\sqrt{2}} \begin{pmatrix} 1 & -1 \\ 1 & 1 \end{pmatrix} \begin{pmatrix} b_k \\ c_k \end{pmatrix}, \tag{C.55}$$

which leads to

$$\det M = (-1)^{n/2} \int db \, e^{-\frac{1}{2} a^T M a} \int dc \, e^{\frac{1}{2} b^T M b}, \tag{C.56}$$

where we assumed n to be even (otherwise $\det M = 0$). As is obvious from the left-hand side of (C.54), the square root of the determinant of an antisymmetric matrix is multilinear in its matrix elements. It is called a Pfaffian.

States $|\psi\rangle$ are represented by Grassmann wavefunctions $\psi(a^+)$ depending only on the a_k^+ (cf. (C.33)). The representatives of operators $\hat{A}$ depend in general also on the a_k:

$$\langle a | \hat{A} | a \rangle =: A(a^+, a). \tag{C.57}$$

In the normal ordered form (C.11), $A(a^+, a)$ is obtained from $\hat{A}$ by replacing everywhere the operators by their Grassmann representative, keeping the same order, and multiplying by $e^{a^+ a}$:

$$A(a^+, a) = e^{a^+ a} \sum_{pq} \frac{1}{p! q!} A_{k_1 \cdots k_p, l_1 \cdots l_q} a_{k_1}^+ \cdots a_{k_p}^+ a_{l_q} \cdots a_{l_1}. \qquad \text{(C.58)}$$

(The $e^{a^+ a}$ just comes from the normalization factor $\langle a | a \rangle$.)

It is now straightforward to derive the following rules:

$$A\psi(a^+) := \langle a | \hat{A} | \psi \rangle$$

$$= \int db^+ \, db \, e^{-b^+ b} A(a^+, b) \psi(b^+), \qquad \text{(C.59)}$$

$$AB(a^+, a) := \langle a | \hat{A}\hat{B} | a \rangle$$

$$= \int db^+ \, db \, e^{-b^+ b} A(a^+, b) B(b^+, a), \qquad \text{(C.60)}$$

$$\hat{A} = A(\hat{a}^\dagger, \hat{a}), \;\; \hat{B} = B(\hat{a}^\dagger), \;\; \hat{C} = C(\hat{a})$$

$$\Rightarrow BAC(a^+, a) = B(a^+) A(a^+, a) C(a). \qquad \text{(C.61)}$$

A useful identity is

$$\hat{A} = \exp\left[\hat{a}_k^\dagger M_{kl} \hat{a}_l\right] \Rightarrow A(a^+, a) = \exp\left[a_k^+ (e^M)_{kl} a_l\right], \qquad \text{(C.62)}$$

This identity can be derived with well-known differentiation/integration tricks. Let $F(t)$ be given by

$$F(t) = \langle a | e^{t \hat{a}^\dagger M \hat{a}} | a \rangle. \qquad \text{(C.63)}$$

To compute $F(1) = A(a^+, a)$ we differentiate with respect to t and subsequently integrate, with the initial condition $F(0) = \exp(a^+ a)$. Differentiation gives

$$F'(t) = \langle a | \hat{a}^\dagger M \hat{a} \, e^{t \hat{a}^\dagger M \hat{a}} | a \rangle = a_k^+ M_{kl} \langle a | \hat{a}_l e^{t \hat{a}^\dagger M \hat{a}} | a \rangle. \qquad \text{(C.64)}$$

The $\hat{a}_l$ needs to be pulled trough the exponential so that we can use $\hat{a}_l | 0 \rangle = a_l | 0 \rangle$. For this we use a similar differentiation trick:

$$\hat{G}_l(t) \equiv e^{-t \hat{a}^\dagger M \hat{a}} \hat{a}_l e^{t \hat{a}^\dagger M \hat{a}},$$

$$\hat{G}_l'(t) = e^{-t \hat{a}^\dagger M \hat{a}} [\hat{a}_l, \hat{a}^\dagger M \hat{a}] e^{t \hat{a}^\dagger M \hat{a}} = M_{lm} \hat{G}_m(t), \;\; \hat{G}(0) = \hat{a}_l,$$

$$\hat{G}_l(t) = (e^{tM})_{lm} \hat{a}_m,$$

$$\hat{a}_l e^{t \hat{a}^\dagger M \hat{a}} = e^{t \hat{a}^\dagger M \hat{a}} (e^{tM})_{lm} \hat{a}_m. \qquad \text{(C.65)}$$

The differential equation for $F(t)$ now reads

$$F'(t) = a_k^+ M_{kl} \langle a | e^{t \hat{a}^\dagger M \hat{a}} | a \rangle (e^{tM})_{lm} a_m = (a^+ e^{tM} a)' F(t), \qquad \text{(C.66)}$$

with the solution

$$F(t) = \exp(a^+ e^{tM} a), \quad A(a^+, a) = F(1) = \exp(a^+ e^M a). \quad \text{(C.67)}$$

Next we derive an important formula for the trace of a fermionic operator. It is usually sufficient to consider only even operators, i.e. operators containing only terms with an even number of fermionic operators or fermionic variables. Such $\hat{A}$ and also their representative $A(a^+, a)$ commute with arbitrary anticommuting numbers, for example $A(a^+, b)c_k = +c_k A(a^+, b)$. The formula reads

$$\text{Tr}\,\hat{A} = \int da^+\, da\, e^{-a^+ a}\, A(a^+, -a), \quad \text{(C.68)}$$

for even $\hat{A}$. This trace formula can be derived as follows:

$$\text{Tr}\,\hat{A} = \sum_{p=0}^{n} \frac{1}{p!} \sum_{k_1 \cdots k_p} \langle k_1 \cdots k_p | \hat{A} | k_1 \cdots k_p \rangle$$

$$= \int (da^+\, da)\,(db^+\, db)\, e^{-a^+ a - b^+ b}$$

$$\sum_{p} \frac{1}{p!} \sum_{k_1 \cdots k_p} \langle k_1 \cdots k_p | a \rangle \langle a | \hat{A} | b \rangle \langle b | k_1 \cdots k_p \rangle$$

$$= \int (da^+\, da)\,(db^+\, db)\, e^{-a^+ a - b^+ b} \sum_p \frac{1}{p!} a_{k_p} \cdots a_{k_1} A(a^+, b) b_{k_1}^+ \cdots b_{k_p}^+$$

$$= \int (da^+\, da)\,(db^+\, db)\, e^{-a^+ a - b^+ b} \sum_p \frac{1}{p!} a_{k_p} \cdots a_{k_1} b_{k_1}^+ \cdots b_{k_p}^+ A(a^+, b)$$

$$= \int (da^+\, da)\,(db^+\, db)\, e^{-a^+ a - b^+ b}\, e^{a_k b_k^+}\, A(a^+, b)$$

$$= (-1)^n \int (da^+\, db)\, e^{a^+ b}\, A(a^+, b)$$

$$= \int (da^+\, db)\, e^{-a^+ b} A(a^+, -b), \quad \text{(C.69)}$$

which is the desired result. We integrated over a and b^+ using $(da^+\, da)$ $\times (db^+\, db) = (-1)^n (da^+\, db)\,(db^+\, da)$ and (C.49). In the last line we made the substitution $b \to -b$ using (C.52).

We note furthermore that omitting the minus sign from $A(a^+, -a)$ in (C.68) leads to

$$\int da^+\, da\, e^{-a^+ a} A(a^+, a) = \text{Tr}(-1)^{\hat{N}} \hat{A}, \quad \text{(C.70)}$$

where $\hat{N}$ is the fermion-number operator (C.12). This formula can be derived from the trace formula (C.68), the operator-product rule (C.60), with $B = \exp(i\pi\hat{N})$, and the application

$$\hat{B} = e^{i\pi\hat{a}^\dagger\hat{a}} \rightarrow B(a^+, a) = e^{-a^+a} \qquad (C.71)$$

of the rule (C.62).

Appendix D
Spinor fields

In this appendix we record the basics of spinor fields. We start with the properties of Dirac matrices in Euclidean space–time. The four Dirac matrices γ_μ, $\mu = 1, 2, 3, 4$, are 4×4 matrices with the properties

$$\gamma_\mu \gamma_\nu + \gamma_\nu \gamma_\nu = 2\delta_{\mu\nu} \mathbb{1}. \tag{D.1}$$

So they anticommute: $\gamma_\mu \gamma_\nu = -\gamma_\nu \gamma_\mu$, $\mu \neq \nu$. They can be chosen Hermitian and unitary, $\gamma_\mu^\dagger = \gamma_\mu = \gamma_\mu^{-1}$. The matrix

$$\gamma_5 \equiv -\gamma_1 \gamma_2 \gamma_3 \gamma_4 \tag{D.2}$$

anticommutes with the γ_μ, $\gamma_\mu \gamma_5 = -\gamma_5 \gamma_\mu$, and it is also Hermitian and unitary, $\gamma_5 = \gamma_5^\dagger$, $\gamma_5^2 = \mathbb{1}$. A realization can be given in terms of tensor products of the 2×2 Pauli matrices σ_k, $k = 1, 2, 3$, and $\sigma_0 \equiv \mathbb{1}_{2 \times 2}$: $\gamma_k = -\sigma_2 \otimes \sigma_k$, $\gamma_4 = \sigma_1 \otimes \sigma_0$, $\gamma_5 = \sigma_3 \otimes \sigma_0$. Usually one does not need a realization as almost all relations follow from the basic anticommutation relations (D.1). Other realizations are related by unitary transformations, which preserve the Hermiticity and unitarity of the Dirac matrices, but not the behavior under complex conjugation or transposition. It can be shown that, in every such realization, there is an antisymmetric unitary 4×4 matrix C, called the charge-conjugation matrix, which relates γ_μ to its transpose:

$$\gamma_\mu^{\mathrm{T}} = -C^\dagger \gamma_\mu C, \quad C^{\mathrm{T}} = -C, \quad C^\dagger C = \mathbb{1}, \tag{D.3}$$
$$\Rightarrow \gamma_5^{\mathrm{T}} = \gamma_5^* = C^\dagger \gamma_5 C. \tag{D.4}$$

In the above realization a possible C is given by $C = \sigma_3 \otimes \sigma_2$. The matrices $\Gamma = \mathbb{1}$, γ_μ, $(-i/2)[\gamma_\mu, \gamma_\nu]$, $i\gamma_\mu \gamma_5$ and γ_5 form a complete set of 16 independent Hermitian 4×4 matrices with the properties $\Gamma^2 = \mathbb{1}$, $\mathrm{Tr}\,\Gamma = 0$ except for $\Gamma = \mathbb{1}$, $\mathrm{Tr}\,(\Gamma\Gamma') = 0$ for $\Gamma \neq \Gamma'$. Useful relations are furthermore $\gamma_5\gamma_\kappa = \epsilon_{\kappa\lambda\mu\nu}\gamma_\lambda\gamma_\mu\gamma_\nu$, with $\epsilon_{\kappa\lambda\mu\nu}$ the completely antisymmetric Levi-Civita tensor, $\epsilon_{1234} = +1$, the trace of an odd number of γ_μ's is zero, $\mathrm{Tr}\,(\gamma_5\gamma_\mu\gamma_\nu) = 0$, and

$$\mathrm{Tr}\,(\gamma_\mu\gamma_\nu) = 4\delta_{\mu\nu}, \tag{D.5}$$

$$\mathrm{Tr}\,(\gamma_\kappa\gamma_\lambda\gamma_\mu\gamma_\nu) = 4(\delta_{\kappa\lambda}\delta_{\mu\nu} - \delta_{\kappa\mu}\delta_{\lambda\nu} + \delta_{\kappa\nu}\delta_{\lambda\mu}), \tag{D.6}$$

$$\mathrm{Tr}\,(\gamma_5\gamma_\kappa\gamma_\lambda\gamma_\mu\gamma_\nu) = -4\epsilon_{\kappa\lambda\mu\nu}. \tag{D.7}$$

More trace relations are given in most textbooks on relativistic field theory.

The Dirac matrices are used to describe covariance under (in our case) Euclidean rotations, which are elements of the group $SO(4)$. A rotation in the μ–ν plane over a small angle $\omega_{\mu\nu}$ can be written as

$$R_{\mu\nu} = \delta_{\mu\nu} + \omega_{\mu\nu} + O(\omega^2), \quad \omega_{\mu\nu} = -\omega_{\nu\mu} \tag{D.8}$$

$$= \delta_{\mu\nu} + i\tfrac{1}{2}\omega_{\kappa\lambda}(M_{\kappa\lambda})_{\mu\nu} + \cdots, \tag{D.9}$$

$$(M_{\kappa\lambda})_{\mu\nu} = -i(\delta_{\kappa\mu}\delta_{\lambda\nu} - \delta_{\kappa\nu}\delta_{\lambda\mu}). \tag{D.10}$$

The antisymmetry of $\omega_{\mu\nu}$ ensures that $R_{\mu\nu}$ is orthogonal, $R_{\kappa\mu}R_{\lambda\mu} = \delta_{\kappa\lambda}$, with $\det R = 1$. The $M_{\kappa\lambda}$ are the generators of $SO(4)$ in the defining representation. The structure constants $C^{\rho\sigma}_{\kappa\lambda\mu\nu}$ defined by $[M_{\kappa\lambda}, M_{\mu\nu}] = C^{\rho\sigma}_{\kappa\lambda\mu\nu}M_{\rho\sigma}$ are easily worked out.

The 4×4 *spinor representation* of these rotations can be written in terms of Dirac matrices as

$$\Lambda = e^{i\frac{1}{2}\omega_{\mu\nu}\Sigma_{\mu\nu}} = \mathbb{1} + i\tfrac{1}{2}\omega_{\mu\nu}\Sigma_{\mu\nu} + \cdots, \tag{D.11}$$

$$\Sigma_{\mu\nu} = -i\tfrac{1}{4}[\gamma_\mu, \gamma_\nu], \tag{D.12}$$

where the $\Sigma_{\mu\nu}$ are the generators in the spinor representation. They satisfy the same commutation relations as the $M_{\mu\nu}$, as follows from the basic relations (D.1). The matrices Λ are unitary,

$$\Lambda^\dagger = \Lambda^{-1}, \quad \text{Euclid.} \tag{D.13}$$

They form a unitary representation up to a sign, e.g. for a rotation over an angle 2π in the 1–2 plane, $\omega_{12} = -\omega_{21} = 2\pi$, and in the realization of the Dirac matrices introduced above, $\Lambda = \exp(\tfrac{1}{4}\omega_{\mu\nu}\gamma_\mu\gamma_\nu) = \exp(i\pi\sigma_0 \otimes \sigma_3) = -\mathbb{1}$.

The representation Λ is reducible, as follows from the fact that Λ commutes with γ_5, $[\Lambda, \gamma_5] = 0$. Introducing the projectors $P_{\mathrm{R,L}}$ onto the eigenspaces ± 1 of γ_5,

$$P_R = \tfrac{1}{2}(\mathbb{1} + \gamma_5), \quad P_L = \tfrac{1}{2}(\mathbb{1} - \gamma_5), \quad P_L^2 = P_L, \quad P_R^2 = P_R,$$
$$P_L P_R = 0, \quad P_L + P_R = \mathbb{1}, \tag{D.14}$$

we can decompose Λ into two components Λ_L and Λ_R as

$$\Lambda = \Lambda P_L + \Lambda P_R \equiv \Lambda_L + \Lambda_R. \tag{D.15}$$

The Λ_L and Λ_R are inequivalent irreducible representations (up to a sign) of $SO(4)$. They are essentially two-dimensional, because the subspace of $\gamma_5 = 1$ or -1 is two-dimensional, but we shall keep them as 4×4 matrices. The Λ's are real up to equivalence,

$$\Lambda^* = e^{\frac{1}{4}\omega_{\mu\nu}\gamma_\mu^*\gamma_\nu^*} = e^{\frac{1}{4}\omega_{\mu\nu}\gamma_\mu^T\gamma_\nu^T} = C^\dagger e^{\frac{1}{4}\omega_{\mu\nu}\gamma_\mu\gamma_\nu} C$$
$$= C^\dagger \Lambda C, \tag{D.16}$$
$$\Lambda_{L,R}^* = C^\dagger \Lambda_{L,R} C. \tag{D.17}$$

The γ_μ are vector matrices in the sense that

$$\Lambda^\dagger \gamma_\mu \Lambda = R_{\mu\nu}\gamma_\nu. \tag{D.18}$$

This follows from the basic anticommutation relations between the γ's, as can easily be checked for infinitesimal rotations. Products $\gamma_\mu \gamma_\nu \cdots$ transform as tensors. Because $\gamma_\mu P_{R,L} = P_{L,R}\gamma_\mu$, the projected relations have the form $\Lambda_R^\dagger \gamma_\mu \Lambda_L = R_{\mu\nu}\gamma_\nu P_L$, and similarly for $L \leftrightarrow R$. It follows that

$$R_{\mu\nu} = \tfrac{1}{2} \operatorname{Tr}\left(\Lambda_R^\dagger \gamma_\mu \Lambda_L \gamma_\nu\right), \tag{D.19}$$

which illustrates the relation

$$SO(4) \simeq SU(2) \times SU(2)/Z_2 \tag{D.20}$$

(interpreted as 2×2 matrices, $\Lambda_{L,R}$ are elements of $SU(2)$, and $Z_2 = \{1, -1\}$ compensates for $\Lambda_{L,R}$ and $-\Lambda_{L,R}$ giving the same R).

We can enlarge $SO(4)$ to $O(4)$ by adding reflections to the set of R's, which have determinant -1. An important one is parity $P \equiv \operatorname{diag}(-1, -1, -1, 1)$. Its spinor representation can be taken as $\Lambda_P = \gamma_4$, which has the expected effect on the γ_μ:

$$\gamma_4 \, \gamma_\mu \, \gamma_4 = P_{\mu\nu} \, \gamma_\nu, \tag{D.21}$$

and it has therefore also the required effect on the generators $\Sigma_{\mu\nu}$, such that we have a representation of $O(4)$. Because $\gamma_4 P_{L,R} \gamma_4 = P_{R,L}$ we have $\gamma_4 \Lambda_{L,R} \gamma_4 = \Lambda_{R,L}$. So we need both irreps L and R in order to be able to incorporate parity transformations.

Vector fields $V_\mu(x)$ transform under $SO(4)$ rotations as

$$V'_\mu(x) = R_{\mu\nu} V_\nu(R^{-1}x), \quad (R^{-1}x)_\mu = R_{\nu\mu} x_\nu, \tag{D.22}$$

which can be understood by drawing a vector field in two dimensions on a sheet of paper and seeing how it changes under rotations. Spinor fields $\psi(x)$ transform according to

$$\psi'_\alpha(x) = \Lambda_{\alpha\beta} \psi_\beta(R^{-1}x), \tag{D.23}$$

where α and β are matrix indices ('Dirac indices'). The fields can be decomposed into irreducible components as

$$\psi_L(x) = P_L \psi(x), \quad \psi_R(x) = P_R \psi(x). \tag{D.24}$$

It is customary to introduce a separate notation $\bar\psi$ for fields transforming with the inverse $\Lambda^\dagger$ as

$$\bar\psi'(x) = \bar\psi(R^{-1}x)\,\Lambda^\dagger \tag{D.25}$$

(so ψ is a column vector and $\bar\psi$ a row vector in the matrix sense). Under parity we have

$$\psi'(x) = \gamma_4 \psi(Px), \quad \bar\psi'(x) = \bar\psi(Px)\gamma_4. \tag{D.26}$$

In general ψ and $\bar\psi$ are independent fields, but with the help of the charge-conjugation matrix C we can make a $\bar\psi$-type object out of ψ and vice-versa:

$$\bar\psi^{(c)} \equiv -(C^\dagger \psi)^T = \psi^T C^\dagger, \quad \bar\psi^{(c)\prime}(x) = \bar\psi^{(c)}(R^{-1}x)\Lambda^{-1}$$
$$\psi^{(c)} \equiv (\bar\psi C)^T = -C\bar\psi^T, \quad \psi^{(c)\prime}(x) = \Lambda\psi^{(c)}(R^{-1}x). \tag{D.27}$$

The fields $\bar\psi^{(c)}$ and $\psi^{(c)}$ are called the charge conjugates of ψ and $\bar\psi$, respectively.

Note the standard notation for the projected $\bar\psi$'s,

$$\bar\psi_L = \bar\psi\, P_R, \quad \bar\psi_R = \bar\psi\, P_L. \tag{D.28}$$

This looks unnatural here but it is natural in the operator formalism where $\hat{\bar\psi}_{L,R} \equiv \hat\psi^\dagger_{L,R}\gamma_4 = \hat{\bar\psi}P_{R,L}$. In the path-integral formalism (in real as well as imaginary time) one introduces independent generators $\psi_\alpha(x)$ and $\psi^+_\alpha(x)$ of a Grassmann algebra, which are related by Hermitian conjugation, such that $\psi_{L,R} = P_{L,R}\psi$ implies $\psi^+_{L,R} = \psi^+ P_{L,R}$, and then $\bar{\psi}_{L,R} \equiv \psi^+_{L,R}\gamma_4$ also gives (D.28). The fields $\bar{\psi}_{L,R}$ transform in representations equivalent to $\Lambda_{R,L}$:

$$\bar{\psi}_L \to \bar{\psi}_L \Lambda^\dagger_R \Rightarrow (\bar{\psi}_L C)^T \to \Lambda_R (\bar{\psi}_L C)^T, \qquad (D.29)$$

$$\bar{\psi}_R \to \bar{\psi}_R \Lambda^\dagger_L \Rightarrow (\bar{\psi}_R C)^T \to \Lambda_L (\bar{\psi}_R C)^T, \qquad (D.30)$$

where we used (D.17) and for clarity used the arrow notation for transformations, while suppressing the space–time index x.

An $O(4)$ invariant action which contains all the types of fields introduced so far with a minimum number (>0) of derivatives is given by

$$S = -\int d^4x\, \bar{\psi}(m + \gamma_\mu \partial_\mu)\psi \qquad (D.31)$$

$$= -\int d^4x\, \left[m(\bar{\psi}_L \psi_R + \bar{\psi}_R \psi_L) + \bar{\psi}_L \gamma_\mu \partial_\mu \psi_L + \bar{\psi}_R \gamma_\mu \partial_\mu \psi_R \right].$$

Finally, we can get corresponding formulas for Minkowski space–time by raising indices in contractions such that there is always a contraction between an upper and a lower index, e.g. $\omega_{\mu\nu}\Sigma_{\mu\nu} = \omega^{\mu\nu}\Sigma_{\mu\nu}$ (we do not make a distinction between upper and lower indices in Euclidean space–time), and substituting $x^4 = x_4 \to ix^0 = -ix_0$, $\omega^{4k} = \omega_{4k} \to i\omega^{0k} = -i\omega_{0k}$. This implies that $\partial_4 \to -i\partial_0$, $\partial_0 = \partial/\partial x^0$. It is then also expedient to use $\gamma^0 = -\gamma_0 = -i\gamma_4$. We have to be careful with Hermiticity properties of Λ, because after the substitution it is no longer unitary:

$$\Lambda^{-1} = \beta\Lambda^\dagger\beta, \quad \beta \equiv i\gamma^0, \quad \text{Minkowski}. \qquad (D.32)$$

In Minkowski space–time $\mu = 0,1,2,3$ and indices are raised and lowered with the metric tensor $\eta_{\mu\nu} = \eta^{\mu\nu} = \text{diag}(-1,1,1,1)$, e.g. $\partial^0 = -\partial_0$, $\partial_k = \partial^k = \partial/\partial x^k$.

Notes

Chapter 1

1 To avoid cluttering of brackets, we use the notation $e^2/4\pi \equiv e^2/(4\pi)$, etc. Furthermore, units $\hbar = 1$, $c = 1$ will be used. Then dimensions are like [mass] = [energy] = [momentum] = $[(\text{length})^{-1}] = [(\text{time})^{-1}]$, etc.

2 As a model for mesons we have to take the spins of the quarks into account. In a first approximation we can imagine neglecting spin-dependent forces. Then the maximum spin is $J = L + S$, with L the orbital angular momentum and $S = 0, 1$ the total spin of the quark–antiquark system. The π has the $q\bar{q}$ spins antiparallel, $S = 0$, the ρ has parallel $q\bar{q}$ spins, $S = 1$. In a second approximation spin-dependent forces have to be added, which split the π and ρ masses. In picking the right particles out of the tables of the Particle Data Group [2], we have to choose quantum numbers corresponding to the same S but changing L. This means that the parity and charge-conjugation parity flip signs along a Regge trajectory. The particles on the ρ trajectory in figure 1.3 are $\rho(769)$, $a_2(1320)$, $\rho_3(1690)$, and $a_4(2040)$, those on the π trajectory are $\pi(135)$, $\pi(135)$, $b_1(1235)$, and $\pi_2(1670)$. The mass m_q used in this model is an effective ('constituent') quark mass, $m_u \approx m_d \approx m_\rho/2 = 385$ MeV, which is much larger than the mass parameters appearing in the Lagrangian (the so-called 'current masses'), which are only a few MeV. In the last chapter we shall arrive at an understanding of this in terms of chiral-symmetry breaking.

Chapter 2

1 The formal canonical quantization of the scalar field in the continuum is done as follows. Given the Lagrangian of the system

$$L(\varphi, \dot{\varphi}) = \int d^3x \, \tfrac{1}{2}(\dot{\varphi})^2 - V(\varphi), \qquad (N.1)$$

the canonical momentum follows from varying with respect to $\dot{\varphi}$,

$$\delta_{\dot{\varphi}} L = \int d^3x \, \dot{\varphi} \, \delta\dot{\varphi} \Rightarrow \pi \equiv \frac{\delta L}{\delta \dot{\varphi}} = \dot{\varphi}. \qquad (N.2)$$

Solving for $\dot{\varphi}$ in terms of π, the Hamiltonian is given by the Legendre transformation

$$H(\varphi, \pi) = \int d^3x\, \pi\dot{\varphi} - L(\varphi, \dot{\varphi}) = \int d^3x\, \tfrac{1}{2}\pi^2 + V(\varphi). \qquad \text{(N.3)}$$

Defining the Poisson brackets as

$$(A, B) = \int d^3x\, \frac{\delta A}{\delta\varphi(x)}\frac{\delta B}{\delta\pi(x)} - A \leftrightarrow B, \qquad \text{(N.4)}$$

the canonical (equal time) Poisson brackets are given by

$$(\varphi(\mathbf{x}), \pi(\mathbf{y})) = \delta(\mathbf{x} - \mathbf{y}), \quad (\varphi(\mathbf{x}), \varphi(\mathbf{y})) = 0 = (\pi(\mathbf{x}), \pi(\mathbf{y})). \qquad \text{(N.5)}$$

The Lagrange (stationary-action) equations of motion are then identical to Hamilton's equations

$$\dot{\varphi} = (\varphi, H), \quad \dot{\pi} = (\pi, H). \qquad \text{(N.6)}$$

The canonically quantized theory is obtained by considering the canonical variables as operators $\hat{\varphi}$ and $\hat{\pi}$ in Hilbert space satisfying the canonical commutation relations obtained from the correspondence principle Poisson bracket $\rightarrow$ commutator:

$$[\hat{\varphi}(\mathbf{x}), \hat{\pi}(\mathbf{y})] = i\delta(\mathbf{x} - \mathbf{y}), \quad [\hat{\varphi}(\mathbf{x}), \hat{\varphi}(\mathbf{y})] = 0 = [\hat{\pi}(\mathbf{x}), \hat{\pi}(\mathbf{y})]. \qquad \text{(N.7)}$$

Observables such as the Hamiltonian become operators (after symmetrizing products of $\hat{\varphi}$ and $\hat{\pi}$, if necessary). The quantum equations of motion then follow from Heisenberg's equations

$$\partial_0\hat{\varphi} = i[\hat{H}, \varphi], \quad \partial_0\hat{\pi} = i[\hat{H}, \hat{\pi}]. \qquad \text{(N.8)}$$

These need not, but often do, coincide with the classical equations of motion transcribed to $\hat{\varphi}$ and $\hat{\pi}$. From (N.7) one observes that the quantum fields are 'operator-valued distributions', hence products like $\hat{\pi}^2$ occuring in the formal Hamiltonian are mathematically ill-defined.

Chapter 4

1 The derivation leading to (4.72) is how I found the lattice gauge-theory formulation in 1972 (cf. [42]). I still find it instructive how a pedestrian approach can be brought to a good ending.

Chapter 8

1 Only Abelian chiral transformations form a group: if V and W are two chiral transformations, then $U = VW = V_L W_L P_L + V_L^\dagger W_L^\dagger P_R$ has $U_L = V_L W_L \neq U_R^\dagger = W_L V_L$, unless V_L and W_L commute.

2 This can be checked here by re-installing the lattice spacing, writing $M_f = m_f + 4r/a$, and $\psi_{fx} = a^{3/2}\psi_f(x)$, etc. with continuum fields $\psi(x)$, $\bar{\psi}(x)$ that are smooth on the lattice scale (the emerging overall factor a^3 must be dropped to get the continuum currents and divergences). Using for convenience the two-index notation for the lattice gauge field

$(U_{\mu x} = U_{x,x+\mu}, U^{\dagger}_{\mu x-\hat{\mu}} = U_{x,x-\mu})$, we may write
$U_{x,x\pm\hat{\mu}a}\psi_g(x \pm \hat{\mu}a) = \psi_g(x) \pm aD_{\mu}\psi_g(x) + \frac{1}{2}a^2 D^2_{\mu}\psi_g(x) + \cdots$, with
$D_{\mu}\psi_g(x) = [\partial_{\mu} - igG_{\mu}(x)]\psi_g(x)$ the continuum covariant derivative, this
gives the expected result.

3 The way Σ is introduced here corresponds to four staggered flavors,
$\Sigma = \sum_{f=1}^{4}\langle\bar{\psi}_f\psi_f\rangle$. Using the $SU(2)$ value $a\sqrt{\sigma} = 0.2634(14)$ [69] and $\sqrt{\sigma}$
$= 420$ MeV, the ratio $(0.00863/4)^{1/3}/0.263 = 0.491$ corresponds to 206
MeV or $\Sigma = 4(206\,\text{MeV})^3$. This number appears somewhat small, but we
have to keep in mind that this is for $SU(2)$, not $SU(3)$, and it also has to
be multiplied by the appropriate renormalization factor.

4 For staggered fermions to be sensitive to topology, quenched $SU(3)$ gauge
couplings need to be substantially smaller than the value $\beta = 6/g^2 = 5.1$
used in [143, 144]. Vink [116, 117] found that values $\beta \gtrsim 6$ were needed in
order to obtain reasonable correlations between the 'fermionic topological
charge' and the 'cooling charge' (cf. figure 8.2). Note that the change
$\beta = 5.1 \to 6$ corresponds to a decrease in lattice spacing by a factor of
about four.

5 Ironically, when the mechanism of canceling the anomalies out between
different fermion species was proposed [148], I doubted that it was
necessary, and this was one of the reasons (apart from a non-perturbative
formulation of non-Abelian gauge theory) why I attempted to put the
electroweak model on the lattice. On calculating the one-loop gauge-field
self-energy and the triangle diagram, I ran into the species-doubling
phenomenon, without realizing that the lattice produced the very
cancellation mechanism I had wanted to avoid.

6 At the time of writing the direct Euclidean approach is considered
suspect and a Lorentzian formulation is being pursued [176]. For an
impression of what is involved in a non-perturbative computation of
gravitational binding energy, see [177].

7 The problem here is that, in order to deal with the oscillating phase
$\exp(iS)$ in the path integral, one has to make approximations right from
the beginning. To incorporate sphalerons, kinks, etc. one needs a lattice
formulation that allows arbitrarily *in*homogeneous field configurations
[178, 179].

References

[1] B.S. DeWitt, *Dynamical Theory of Groups and Fields*, Blackie, London 1965.

[2] http://pdg.lbl.gov

[3] T-P. Cheng and L-F. Li, *Gauge Theory of Elementary Particle Physics*, Oxford University Press, Oxford 1988.

[4] M. Creutz, *Quarks, Gluons and Lattices*, Cambridge University Press, Cambridge 1983.

[5] C. Rebbi, *Lattice Gauge Theories and Monte Carlo Simulations*, World Scientific, Singapore 1983.

[6] C. Itzykson and J.M. Drouffe, *Statistical Field Theory I & II*, Cambridge University Press, Cambridge 1989.

[7] G. Parisi, *Statistical Field Theory*, Addison-Wesley, New York 1988.

[8] J. Zinn-Justin, *Quantum Field Theory and Critical Phenomena*, Clarendon, Oxford 1989.

[9] H.J. Rothe, *Introduction to Lattice Gauge Theories*, World Scientific, Singapore 1992.

[10] G. Münster and I. Montvay, *Quantum Fields on a Lattice*, Cambridge University Press, Cambridge 1994.

[11] M. Le Bellac and G. Barton, *Quantum and Statistical Field Theory*, Clarendon, Oxford 1992.

[12] M. Nakahara, *Geometry, Topology and Physics*, Adam Hilger, Bristol 1991.

[13] R.B. Dingle, *Asymptotic Expansions, their Derivation and Interpretation*, Academic Press, London and New York 1973.

[14] S.R. Sharpe, Phenomenology from the lattice, TASI 94, hep-ph/9412243.

[15] R. Gupta, Introduction to lattice QCD, Les Houches 1997, hep-lt/9807028.

[16] M. Lüscher, Advanced lattice QCD, Les Houches 1997, hep-lat/9802029.

[17] J. Ambjørn, Quantization of geometry, Les Houches 1994, hep-th/9411179.

[18] P. van Baal, QCD in a finite volume, in Festschrift in honor of B.L. Ioffe, ed. M. Shifman, hep-ph/0008206.

[19] H. Leutwyler, Chiral dynamics, in Festschrift in honor of B.L. Ioffe, ed. M. Shifman, hep-ph/0008124; J. Gasser and H. Leutwyler, *Ann. Phys.* **158** (1984) 142.

[20] M. Lüscher and P. Weisz, *Nucl. Phys.* **B290** [FS20] (1987) 25.

[21] M. Lüscher and P. Weisz, *Nucl. Phys.* **B295** [FS21] (1988) 65.

[22] M. Lüscher and P. Weisz, *Nucl. Phys.* **B300** [FS22] (1988) 325.

[23] M. Lüscher and P. Weisz, *Nucl. Phys.* **B318** (1989) 705; *Phys. Lett.* **B212** (1988) 472.

[24] C. Frick, K. Jansen, J. Jersák, I. Montvay, P. Seuferling and G. Münster, *Nucl. Phys.* **B331** (1990) 515.

[25] J. Kuti and Y. Shen, *Phys. Rev. Lett.* **60** (1988) 85.

[26] L. O'Raifeartaigh, A. Wipf and H. Yoneyama, *Nucl. Phys.* **B271** (1986) 653.

[27] A. Hasenfratz, K. Jansen, J. Jersák, H.A. Kastrup, C.B. Lang, H. Leutwyler and T. Neuhaus, *Nucl. Phys.* **B356** (1991) 332.

[28] M. Göckeler, H.A. Kastrup, T. Neuhaus and F. Zimmermann, *Nucl. Phys.* **B404** (1993) 517.

[29] K.G. Wilson, *Phys. Rev.* **B4** (1971) 3174; B4 (1971) 3184; *Phys. Rev.* **D7** (1973) 2911.

[30] J.B. Kogut and K.G. Wilson, *Phys. Rep.* **12** (1974) 75.

[31] R. Dashen and H. Neuberger, *Phys. Rev. Lett.* **50** (1983) 1897.

[32] H. Neuberger, *Nucl. Phys. B (Proc. Suppl.)* **17** (1992) 17; *Nucl. Phys. B (Proc. Suppl.)* **29B,C** (1992) 19, and references therein.

[33] J. Kuti, *Nucl. Phys. B (Proc. Suppl.)* (1993), talk at Lattice '92.

[34] P. Hasenfratz, Lattice '88.

[35] J. Jersák, Lattice studies of the Higgs system, Proceedings of the Eloisatron Project Workshop on Higgs Particles, Erice 1989.

[36] W. Bock, J. Smit and J.C. Vink, *Phys. Lett.* **B291** (1992) 297; W. Bock, C. Frick, J. Smit and J.C. Vink, *Nucl. Phys.* **B400** (1993) 309.

[37] H. Neuberger, *Nucl. Phys. B (Proc. Suppl.)* **29B,C** (1992) 19.

[38] U.M. Heller, *Nucl. Phys. B (Proc. Suppl.)* **34** (1994) 101.

[39] K.G. Wilson, *Phys. Rev.* **D10** (1974) 2445.

[40] J.B. Kogut and L. Susskind, *Phys. Rev.* **D11** (1975) 395.

[41] A.M. Polyakov, *Phys. Lett.* **B59** (1975) 82.

[42] K.G. Wilson, Future directions in particle theory, Proceedings of the Lepton and Photon Interactions at High Energies Workshop, Ithaca, 1983, pp. 812–834, p. 827.

[43] J.B. Kogut, *Rev. Mod. Phys.* **55** (1983) 775.

[44] R. Balian, J.M. Drouffe and C. Itzykson, *Phys. Rev.* **D10** (1974) 3376; **D11** (1975) 2098; **D11** (1975) 2104, Erratum **D19** (1979) 2514.

[45] A. Hasenfratz and P. Hasenfratz, *Phys. Lett.* **B93** (1980) 165.

[46] R. Dashen and D.J. Gross, *Phys. Rev.* **D23** (1981) 2340.

[47] P. Weisz, *Phys. Lett.* **B100** (1981) 331.

[48] A. Billoire, *Phys. Lett.* **B104** (1981) 472.

[49] Y. Nambu, *Phys. Rev.* **D10** (1974) 4262.

[50] G. 't Hooft, in *High Energy Physics*, Proceedings of the EPS International Conference, Palermo 1975, ed. A. Zichichi (Editrice Compositori, Bologna, 1976); *Nucl. Phys.* **B190** [FS3] (1981) 455; *Phys. Scr.* **25** (1982) 133.

[51] G. 't Hooft, *Nucl. Phys.* **B138** (1978) 1.

[52] G. Mack, in *Recent Progress in Gauge Theories*, Cargèse 1979, eds.

G. 't Hooft *et al.*, Plenum Press, New York 1980; *Phys. Rev. Lett.* **45** (1980) 1378.

[53] J. Smit and A.J. van der Sijs, *Nucl. Phys.* **B355** (1991) 603; **B422** (1994) 349.

[54] A. di Giacomo, *Nucl. Phys. B (Proc. Suppl.)* **47** (1996) 136; M.I. Polikarpov, *Nucl. Phys. B (Proc. Suppl.)* **53** (1997) 134; L. del Debbio, M. Faber, J. Greensite and S. Olejnik, *Nucl. Phys. B (Proc. Suppl.)* **53** (1997) 141; P. van Baal, *Nucl. Phys. B (Proc. Suppl.)* **63** (1998) 126; M. García Pérez, *Nucl. Phys. B (Proc. Suppl.)* **94** (2001) 27.

[55] P. van Baal, hep-th/0109148.

[56] M. Faber, J. Greensite and S. Olejnik, hep-lat/0109032

[57] J. Ambjørn, J. Giedt and J. Greensite, JHEP 0002 (2000) 033.

[58] Y. Koma, E.-M. Ilgenfritz, H. Toki and T. Suzuki, *Phys. Rev.* **D64** (2001) 011501.

[59] F.V. Gubarev, E.-M. Ilgenfritz, M.I. Polikarpov and T. Suzuki, *Phys. Lett.* **B468** (1999) 134.

[60] G.P. Lepage, P.B. Mackenzie, *Phys. Rev.* **D48** (1993) 2250.

[61] R. Sommer, *Nucl. Phys.* **B411** (1994) 839.

[62] S. Necco and R. Sommer, hep-lat/0108008.

[63] S. Capitani, M. Lüscher, R. Sommer and H. Wittig, *Nucl. Phys.* **B544** (1999) 669.

[64] S. Necco and R. Sommer, hep-lat/0109093.

[65] G. Bali, *Phys. Rep.* **343** (2001) 1.

[66] P. Weisz, *Nucl. Phys. (Proc. Suppl.)* **47** (1996) 71, hep-lat/9511017.

[67] G. 't Hooft, *Nucl. Phys.* **B72** (1974) 461.

[68] B. de Wit and G. 't Hooft, *Phys. Lett.* **B69** (1977) 61.

[69] B. Lucini and M. Teper, JHEP 0106 (2001) 050.

[70] L.H. Karsten and J. Smit, *Nucl. Phys.* **B183** (1981) 103.

[71] K.G. Wilson, in *New Phenomena in Subnuclear Physics*, ed. A. Zichichi, Plenum, New York 1977 (Erice 1975).

[72] L. Susskind, *Phys. Rev.* **D16** (1977) 3031; Coarse grained quantum chromodynamics, in *Weak and Electromagnetic Interactions at High Energies*, North-Holland, Amsterdam 1977 (Les Houches 1976).

[73] M.F.L. Golterman and J. Smit, *Phys. Lett.* **B140** (1984) 392.

[74] M.F.L. Golterman and J. Smit, *Nucl. Phys.* **B245** (1984) 64.

[75] M.F.L. Golterman and J. Smit, *Nucl. Phys.* **B255** (1985) 328.

[76] M.F.L. Golterman, *Nucl. Phys.* **B273** (1986) 666.

[77] M.F.L. Golterman, *Nucl. Phys.* **B278** (1986) 417.

[78] H.S. Sharatshandra, H.J. Thun and P. Weisz, *Nucl. Phys.* **B192** (1981) 205.

[79] C. van den Doel and J. Smit, *Nucl. Phys.* **B228** (1983) 122.

[80] J. Smit, *Nucl. Phys. B (Proc. Suppl.)* **29B,C** (1992) 83.

[81] P. Becher and H. Joos, *Z. Phys.* **C15** (1982) 343. M. Göckeler and H. Joos, in *Progress in Field Theory*, eds. G. 't Hooft *et al.* Plenum, New York 1984.

[82] J. Hoek and J. Smit, *Nucl. Phys.* **B263** (1986) 129.

[83] N. Kawamoto and J. Smit, *Nucl. Phys.* **B192** (1981) 100.

[84] J. Hoek, N. Kawamoto and J. Smit, *Nucl. Phys.* **B199** (1982) 495.

[85] J. Hoek, *J. Comput. Phys.* **49** (1983) 265; **54** (1984) 245.

[86] K.G. Wilson, in *New Phenomena in Quantum Field Theory and*

Statistical Mechanics, ed. M. Lévy and P. Mitter, Plenum, New York 1977 (Cargèse 1976).

[87]   M. Lüscher, *Commun. Math. Phys.* **54** (1977) 283.

[88]   M. Creutz, *Phys. Rev.* **D35** (1987) 1460.

[89]   J. Smit, *Nucl. Phys. B (Proc. Suppl.)* **20** (1991) 542.

[90]   W. Bock, J.E. Hetrick and J. Smit, *Nucl. Phys.* **B437** (1995) 585.

[91]   G. Münster, *Nucl. Phys.* **B190** (1981) 439; Erratum **B205** (1982) 648.

[92]   K. Seo, *Nucl. Phys.* **B209** (1982) 200.

[93]   G. Bali and K. Schilling, *Phys. Rev.* **D47** (1993) 661.

[94]   M. Luscher, R. Sommer, P. Weisz and U. Wolff, *Nucl. Phys.* **B413** (1994) 481.

[95]   J. Jersák, T. Neuhaus and P.M. Zerwas, *Nucl. Phys.* **B251** (1985) 299.

[96]   E. Witten, *Nucl. Phys.* **B160** (1979) 57.

[97]   CP-PACS collaboration, A. Alikhan *et al.*, hep-lat/0105015.

[98]   CP-PACS collaboration, S. Aoki *et al.*, *Phys. Rev. Lett.* **84** (2000) 238.

[99]   UKQCD collaboration, C.R. Allton *et al.*, hep-lat/0107021.

[100]   S.R. Coleman and E. Witten, *Phys. Rev. Lett.* **54** (1980) 100.

[101]   G. 't Hooft, *Phys. Rep.* **142** (1986) 357.

[102]   E. Witten, *Nucl. Phys.* **B156** (1979) 269.

[103]   G. Veneziano, *Nucl. Phys.* **B159** (1979) 213.

[104]   J. Smit and J.C. Vink, *Nucl. Phys.* **B284** (1987) 234.

[105]   P. di Vecchia, K. Fabricius, G.C. Rossi and G. Veneziano, *Nucl. Phys.* **B192** (1981) 392; *Phys. Lett.* **B108** (1982) 323.

[106]   S.L. Adler, *Phys. Rev.* **177** (1969) 2426.

[107]   J. Smit and J.C. Vink, *Nucl. Phys.* **B286** (1987) 485.

[108]   M.L. Laursen, J. Smit and J.C. Vink, *Nucl. Phys.* **B343** (1990) 522.

[109]   R. Groot, J. Hoek and J. Smit, *Nucl. Phys.* **B237** (1984) 111.

[110]   G.C. Rossi, *Nucl. Phys. B (Proc. Suppl.)* **53** (1997) 3, hep-lat/9609038.

[111]   B. Allés, M. Campostrini, A. di Giacomo, Y. Gündüc and E. Vicari, *Phys. Rev.* **D48** (1993) 2284; B. Allés, M. D'Elia, A. di Giacomo and R. Kirchner, *Phys. Rev.* **D58** (1998) 114506.

[112]   M. Teper, *Nucl. Phys. B (Proc. Suppl.)* **83–84** (2000) 146; M. Teper, *Phys. Lett.* **B202** (1988) 553; J. Hoek, M. Teper and J. Waterhouse, *Nucl. Phys.* **B288** (1987) 589.

[113]   J. Smit and J.C. Vink, *Nucl. Phys.* **B298** (1988) 557.

[114]   J. Smit and J.C. Vink, *Nucl. Phys.* **B303** (1988) 36.

[115]   J.C. Vink, *Nucl. Phys.* **B307** (1988) 549.

[116]   J.C. Vink, *Phys. Lett.* **B210** (1988) 211.

[117]   J.C. Vink, *Phys. Lett.* **B212** (1988) 483.

[118]   M. Bochicchio, L. Maiani, G. Martinelli, G. Rossi and M. Testa, *Nucl. Phys.* **B262** (1985) 331.

[119]   J. Smit, *Nucl. Phys.* **B175** (1980) 307.

[120]   J. Greensite and J. Primack, *Nucl. Phys.* **B189** [FS2] (1981) 170.

[121]   H. Kluberg-Stern, A. Morel, O. Napoly and B. Petersson, *Nucl. Phys.* **B190** [FS3] (1981) 504.

[122]   S. Aoki, *Phys. Rev.* **D30** (1984) 2653; *Nucl. Phys.* **B314** (1989) 79.

[123]   J. Fröhlich and C. King, *Nucl. Phys.* **B290** (1987) 157.

[124]   P.H. Ginsparg and K.G. Wilson, *Phys. Rev.* **D25** (1982) 2649.

[125]   P. Hasenfratz, The theoretical background and properties of perfect actions, hep-lat/9803027; *Nucl. Phys.* **B525** (1998) 401.

[126] H. Neuberger, *Phys. Lett.* **B417** (1998) 141.

[127] R. Narayanan and H. Neuberger, *Phys. Lett.* **B302** (1993) 62; *Nucl. Phys.* **B412** (1994) 574; *Nucl. Phys.* **B443** (1995) 305.

[128] D.B. Kaplan, *Phys. Lett.* **B288** (1992) 342.

[129] Y. Shamir, *Nucl. Phys.* **B406** (1993) 90.

[130] P. Hasenfratz, *Nucl. Phys.* **B525** (1988) 401.

[131] M. Lüscher, *Phys. Lett.* **B428** (1998) 342.

[132] K. Fujikawa, *Phys. Lett.* **42** (1979) 1195; *Phys. Rev.* **D21** (1980) 2848; **D22** (1980) 1409.

[133] W. Kerler, *Phys. Rev.* **D23** (1981) 2384.

[134] E. Seiler and I.O. Stamatescu, *Phys. Rev.* **D25** (1982) 2177; **D26** (1982) 534.

[135] F. Niedermayer, *Nucl. Phys. B (Proc. Suppl.)* **73** (1999) 105.

[136] R.G. Edwards, U.M. Heller and R. Narayanan, *Nucl. Phys.* **B535** (1998) 403.

[137] L. Giusti, C.G. Rossi, M. Testa and G. Veneziano, hep-lat/0108009.

[138] T. Banks and A. Casher, *Nucl. Phys.* **169** (1980) 103.

[139] J.J.M. Verbaarschot, *Nucl. Phys. B (Proc. Suppl.)* **53** (1997) 88.

[140] P.H. Damgaard, hep-lat/0110192.

[141] E.V. Shuryak and J.J. Verbaarschot, *Nucl. Phys.* **A560** (1993) 306.

[142] M.E. Berbenni-Bitsch, S. Meyer, A. Schäfer, J.J.M. Verbaarschot and T. Wettig, *Phys. Rev. Lett.* **80** (1998) 1146.

[143] P.H. Damgaard, U.M. Heller, R. Niclasen and K. Rummukainen, *Phys. Rev.* **D61** (2000) 014501.

[144] P.H. Damgaard, U.M. Heller, R. Niclasen and B. Svetitsky, hep-lat/0110028.

[145] F. Farchioni, Ph. de Forcrand, J. Hip, C.B. Lang and K. Splittorff, *Phys. Rev.* **D62** (2000) 014503.

[146] P. Hernandez, K. Jansen and L. Lellouch, *Phys. Lett.* **B469** (1999) 198.

[147] P. Hernandez, K. Jansen and L. Lellouch, JHEP 0107 (2001) 018.

[148] C. Bouchiat, J. Iliopoulos and Ph. Meyer, *Phys. Lett.* **B38** (1972) 519. D.J. Gross and R. Jackiw, *Phys. Rev.* **D6** (1972) 477.

[149] J. Smit, *Acta Phys. Polon.* **B17** (1986) 531.

[150] P.V.D. Swift, *Phys. Lett.* **145B** (1984) 256.

[151] E. D'Hoker and E. Farhi, *Nucl. Phys.* **B248** (1984) 59, 77.

[152] T. Banks and A. Dabholkar, *Phys. Rev.* **D46** (1992) 4016; *Nucl. Phys. B (Proc. Suppl.)* **29B,C** (1992) 46.

[153] W. Bock, A.K. De, K. Jansen, J. Jersák, T. Neuhaus and J. Smit, *Nucl. Phys.* **B344** (1990) 207–237.

[154] W. Bock, A.K. De and J. Smit, *Nucl. Phys.* **B388** (1992) 243.

[155] M.F.L. Golterman, D.N. Petcher and J. Smit, *Nucl. Phys.* **B370** (1992) 51.

[156] E. Eichten and J. Preskill, *Nucl. Phys.* **B268** (1986) 179.

[157] M. Creutz, M. Tytgat, G. Rebbi and S.-S. Xue, *Phys. Lett.* **B402** (1997) 341.

[158] M.F.L. Golterman, D.N. Petcher and E. Rivas, *Nucl. Phys.* **B395** (1993) 596.

[159] I. Montvay, *Phys. Lett.* **B199** (1987) 89; *Nucl. Phys. B (Proc. Suppl.)* **29B,C** (1992) 159.

[160] H.B. Nielsen and M. Ninomiya, *Nucl. Phys.* **B185** (1981) 20; **B193** (1981) 173.

[161] L.H. Karsten, *Phys. Lett.* **B104** (1981) 315.

[162] J. Smit, *Nucl. Phys. B (Proc. Suppl.)* **4** (1988) 415.

[163] A. Borrelli, L. Maiani, G.C. Rossi, R. Sisto and M. Testa, *Phys. Lett.* **B221** (1989) 360; *Nucl. Phys.* **B333** (1990) 355.

[164] G.C. Rossi, R. Sarno and R. Sisto, *Nucl. Phys.* **B398** (1993) 101.

[165] W. Bock, J.E. Hetrick and J. Smit, *Nucl. Phys.* **B437** (1995) 585.

[166] M.F.L. Golterman and Y. Shamir, *Phys. Lett.* **B399** (1997) 148.

[167] D. Förster, H.B. Nielsen and M. Ninomiya, *Phys. Lett.* **B94** (1990) 135.

[168] M. Golterman, *Nucl. Phys. B (Proc. Suppl.)* **94** (2001) 189.

[169] J. Smit, *Nucl. Phys. B (Proc. Suppl.)* **29B,C** (1992) 83.

[170] W. Bock, J. Smit and J.C. Vink, *Nucl. Phys.* **B414** (1994) 73; **B416** (1994) 645.

[171] J. Smit, *Nucl. Phys. B (Proc. Suppl.)* **17** (1990) 3; J. Shigemitsu, *Nucl. Phys. B (Proc. Suppl.)* **20** (1991) 515; M. Testa, *Nucl. Phys. B (Proc. Suppl.)* **26** (1992) 228; D.N. Petcher, *Nucl. Phys. B (Proc. Suppl.)* **30** (1993) 50; M. Creutz, *Nucl. Phys. B (Proc. Suppl.)* **42** (1995) 56; Y. Shamir, *Nucl. Phys. B (Proc. Suppl.)* **47** (1996) 212.

[172] M. Lüscher, *Nucl. Phys.* **B549** (1999) 295; **B568** (2000) 162; JHEP 0006 (2000) 028.

[173] R. Narayanan, *Nucl. Phys. B (Proc. Suppl.)* **34** (1994) 95.

[174] H. Neuberger, *Nucl. Phys. B (Proc. Suppl.)* **83–84** (2000) 67.

[175] M. Lüscher, *Nucl. Phys. B (Proc. Suppl.)* **83–84** (2000) 34; Chiral gauge theories revisited, hep-th/0102028.

[176] J. Ambjørn, J. Jurkiewicz and R. Loll, hep-th/0105267.

[177] B.V. de Bakker and J. Smit, *Nucl. Phys.* **B484** (1997) 476.

[178] G. Aarts and J. Smit, *Nucl. Phys.* **B555** (1999) 355.

[179] M. Salle, J. Smit and J.C. Vink, *Phys. Rev.* **D64** (2001) 025016.

Index

Abelian gauge theory, 97–98
Abelian group, 88
Abrikosov flux tubes, 139
action
  continuum, 11
  discretized, 10
  functional, 9
  plaquette, 140
  Symanzik-improved, 66–67
action density, *see* Lagrange function
anharmonic oscillator, 6, 12
anticommuting numbers, 149
antiferromagnetic phase, 38
antiquarks, 179, 181
asymptotic freedom, 3, 121–125, 189
asymptotic scaling, 140–144
Atiyah–Singer index theorem, 204
averaging function, 68
axial vector current, 202
axis-reversal symmetry, 74

bare coupling, 141, 180
  constant, 50
baryons, 2, 175, 179–180, 184–185
  masses, 184
  propagators, 173–177
  three-quark hopping, 177
Berezin integral, 246
beta function, 49, 55–56, 58, 122
  coefficient, 59
  trick, 49
Born approximation, 59
Bose-field denominator, 154
bosons, 2, 26, 77, 158

Callan–Symanzik
  beta function, 55
  coefficient, 57
  equations, 49, 56
canonical quantization, 106

Casimir operator, 232
  quadratic, 113, 120
character expansion, 128
charge quantization, 85
charge-conjugation matrix, 174
Chern classes, 203
chiral
  anomaly, 202–204, 227
  gauge theory, 217–223
  perturbation theory, 63, 195
  symmetry, 193–228
  symmetry breaking, 182, 212
  transformation, 207
class function, 104
classical
  action, 65, 95
  continuum limit, 18
  field theory, 6
  ground state, 195
  Hamilton equations, 4
  QCD action, 200
Coleman's theorem, 32
color index, 87, 171
commutation relations, 14, 29, 231, 242
complex Higgs doublet, 75
composite scalar fields, 5
confinement
  mechanisms, 138–140
  phase, 136–138
contact terms, 208
contour integration, 24–25
coordinate representation, 15
correlation function, 27, 30, 38, 61–62,
  68, 173
correlation length, 38, 69, 140
Coulomb
  phase, 136–138
  potential, 119
  self-mass, 139
coupling

renormalized, 46, 70, 190
    tuning, 74
covariant derivative, 85, 87
creation and annihilation operators,
    13–14, 25–26, 29
critical
    exponent, 38
    fixed point, 69, 71
    lines, 37
    phenomena, 38
    surface, 69–70
current algebra, 211

Dirac
    action, naive discretization, 149–151
    fermions, 161, 220
    fields, 155
    index, 87, 149, 171, 213, 256
    matrices, 150, 162, 253, 254
    operator eigenvalues, 214
    particles, 151
    spinor, 160
domain-wall fermions, 212
dressed
    particle states, 26
    propagator, 62

effective action, 68, 70, 190, 197
effective interaction, 2–3, 6, 62
electroweak gauge fields, 75
elementary scalar fields, 5
energy
    density, 6–7
    spectrum, 29
    zero point, 141
equation of motion, 6, 33
Euclidean action, 32
Euclidean formalism, 16
evolution operator, 16
expectation values, 54

Faddeev–Popov
    factor, 147
    measure, 146
fermion, 2, 151, 180
    free, 225
    masses, 76
    staggered, 149, 160–161, 169, 213–214
fermion action, 156
fermion doubling, 92, 155
fermion fields
    integration over, 170–171
fermion Hamiltonian, 165
fermion propagator, 152
fermion–antifermion correlation
    function, 176
fermionic coherent states, 242–252
Feynman diagrams, 39

Feynman gauge, 116
    self-energy, 120
Feynman propagator, 27
field configurations
    non-perturbative, 139
fine-structure constant, 1, 119, 190
finite renormalization constants, 208
finite-size effects, 146
fixed point
    infrared stable, 51
    ultraviolet stable, 51
flavor index, 87, 171, 213
flux lines
    adjoint representation, 132
    fundamental representation, 132
Fokker–Planck Hamiltonian, 82
formal continuum Hamiltonian, 106
formal continuum limit, 106
four-point function, 46
four-point vertex, 52
Fourier transform, 20–21
free scalar field, 22–25
    action, 22

gamma matrices, 160
gauge coupling, 76, 78
    compution from masses, 190
gauge fixing, 144, 146
gauge potential, 97
gauge theory
    bare coupling, 123
    dimensional transmutation, 124
gauge transformations, 97
gauge-field quantization
    temporal gauge, 239
gauge-fixing action, 147
gauge-Higgs systems
    lattice formulation, 78
gauge-invariant
    action, 95, 145
    correlation function, 173
    measure, 98
    observable, 145
    staggered-fermion action, 160
Gauss-law condition, 102
Gaussian
    fixed point, 70
Gell-Mann matrices, 86
generalized Ising model, 35
Ginsparg–Wilson
    Dirac operator, 223
    fermions, 223, 226
    relation, 211
glueball, 2–3, 97, 115, 135–136, 140,
    145, 180
    mass, 114, 144
    mass ratio, 140
gluon, 1–2, 120, 135

Goldstone bosons, 32, 34, 58, 63, 77, 198
  finite-size effects, 63
  massless, 58
Goldstone's theorem, 34
Grand Unified Theories, 85
Grassmann algebra, 244–245, 256
  independent generators, 150
Grassmann representation, 162
Grassmann variables, 149, 242, 245
Gribov copies, 147
ground state, 6–7, 15, 31, 33, 114
group velocity, 7

Haar measure, 96, 169
hadrons, 1–3, 170, 178–180, 185, 190
Hamilton
  function, 8
  operator, 8, 11
Hamiltonian, 4, 20, 105, 156, 214
  continuous time, 105–106
  operator, 13
harmonic oscillator, 13
  discretized, 14
Heisenberg
  model, 35
  operators, 30
Hermitian, 86
  conjugate, 149, 244
  Hamiltonian, 11, 102
  matrix, 153
  operator, 82, 99, 102, 242
Higgs
  field, 75
  mass, 76–77
  particle, 77
Hilbert space, 8, 99–100, 243
  fermion, 155
hopping
  expansion, 51–55, 57, 62, 66–67,
    171–173
  matrix, 53
  result, 80
Hurewicz measure, 96

index theorem, 217, 225–226
infrared divergencies, 58
intermediate-distance regime, 142
irreducible representations, 103

kinetic-energy operator, 102–105
Kogut–Susskind Hamiltonian, 106

Lagrange function, 6, 8, 10
  discretized, 10
Lagrangian, 6, 239
lambda scale, 140
Landau pole, 49
Langevin algorithm, 60

Laplace–Beltrami operator, 236
lattice
  artifact region, 167
  derivatives, 90
  finite-size effects, 61
  formulation, 209
  gauge field, 90–95, 105, 145
  path integral, 95–97
  regularization, 8–31, 34, 76, 167, 227
  scaling violations, 61
  symmetry, 73, 168
Levi-Civita tensor, 254
link variable, 134
  integration, 131
Lorentz covariant expression, 25
Lüscher–Weisz solution, 55–60

magnetic monopoles, 139
magnetization observable, 63
Majorana fermions, 161
Maxwell theory, 107
mean-field approximation, 37–38, 54
Merwin–Wagner theorem, 32
mesons, 2, 175–177, 179–182, 185, 199,
    201
  propagators, 173–177
Metropolis algorithm, 60–61
Mexican hat potential, 33–34
microscopic spectral density, 216
minimal subtraction, 80, 191
Minkowski-space propagator, 27
Minkowski space–time, 149
Monte Carlo methods, 60–61
MS-bar scheme, 124, 213

Nambu–Goldstone bosons, 196–197,
    199, 213
Neuberger's Dirac operator, 217
no-go theorem, 221
Noether currents, 205
non-Abelian gauge theory, 90
  quadratic divergencies, 169

$O(4)$ model, 55, 59–60, 74–77
$O(n)$ model, 32–82
occupation number, 29
one-loop diagram, 46
operator mixing, 208

parallel transport, 92, 94
paramagnetic strong coupling, 220
partition function, 13, 22–23, 35, 51,
    97–98
  canonical, 13
path integral, 8–31, 38, 67, 98–99, 102,
    146, 149–151, 202, 204, 256
  discretized, 10
  expectation value, 39

imaginary-time, 12
measure, invariant, 207
path-ordered product, 92
Pauli matrices, 75, 86, 253
perfect actions, 211
periodic boundary conditions, 13, 17
periodic delta function, 148
perturbation theory, 1, 3, 48
  renormalized, 46, 56, 63
phase transition, 37–38
φ^4 model, 6, 16, 33, 71–72
  one-loop approximation, 79
photons, 1, 135, 139
  self-energy, 166
pion–nucleon
  mass ratio, 184
  sigma model, 224
pions, 1, 5, 34, 224
  low-energy action, 32
Planck mass, 76
plaquette, 102, 125–127, 131–133
  correlator, 136
  field, 97
  hypercubic lattice, 94
  spacelike, 106, 127
  timelike, 127
  tube, 132
  Wilson loop, 108
Poincaré–Hopf theorem, 222
Polyakov line, 107, 111–112
polymers, 128
polynomials
  contact terms, 74
Pontryagin index, 203
propagators, 52
pseudoscalar masses, 182, 199–202
pseudoscalar mesons, 180

QCD, 1, 2, 5, 26, 76–77, 83, 85, 90, 95,
  138, 143, 159–160, 165, 173, 186,
  188, 190, 193, 195, 197, 200, 202,
  212, 217–218, 241
  action, 85–90, 193, 226
  confinement, 139
  continuum limit, 160
  lambda scales, 124
  low-mass hadrons, 170–192
  multicolor, 214
  parameters, 188–189
  scaling region, 178
  theta parameter, 211, 226
  weak-coupling expansion, 123
QED, 1, 83, 176, 217–220, 241
  action, 83–85
  electrons, 83
  fermion propagation, 177
  gauge invariance, 84

quantum chromodynamics, *see* QCD
quantum electrodynamics, *see* QED
quark, 1–2, 135, 160, 170, 180–182, 187,
  193
quark–antiquark potential, 2, 190
quark–gluon theory, 224
quenched approximation, 209
quenched simulation, 187

random-walk approximation, 53–54
real-space renormalization-group
  method, 67, 71
Regge slope, 193
Regge trajectories, 3
regularization
  Pauli–Villars, 77
  space, 77
renormalization-group
  beta functions, 48–51, 121
  Callan–Symanzik, 55
  equation, 48, 123
  transformation, 68
renormalized mass parameter, 46–47
representation
  n-ality, 135
  ρ mesons, 34
  ρ resonance, 78
rotation invariance, 28
rotation method, 65
running coupling, 47, 49, 51

saddle-point
  approximation, 65
  expansion, 71
scalar field, 5, 15, 32, 75, 125, 174, 194,
  224
  propagator, 23, 28
scale factor, 68
scaling
  arguments, 216
  numerical results, 140–144
Schur's lemma, 104
Schwinger's Source Theory, 56
sea-quark
  loops, 176, 180
semiclassical expansion, *see*
  weak-coupling expansion
σ model, 34
  action, 224
small-distance regime, 142–143
space–time, xi, 32, 67, 90, 156, 209, 257
  domain, 6
  index, 171
  Minkowski, 257
  symmetry, 16, 74
  volume, 146, 151
spatial links, 106

species doubling, 149, 151–156
spectral density, 215
spectral representation, 26, 31, 38
spin models generalized, 35
spinor fields, 253–257
spinor representation, 254
Standard Model, 32, 58, 76, 218–220
  anomaly-free, 219
  Higgs–Yukawa sector, 161
  lattice formulation, 78
  triviality, 74–78
stationary points, 33
stationary-phase approximation, 9
string tension, 1–3, 114, 131, 193
  non-zero, 140
strong-coupling, 134
  diagrams, 136
  expansion, 125–129, 144, 178
  hadron masses, 177–178
  region, 132
$SU(n)$, 229–238
  action, 239
  adjoint representation, 231–234
  fundamental representation, 229–231
  gauge theory, 90, 112, 115–148
  left and right translations, 234–236
  phase diagram, 138
  tensor method, 236–238
sum over representations, 95
superconductor, 139

topological charge density, 203
topological susceptibility, 204, 212
trajectory
  physical, 70
  renormalized, 70
transfer operator, 10–11, 13, 18–20, 26, 29
transfer-matrix
  elements, 163
  formalism, 135
transformation of variables, 207
translation operator, 31, 102
tree-graph, 75
triangle diagram, 203, 207, 218
two-point correlation function, 53
two-point function, 52–53

$U(1)$, 85
  compact gauge theory, 98
  gauge theory, 115–148
  gauge-fixed model, 222
  problem, 199–202
unitarity bound, 59
unitary matrix, 153
universality, 67, 71
  classes, 67

vacuum, 26
  bubbles, 41
  expectation value, 75
  polarization tensor, 119
valence-quark
  lines, 176
  propagators, 180–181
variational methods, 144
vector bosons, 75
vertex function, 41–42, 45–47, 52–53, 70, 73, 79, 165–166
vortex configurations, 139

Ward identity, 78, 167
  constraints, 168
Ward–Takahashi identities, 166, 207, 209, 211–212
wave vector, 7, 20, 156
wave-function renormalization constant, 58, 62
weak coupling, 71, 139
  expansion, 39–46, 56–58, 72, 79, 116, 135, 139, 144–147, 165
  perturbation theory, 160
  potential, 115–121
Weyl fermions, 161
Wick rotation, 27
  inverse, 154
Wilson action, 95, 187
  parameters, 188
Wilson fermions, 149, 170–171, 212, 214, 223
  free, 225
  method, 156–160, 165, 204, 227
  propagator, 172
  transfer operator, 161–165
Wilson line, *see* Polyakov line
Wilson loop, 107–108, 110–111, 115–116, 119–120, 129–133, 135–136, 138, 175, 241
  confinement phase, 137
  Coulomb phase, 137
  plaquette, 108
  rectangular, 137
  spacelike, 135
Wilson's hopping parameter, 159, 182
Wilson's renormalization theory, 56
Wilson–Dirac Hamiltonian, 165
Wilson–Dirac operator, 214
Wilson–Yukawa framework, 220
Witten–Veneziano formula, 204, 209, 211

Yukawa couplings, 75–76, 78, 220

$Z(n)$, 136
  vortices, 140

Printed in the United States
by Baker & Taylor Publisher Services